W9-CQQ-977

102 Springer Series in Solid-State Sciences

Edited by Hans-Joachim Queisser

Springer Series in Solid-State Sciences

Editors: M. Cardona   P. Fulde   K. von Klitzing   H.-J. Queisser

Managing Editor: H. K. V. Lotsch   Volumes 1–89 are listed at the end of the book

H. G. Kiess (Ed.)

Conjugated Conducting Polymers

With 118 Figures

With Contributions by
D. Baeriswyl, D. K. Campbell, G. C. Clark,
G. Harbeke, P. K. Kahol, H. G. Kiess,
S. Mazumdar, M. Mehring, W. Rehwald

Springer-Verlag
Berlin Heidelberg New York
London Paris Tokyo
Hong Kong Barcelona
Budapest

Dr. rer. nat. Helmut G. Kiess
Paul Scherrer Institut Zürich
Badenerstrasse 569
CH-8048 Zürich, Switzerland

Series Editors:

Professor Dr., Dres. h.c. Manuel Cardona
Professor Dr., Dr. h.c. Peter Fulde
Professor Dr., Dr. h.c. Klaus von Klitzing
Professor Dr. Hans-Joachim Queisser

Max-Planck-Institut für Festkörperforschung, Heisenbergstrasse 1
W-7000 Stuttgart 80, Fed. Rep. of Germany

Managing Editor:

Dr. Helmut K. V. Lotsch

Springer-Verlag, Tiergartenstrasse 17
W-6900 Heidelberg, Fed. Rep. of Germany

ISBN 3-540-53594-2 Springer-Verlag Berlin Heidelberg New York
ISBN 0-387-53594-2 Springer-Verlag New York Berlin Heidelberg

Library of Congress Cataloging-in-Publication Data
Conjugated conducting polymers / H. Kiess (ed.); with contributions
by H. Kiess...[et al.]. p. cm.—(Springer series in solid state sciences; vol. 102)
ISBN 3-540-53594-2 (Berlin: acid-free paper)
ISBN 0-387-53594-2 (New York: acid-free paper)
1. Polymers—Electric properties. 2. Polymers—Optical properties. I. Kiess, H. (Helmut),
1931–   II. Series: Springer series in solid state sciences; 102.
QD381.9.E38C664 1992 530.4'13—dc20   91-25652 CIP

This work is subject to copyright. All rights are reserved, whether the whole or part of the material is concerned, specifically the rights of translation, reprinting, reuse of illustrations, recitation, broadcasting, reproduction on microfilms or in any other way, and storage in data banks. Duplication of this publication or parts thereof is permitted only under the provisions of the German Copyright Law of September 9, 1965, in its current version, and permission for use must always be obtained from Springer-Verlag. Violations are liable for prosecution under the German Copyright Law.

© Springer-Verlag Berlin Heidelberg 1992
Printed in the United States of America

The use of general descriptive names, registered names, trademarks, etc. in this publication does not imply, even in the absence of a specific statement, that such names are exempt from the relevant protective laws and regulations and therefore free for general use.

Typesetting: Thomson Press (India) Ltd., New Delhi
54/3020-5 4 3 2 1 0 – Printed on acid-free paper

Preface

This book reviews the current understanding of electronic, optical and magnetic properties of conjugated polymers in both the semiconducting and metallic states. It introduces in particular novel phenomena and concepts in these quasi-one-dimensional materials that differ from the well-established concepts valid for crystalline semiconductors.

After a brief introductory chapter, the second chapter presents basic theoretical concepts and treats in detail the various models for π-conjugated polymers and the computational methods required to derive observable quantities. Specific spatially localized structures, often referred to as solitons, polarons and bipolarons, result naturally from the interaction between π-electrons and lattice displacements. For a semi-quantitative understanding of the various measurements, electron–electron interactions have to be incorporated in the models; this in turn makes the calculations rather complicated. The third chapter is devoted to the electrical properties of these materials. The high metallic conductivity achieved by doping gave rise to the expression *conducting polymers*, which is often used for such materials even when they are in their semiconducting or insulating state. Although conductivity is one of the most important features, the reader will learn how difficult it is to draw definite conclusions about the nature of the charge carriers and the microscopic transport mechanism solely from electrical measurements. Optical properties are discussed in the fourth chapter. Measurements on dopant- and light-induced changes in the optical spectra help to clarify many controversial aspects concerning the nature of the charge carriers and the question of electron–phonon and electron–electron interactions. It is important to note also that the nonlinear optical coefficients of these materials are high, so they could conceivably become useful in optical processing. The final chapter gives an account of the magnetic properties of these polymers. Nuclear magnetic resonance (NMR) and electron spin resonance (ESR) measurements allow direct probing of properties on a microscopic scale and can thus give detailed information which is otherwise not accessible, for example, we can establish whether the charge carriers induced by doping carry spin (i.e. are bipolarons, solitons or polarons). Data on spin dynamics also provide information on the mobility of the carriers, the spin distribution in defects and electron–electron interactions, which is otherwise difficult to obtain.

This book is dedicated to the late Prof. Günther Harbeke, who contributed to one of the chapters. His interest in the optical properties of condensed matter stimulated much of the work which was followed up experimentally in the former Laboratories RCA Ltd., Zürich, and which finally helped to shape the chapter on optics. It is therefore with deep respect that the coauthors dedicate this book to his memory.

Zürich, February 1992   Helmut Kiess

Contents

Contributors

Baeriswyl, Dionys
Institut de Physique Théorique, Université de Fribourg,
CH-1700 Fribourg, Switzerland

Campbell, David K.
Center for Nonlinear Studies, Los Alamos National Laboratory,
Mail Stop B 262, Los Alamos, NM 87545, USA

Clark, Gilbert C.
Department of Physics, University of California,
Los Angeles, CA 90024, USA

Harbeke, Günther (deceased)
Formerly at Paul Scherrer Institut, c/o Laboratories RCA Ltd.,
Badenerstrasse 569, CH-8048 Zürich, Switzerland

Kahol, Pawan K.
II. Physikalisches Institut, Universtät Stuttgart,
Pfaffenwaldring 57, W-7000 Stuttgart 80,
Federal Republic of Germany

Kiess, Helmut G.
Paul Scherrer Institut, c/o Laboratories RCA Ltd.,
Badenerstrasse 569, CH-8048 Zürich, Switzerland

Mazumdar, Sumitendra
Department of Physics, University of Arizona,
Tucson, AZ 85721, USA

Mehring, Michael
II. Physikalisches Institut, Universität Stuttgart,
Pfaffenwaldring 57, W-7000 Stuttgart 80,
Federal Republic of Germany

Rehwald, Walther
Paul Scherrer Institut, c/o Laboratories RCA Ltd.,
Badenerstasse 569, CH-8048 Zürich, Switzerland

1. Introduction

HELMUT G. KIESS

The usefulness of conventional polymers is largely due to their special properties such as low specific weight, mechanical strength, plastic deformability and their unusually high resistivity. These features have allowed plastic materials to intrude into most sectors of our civilization, be it in the form of low cost and simple utilities or in highly sophisticated technologies like in aircraft construction or modern electronic circuits. Accordingly synthetic chemists have directed much effort towards creating new materials with improved properties for these particular applications. Physicists meanwhile, were confronted with a series of problems related to the special features of plastics, and their task was to understand the rheological, mechanical and resistive behaviour. However, it has been obvious ever since polymers became important in technology that their usefulness could be significantly enhanced if their conductivity could be raised, possibly to match that of metals. Metals could then be replaced in many cases by light and easily processible materials.

This challenge was recognized by many chemists active in the field and numerous syntheses were made with the aim of obtaining polymers with semiconducting properties or metallic conductivity [1.1]. This effort was boosted when *Little* [1.2] proposed the possibility of room temperature superconductivity in materials consisting of a conducting backbone to which polarizable side groups are attached. In the years after this proposal many organic crystals with high conductivity were discovered, some of which showed superconductivity at low temperatures. These crystals often consist of stacks of donor and acceptor molecules and they display metallic conductivity when the separation between the molecules along the stack is small. Perpendicular to the stack the crystals are semiconducting. This indicates that the intermolecular overlap of the wavefunctions is important. Polymers with a delocalized π-electron system would ideally fulfil this condition. Indeed, many chemists attempted to prepare polymers with extended π-electron systems in the hope that the delocalisation of the π-electrons would yield a high free carrier concentration and hence high conductivity. But the observed conductivities were low. It was often argued that the strong deviations from the expected values were not intrinsic but caused by impurities in the polymer, breaks in the chain and conformational defects. However, optical spectra of polyenes showed that the threshold for absorption decreases with increasing chain length but tends to saturate at about $2\,eV$. Hence, *Kuhn* [1.3] argued that the optical gap is fundamental and is due to bond alternation, whereas *Ovchinnikov* et al. [1.4] promoted the idea that electron correlation is the main reason for the gap. Irrespective of the physical mechanism, the conductivity is then expected to remain low since excitation of carriers across a gap of 2 eV is highly unlikely. Under these circumstances it is not surprising that a combination of ingeneous preparational techniques and knowledge of semiconductor and solid state physics was needed to produce a breakthrough in this field leading to the invention and preparation of conducting polymers [1.5]. The conductivity of semiconducting polymers was shown to become metallic if materials with extended π-electron systems were doped with strong oxidizing or reducing agents. Indeed, it was experimentally demonstrated that the conductivity could be changed over more than 14 orders in magnitude extending from the insulating to the metallic state.

It soon emerged that the change from the insulating to the metallic state is accompanied by a number of peculiar transport, optical and magnetic phenomena which are not explicable in terms of conventional semiconductor physics. They are induced by the quasi one-dimensionality of these materials, which enhances many effects such as nonlinearity, disorder and fluctuations. It also became clear that polyacetylene $(CH)_n$ is the polymer best suited to studies of these phenomena because it is the simplest of all the conducting polymers. Therefore much of the work done in the field of conducting polymers relates to $(CH)_n$.

Once it was established that nonlinearities play a decisive role for the description of the observed phenomena, it was felt that in turn these polymers might be used 'paradigmatically' to study nonlinearities. The thermal, electronic and optical properties of condensed matter can usually be described by assuming the existence of a stable ground state and of excited states at higher energies. The energy of the excited state is then a sum of the energies of the so-called elementary excitations, phonons, excitons etc. These excitations can often be obtained from the ground state as a small perturbation as in the case of phonons; for others, perturbation theory is insufficient to calculate the new state. The latter applies for example to domain walls in magnetic materials. These so-called nonlinear excitations are localized in space and are stable even in the presence of other excitations. However, the theoretical modeling of the phenomena observed in conducting polymers turned out to be more intricate than originally expected and the experiments also required a high level of sophistication in order to pin down the exact physical processes responsible for the phenomena.

Although a great many reviews exist in the area of conducting polymers (see references quoted at the beginning of the various chapters), the editor and authors perceived a need for a presentation summarizing the physics of the field in a way suited to the solid state physicist as well as the interested polymer chemist and engineer, and describing the unexpected phenomena found in this new class of materials. The field is a complex one requiring a special effort to introduce it to the non-expert with sufficient clarity, but at the same time retaining accuracy and a reasonable degree of completeness. With this aim in mind we decided to divide the book into chapters on theory (Chap. 2), on electrical charge transport (Chap. 3), on optical properties (Chap. 4) and finally on magnetic properties (Chap. 5) of conducting polymers. The sequence of the chapters is deliberate. First, the theoretical concepts are presented in order to convey to the reader the new ideas which had to be introduced in order to describe the properties of these polymers. Then the conductivity and related phenomena are reviewed, with emphasis on the various aspects of the high conductivity which gave this new class of polymers its name. Optical experiments are powerful probes of the material on a more microscopic level and, indeed, they have helped to clarify many of the new phenomena. This is particularly true for investigations of the magnetic properties which have yielded precise information about specific features of the material.

Chapter 2, devoted to the theory of π-conjugated polymers, attempts to differentiate clearly between the models used to describe the physical properties and the computational methods required to perform the calculations. Thus, after an introductory section on theoretical concepts, models, and methods, further sections of this chapter are devoted to the various models (including computational methods) which treat linear π-conjugated polymers exclusively in terms of electron–phonon or electron–electron coupling, or combined electron–phonon and electron–electron interactions. These theoretical models, despite their complexity, are still idealizations of the real world in the sense that they treat ideal, infinitely long chains. Therefore a final section of Chap. 2 discusses briefly the effects of disorder, interchain coupling, three-dimensional effects, etc., as encountered in real polymers.

The next three chapters give an account of the most important experimental results and their interpretation.

Chapter 3 discusses the results of various electrical transport measurements and their interpretation. Electrical transport measurements are in principle quite simple, but they do not yield straightforward information about the microscopic nature of the charge carriers involved. This is particularly true for the conducting polymers that deviate significantly from the theoretical model of the idealized linear chain: these have various degrees of crystallinity, amorphous regions, weak interchain interactions, chain defects and chain breaks, all of which depend on the preparation technique and influence the transport properties differently. Since high metallic conductivity has been achieved with stretch-aligned $(CH)_n$ with low defect concentration, a trend towards the production of better oriented, defect-free material is observed. Depending on the importance attributed to the deviations from the idealized linear chain, the transport measurements have to be interpreted differently. The pertinent models and the related problems are discussed in detail. Obviously, conclusions about the microscopic nature of the charge carriers have to be treated with caution and are often not realistic if they are based solely on transport measurements. Thus, the question of whether electrical transport is due to solitons or electrons in $(CH)_n$ is still controversial. One of the most gnawing questions of conducting polymers, namely the mechanism of the insulator–metal transition, also remains to be answered.

Chapter 4 reviews the optical properties including optical excitations across the gap and high up into the bands, the interaction of electromagnetic radiation with the lattice vibrations and the study of the nonlinear elementary excitations in conjugated polymers. Particular emphasis is placed on those investigations which helped to clarify the nature of the quasi-particles such as solitons, polarons and bipolarons in these materials. Photoinduced absorption experiments performed in polyacetylene are discussed in detail because they are essential for exploring the solitonic state and the relative strength of the electron–phonon and electron–electron coupling in polyacetylene. The nonlinear properties are also mentioned briefly because conjugated polymers exhibit high coefficients for third harmonic generation and also because basic information on the coupling is obtainable from such results. Finally experimental results for polymers in the highly conducting state are presented.

The chapter on magnetic properties (Chap. 5) begins with an overview of the magnetic properties of materials in general and of the measuring techniques. There are, indeed, a wide variety of phenomena and observation techniques which deserve introduction and explanation to the reader not experienced in the field of magnetic properties of materials. These comprise nuclear and electronic spin measurements under various dynamic or static conditions. It is shown that these measurements allow one to probe local properties, thus offering important support in the interpretation of the results of optical and electrical measurements. After this introductory section the chapter focuses on fundamental aspects of conducting polymers such as structural and lattice-dynamical properties of the polymers themselves, the structure and dynamics of solitons and polarons, and of the conjugational defects in general, and on structural and dynamical effects induced by doping. These results allowed definitive conclusions to be reached on otherwise very controversial items.

Thus, as already mentioned, the book focuses on the physics of conducting polymers. It explains and reviews the exciting new ideas and concepts which had to be introduced to describe these materials, but are not required in ordinary condensed matter physics. In addition, it gives an overview of experimental observations of the electrical, optical and magnetic properties and their interpretation. Since unexpected physics is often connected with these polymers, it is certainly legitimate to ask whether still more new applications may emerge. Indeed, many proposals have been made. However, these are not discussed in this book since the only products on the market at present are long life batteries based on polyaniline and polypyrrole, other suggestions not yet having gone beyond experimental tests. This is due to problems of stability and processibility of these materials, though good progress has been achieved recently. It is expected that these difficulties will be overcome in the near future through the combined efforts of chemists, physicists and engineers. A contribution towards this goal is made by investigations that lead to a deeper understanding of the underlying physical processes. We hope that this book will help to spread this knowledge and to stimulate progress in this exciting and rapidly moving field of materials science.

References

1. A. Rembaum, J. Moacanin, H.A. Pohl: In *Progress in Dielectrics*, ed. by J. Birks (Temple, London 1965) Vol. 6, p. 41
2. W.A. Little: Phys. Rev. **134A**, 1416 (1964)
3. H. Kuhn: J. Chem. Phys. **17**, 1189 (1949)
4. A.A. Ovchinnikov, I.I. Ukrainskii, G.V. Kventsel: Sov. Phys. Usp. **15**, 575 (1973)
5. C.K. Chiang, C.R. Fincher, Y.W. Park, A.J. Heeger, H. Shirakawa, E.J. Louis, S.C. Gau, A.G. MacDiarmid: Phys. Rev. Lett. **39**, 1098 (1977)

2. An Overview of the Theory of π-Conjugated Polymers

Dionys Baeriswyl, David K. Campbell, and Sumit Mazumdar

We review recent advances in the theoretical modelling of π-conjugated polymers. Our emphasis is on quasi-one-dimensional π-electron models that include both electron–phonon and electron–electron interactions. We use the widely studied Peierls–Hubbard Hamiltonian as a prototype model, since this contains both the pure electron–phonon (Hückel and SSH) limits and the pure electron–electron (Hubbard and PPP) limits. We attempt to present an integrated perspective by explaining the essential concepts in both chemical language (valence bonds, resonance, bond alternation defects, etc.) and solid-state physics terms (band structure, localized states, broken symmetry, solitons). We argue that modelling π-conjugated polymers is a true many-electron problem requiring advanced techniques that give reliable answers to precise questions, especially for excited states. Among the techniques we discuss are mean-field, perturbative, and variational approximations for infinite polymer chains and (numerically) exact computations (Lanczos and quantum Monte Carlo methods) for finite chains (oligomers). We compare critically the theoretical results obtained by these various methods with experimental observations, in particular with optical spectroscopy and nuclear magnetic resonance, both for the archetypal π-conjugated system polyacetylene and for other conjugated polymers including polydiacetylenes and polythiophene. Our goal is to find a model consistent with the broad range of experimental data. This analysis establishes that electron–phonon and electron–electron interactions are likely to be of equal importance in determining the properties of these materials. Hence neither the Hückel/SSH nor the PPP/Hubbard models are sufficient for a complete description of the observed behaviour of π-conjugated polymers. We add an analysis of factors that go beyond the idealized models, including disorder, inter-chain coupling and quantum lattice fluctuations, and conclude with a brief discussion of the doping-induced insulator-metal transition and of the possible mechanisms of charge transport.

2.1 Synopsis

In any attempt to survey the theoretical perspective on π-conjugated polymers, one initially encounters several problems of presentation. We adopt here the somewhat unusual tactic of discussing these problems explicitly, for we feel that this not only helps to define the scope and form of our overview but also serves to indicate the present limitations on our detailed theoretical understanding of these fascinating but difficult materials.

Perhaps the most obvious problem, common to all such survey articles, is reconciling the demands of completeness and technical detail with those of clarity and pedagogy. We have attempted here to focus on identifying and illustrating the broad theoretical issues and central concepts in a didactic manner. Fortunately, we can refer readers interested in further details to a large number of reviews dealing in depth with particular aspects of the theory [2.1–12 etc].

A second problem, which amplifies the first, arises chiefly because π-conjugated polymers lie at the interface between organic chemistry and solid-state physics and have generated substantial interest in both disciplines. As a consequence, a wide range of theoretical models and an even wider variety of calculational methods have been applied to the study of conducting polymers. Often, similar models are couched in dramatically different jargon, and related concepts are obscured by linguistic barriers. We have attempted to bridge the gulf between the respective approaches of the chemists and the physicists by stressing the relations between various models and, within reason, explaining all basic concepts in both chemical and physical terms. We have further tried to link the studies of π-conjugated polymers with the extensive chemical literature on finite polyenes and other conjugated oligomers. Also, we have tried to delineate consistently the essential distinction between the *models* and the (typically approximate) *methods* used to solve them.

A third problem stems from our natural desire as theorists to develop a broad understanding of classes of related phenomena. As materials, conjugated polymers can indeed be viewed as members of the class of quasi-one-dimensional systems, with other members including (1) organic charge transfer salts (e.g. TTF-TCNQ, TTF-Chloranil, etc.), (2) inorganic charge-density wave compounds (e.g. TaS_3, $(TaSe_4)_2I$, etc.), (3) metal chain compounds (KCP, halogen-bridged metallic chains and $Hg_{3-\delta}AsF_6$) and (4) σ-bonded electronically active polymers (e.g. polysilanes). Furthermore, these one-dimensional systems share many common features with certain quasi-two-dimensional materials, including the organic superconductors (e.g. $(BEDT-TTF)_2X$) and the high T_c superconducting copper oxides. An essential conceptual link between these materials is the idea of *broken symmetry* ground states in electronic systems in reduced dimensions. From a theoretical perspective, analysis of the possible broken symmetries— bond-order waves (BOW), charge-density waves (CDW), spin-density waves (SDW) and superconductivity—as functions of effective dimensionality, the strengths of various interactions, and band fillings allows for a deep unification of this whole class of related novel electronic materials. Although we shall mention in passing many of these theoretical concepts and the consequent relations between these materials, limitations of space and the stated focus of this volume lead us to concentrate primarily on the familiar π-conjugated polymers.

The polymers themselves present a fourth and serious problem for theoretical modelling. In Figs. 2.1–6 we illustrate the idealized structures of several of the most widely studied conducting polymers. The complexity of the chemical moieties in the monomers changes dramatically from the simple units in *trans-* and *cis-*polyacetylene ((CH)$_x$, Figs. 2.1 and 2.2) through polymers involving heteroatoms (polythiophene and polypyrrole, Figs. 2.3a and 2.3b) and those involving aromatic units (poly-(para)-phenylene, Fig. 2.4a and poly-(para)-phenylene vinylene, Fig. 2.4b) to polymers involving both heteroatoms and aromatic subunits (polyaniline, Fig. 2.5). Furthermore, the polydiacetylenes (Fig. 2.6) are typically stabilized by large side groups, which may affect the properties of the conjugated backbone. As indicated by several other studies in this volume, real conducting polymers are not obviously related to the theorists' isolated, infinite, defect-free one-dimensional chains. Limited conjugation lengths, subtle solid-state inter-chain effects, direct inter-chain chemical bonding,

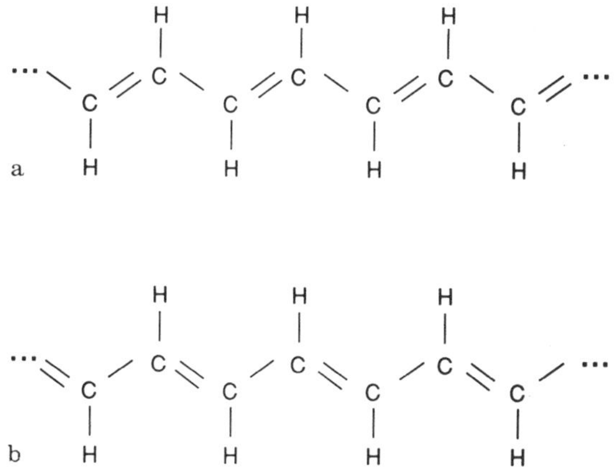

Fig. 2.1 a, b. The two degenerate bond alternation patterns in an idealized quasi-one-dimensional *trans*-(CH)$_x$ polymer chain

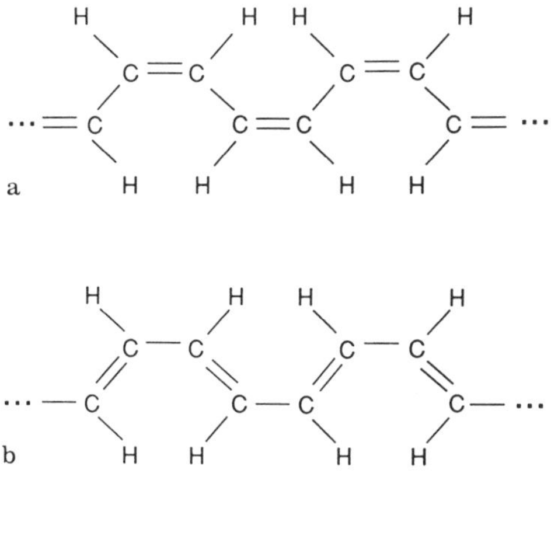

Fig. 2.2 a, b. The two possible (*non*-degenerate) bond alternation patterns in an idealized quasi-one-dimensional *cis*-(CH)$_x$ polymer chain: a *cis-transoid* form; b *trans-cisoid* form

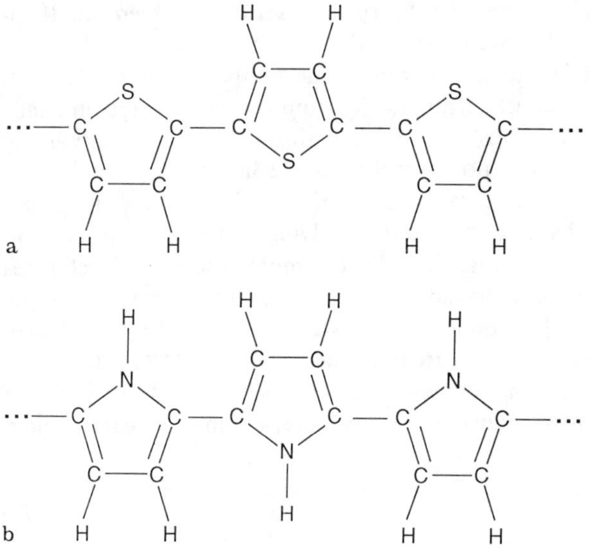

Fig. 2.3 a, b. The idealized structures of **a** polythiophene and **b** polypyrrole

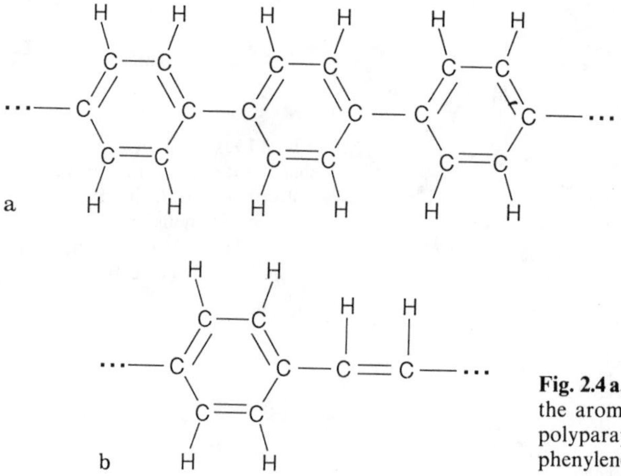

Fig. 2.4 a, b. The idealized structures of the aromatic π-conjugated polymers (**a**) polyparaphenylene and (**b**) polyparaphenylene-vinylene

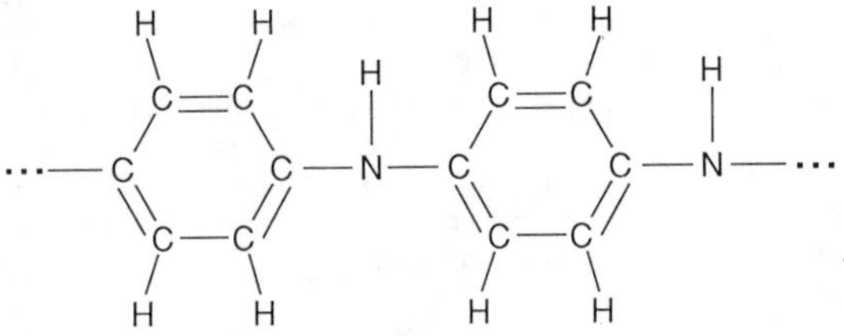

Fig. 2.5. The idealized structure of polyaniline (in its leuco emeraldine base polymeric form) [2.425]

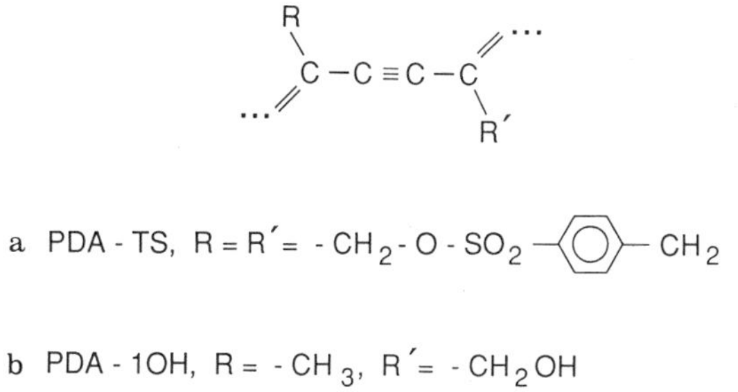

$$\begin{array}{c} R \\ \diagdown \\ \diagup \\ \cdots \end{array} C - C \equiv C - C \begin{array}{c} \diagup \cdots \\ \\ \diagdown \\ R' \end{array}$$

a PDA - TS, $R = R' = - CH_2 - O - SO_2 - \langle O \rangle - CH_2$

b PDA - 1OH, $R = - CH_3$, $R' = - CH_2 OH$

Fig. 2.6 a, b. The idealized chemical structure of polydiacetylene (PDA). The side groups, R and R', are as indicated in the figure for (**a**) PDA-toluene-sulfonate (PTS) and (**b**) for PDA-1OH

and impurities and defects complicate substantially the present real materials. Furthermore, the same polymer prepared by different synthesis procedures can have, in many cases, quite distinct morphologies and properties. Eventually, one hopes that clever experimental techniques (such as solid-state polymerization within a host matrix or the use of soluble co-polymers) will permit the idealized limit of isolated infinite chains to be more closely approached experimentally. Indeed, the synthetic route to poly-(para)-phenylene vinylene that allows the casting of films that are both dense and very highly oriented is a good example of recent improvements in materials processing [2.13]. Reflecting both this hope and the existing theoretical literature, we shall concentrate primarily on models for single, effectively one-dimensional chains and only briefly examine the limited attempts to model in detail the present real materials.

Finally, by focusing on the *theory* of conjugated polymers we do not explore the primary origins of considerable recent interest in these materials and the related synthetic metals mentioned above. Much of this interest stems from the hope that one can use the richness of organic chemistry to tune molecular parameters to achieve particular macroscopic properties. Originally this interest focused on the prospects for making "plastics that conduct"; indeed, this prospect is sufficiently intriguing that although the pristine conjugated polymers are in fact insulators or, at best, semiconductors, the near-metallic conductivities that they achieve upon chemical doping has led to the terminology *conducting polymers*. More recently, the possibility of using the *semi*-conducting versions of these materials as nonlinear optical devices has attracted considerable attention. The exciting potential technological applications of these conducting polymers and related synthetic metals promise to keep the field of π-conjugated polymers active and vital, independent of the theorists' substantial interest in understanding the basic phenomena.

Having made clear some of the limitations and omissions of our theoretical overview of π-conjugated polymers, we can now describe the remaining six sections of this chapter. In Sect. 2.2, we introduce the conceptual aspects of theoretical models and methods, and try to motivate the simplifications required to produce theoretically tractable models. In particular, after a brief discussion of more general models, we focus on models that treat only the π electrons as dynamical entities. We indicate how these models can be 'derived' from more microscopic theories and discuss the essential roles played by both electron–phonon (e–p) and electron–electron (e–e) interactions among the π electrons. This leads to the crucial distinction between independent-electron theories and interacting-electron theories. If e–p interactions dominated in real π-conjugated polymers, these systems could be treated accurately within independent-electron theories. Chemists would then say that Hückel molecular orbital theory is adequate, whereas physicists would call the polymers Peierls semiconductors. If, in contrast, e–e interactions dominated, then—particularly since the polymers are quasi-one-dimensional so that mean-field or Hartree–Fock methods that ignore quantum fluctuation effects are especially suspect—accurate many-electron methods must be used. In this limit chemists would refer to substantial correlation effects, whereas physicists would refer to the polymers as 'Mott–Hubbard insulators'. To illustrate both cases, Sects. 2.3 and 2.4 review paradigms for, respectively, pure e–p and pure e–e interactions. In Sect. 2.3, we summarize the independent-electron theories, discussing the general Hückel approach and examining in some detail the celebrated Su-Schrieffer-Heeger (SSH) model of $trans$-(CH)$_x$. The explicitness and relative simplicity of the SSH model allows one to make extensive predictions for the properties of $trans$-(CH)$_x$, and we survey these predictions and compare them briefly to experimental results. In Sect. 2.4 we examine the one-dimensional Hubbard model as a paradigm of many-electron theories. Although the pure Hubbard model cannot be viewed as a realistic model of the conjugated polymers, the existence of exact, analytic solutions (a distinct rarity for strongly correlated many-body Hamiltonians) for various observables within this model provides important insight into the possible effects and signatures of e–e interactions in real conjugated polymers.

In Sect. 2.5 we confront the more likely prospect that $both$ e–p and e–e interactions play significant roles in conjugated polymers. We introduce a class of models known to physicists as the Peierls–Hubbard models and to chemists as (or related to) the Pariser–Parr–Pople (PPP) models, which incorporate both types of interactions in a flexible manner. We discuss the interpretation and the expected values of the parameters in these models and then examine the methods used to study the models. In this discussion we focus on numerically exact results (obtained primarily for finite-sized systems using quantum Monte Carlo or exact diagonalization techniques) for the anticipated range of parameters relevant to conjugated polymers. This approach has the dual virtues of providing $unambiguous$ and $definitive$ predictions from the models and of validating, or in some cases invalidating, the results of various $approximate$ solution methods, including perturbation theory and various variational wave-functions. In addition to its virtues, however, our approach in this section has the vice of being somewhat technical; readers who are not interested in this level of technical detail can skip the section without losing the essential thread of the review.

In Sect. 2.6 we compare the theoretical predictions of Peierls–Hubbard models with experimental data on π-conjugated polymers. Our approach is to demand broad, semi-quantitative agreement with data on a variety of experimental observables in several different classes of polymers. The polymers we consider include trans- and cis-polyacetylene, polythiophene, polypyrrole, and polydiacetylene. Among the experimental observables on which we focus are the extent of bond alternation, the charge and spin densities (as measured by XPS and ENDOR), optical gaps, optically forbidden (e.g. A_g) states, photo-induced absorption, bleaching and luminescence, Raman and infrared absorption, and properties of intrinsic bond-alternation defects (solitons, polarons, and bipolarons). Although the experimental situation is far from being completely clarified, we believe it suggests that both e–p and e–e interactions play essential, non-perturbative roles in conjugated polymers and that exact many-body methods, or faithful approximations to such methods, will be necessary to obtain both definitive predictions from the models and detailed fits to experiment.

Finally, in Sect. 2.7 we examine critically the most obvious limitations of the standard theoretical approaches to conducting polymers. We discuss briefly the role of disorder, including intrinsic defects (e.g. cross links, sp^3 hybridized bonds and chain ends) and extrinsic impurities which, in view of the organo-metallic catalysts used in many syntheses, are a particular bane of these systems. We also examine in more detail recent results that suggest the importance of three-dimensional effects in high purity, crystalline trans-$(CH)_x$ and the possible breakdown of the quasi-one-dimensional picture of localized, nonlinear excitations. In addition, we discuss the (fairly incomplete) current understanding of what remain perhaps the two most interesting properties of conjugated polymers: the origin of the insulator-metal transition that gives rise to the high conductivity of heavily doped polymers and the nature of electronic transport in the real materials in both the insulating and metallic regimes.

Throughout this review we attempt to develop a forward-looking perspective, making clear what is *not*—as well as what is—known and indicating areas for future research. We hope that this approach will prove successful and will help to stimulate further research in this often frustrating but always exciting field.

2.2 Theoretical Concepts, Models and Methods

In view of their long history (see, e.g., [2.14]), it is hardly surprising that the related problems of bond alternation and electronic structure in finite polyenes, polyacetylene and, more generally, π-conjugated polymers have been studied using a wide range of approximate models and virtually the full panoply of known many-body methods. In this section, we review the steps leading from the fundamental microscopic Hamiltonian through a series of approximations to models of increasing simplicity. This will allow us to focus on the conceptual aspects of these models and to discuss at a general level their respective virtues and limitations. From our remarks in the introduction it follows that we should adopt a perspective sufficiently general to capture the essence of both the chemists' and physicists' views of conducting polymers. To avoid vagueness, however, we shall often refer to the specific context of (even) *polyenes*, $C_{2n}H_{2n+2}$, and *polyacetylene*, the idealized $n \to \infty$ limit of these systems.

2.2.1 The Born–Oppenheimer Approximation

In its full complexity, the theoretical problem of modelling the even polyene $C_{2n}H_{2n+2}$ at a truly microscopic level is described by a quantum-mechanical Hamiltonian, H, which one can formally separate into terms involving (1) the $4n+2$ three-dimensional coordinates, called R_α, of the carbon and hydrogen nuclei, (2) the $(6 \times 2n) + (2n+2) = 14n+2$ three-dimensional coordinates, called r_i, of the electrons and (3) a remainder involving both sets of coordinates. Using the subscripts n for nuclear and e for electron and using braces $\{\cdots\}$ to indicate functional dependence on a set of coordinates, we can write

$$H = H_{n-n}\{R\} + H_{e-e}\{r\} + V_{n-e}\{r, R\} \tag{2.1}$$

where

$$H_{n-n}\{R\} = \sum_\alpha \frac{P_\alpha^2}{2M_\alpha} + \sum_{\alpha > \beta} \frac{Z_\alpha Z_\beta e^2}{|R_\alpha - R_\beta|} \tag{2.2a}$$

$$H_{e-e}\{r\} = \sum_i \frac{p_i^2}{2m_e} + \sum_{j > i} \frac{e^2}{|r_j - r_i|} \tag{2.2b}$$

and

$$V_{n-e}\{r, R\} = -\sum_{\alpha, i} \frac{Z_\alpha e^2}{|R_\alpha - r_i|}. \tag{2.2c}$$

Here Z_α and M_α are, respectively, the charge and mass of the αth nucleus and m_e is the electron mass. The complete Schrödinger equation for the $3 \times (18n + 4) = 54n + 12$ degrees of freedom clearly poses a quantum-mechanical many-body problem which is impossible to solve exactly even numerically. We are thus immediately confronted with the need both to invent approximation *methods* for the full equation and, very importantly, to develop simplified, approximate *models* that capture certain essential features without invoking the full complexity of (2.1) and (2.2). It is important, although sometimes difficult, to keep the distinction between models and methods clear, particularly since a given approximation method, for example Hartree–Fock, can be applied to models at many different levels of sophistication.

An obvious first step for either of these efforts is to recognize the substantial separation of time scales between the nuclear and the electronic motions and the consequent utility of an approximation that focuses on the electrons as the primary dynamical objects determining the chemical bonding and transport properties of interest. The formal embodiment of this observation is provided by the Born–Oppenheimer approximation, which involves studying in a first step the dynamics of the electronic system for a *fixed* geometry of the nuclei. The total energy of the electrons thus is a *functional* of the nuclear coordinates; when this energy is added as a potential energy to H_{n-n}, the resulting Hamiltonian then depends only on the nuclear coordinates R_α. In a second step this Hamiltonian is used for calculating the ground-state configuration of the nuclei and the (adiabatic) dynamics around this ground state. The Born–Oppenheimer approximation is justified if the electronic motion is much more rapid than the relevant nuclear motions. Within the Born–Oppenheimer approximation one must first diagonalize the electronic Hamiltonian

$$H_e\{r, R\} \equiv H_{e-e}\{r\} + V_{n-e}\{r, R\} \tag{2.3}$$

for fixed nuclear coordinates R. The Coulomb interaction between electrons and nuclei (2.2c) is a sum of terms each of which depends on a single electron coordinate, r_i. In contrast, the electron–electron Coulomb repulsion in (2.2b) involves the coordinates of two electrons, r_j and r_i, in each term and thus obviously correlates the dynamics of the electrons beyond the constraints of the Pauli principle.

The problem of solving the Schrödinger equation for the Hamiltonian (2.3) for fixed $\{R_\alpha\}$ still represents a formidable task and hence requires further approximations, These can be divided roughly into two categories: (1) ab initio calculations, which start from (2.3) and solve this many-body problem approximately and (2) model calculations which replace (2.3) by a simpler, parametrized Hamiltonian, the properties of which can hopefully be obtained in an accurate and controlled manner.

2.2.2 Ab Initio Calculations

Two approximate schemes are used extensively for performing ab initio calculations on the full microscopic Hamiltonian (2.3). Quantum chemists have developed the Hartree–Fock (HF) approximation and related techniques to a high degree of sophistication. Condensed matter physicists tend to favour density-functional theory, which is usually treated in the so-called local density approximation (LDA). Both approximations are effective single-particle theories, as we shall explain.

a) The Hartree–Fock Approximation

Much of the theoretical effort in quantum chemistry in the past several decades has been devoted to developing approximate independent-electron theories for a wide range of molecules and chemical compounds. With the advent of large scale digital computers, it has become increasingly possible to apply these independent-electron methods to the full Hamiltonian (2.3) (for several perspectives, see [2.4, 15–18]). These methods can be applied to fairly large finite clusters: in the case of π-conjugated polymers, oligomers involving up to thirty or forty monomer units can be studied. The approach exists in many variants which differ in subtle but potentially important ways. Here, we shall attempt to give an impressionistic perspective; readers unfamiliar with the method but interested in the details should refer to the articles cited above.

We first rewrite the Hamiltonian (2.3) as

$$H_e\{r, R\} = H_e^0\{r, R\} + V_{e-e}\{r\} \tag{2.4a}$$

where

$$H_e^0\{r, R\} = \sum_i \left[\frac{p_i^2}{2m_e} + V_{n-e}(r_i, \{R\}) \right] \tag{2.4b}$$

is a sum of single-particle terms with single-particle potential

$$V_{n-e}(r_i, \{R\}) = - \sum_\alpha \frac{Z_\alpha e^2}{|R_\alpha - r_i|} \tag{2.4c}$$

and

$$V_{e-e}\{r\} = \sum_{j>i} \frac{e^2}{|r_i - r_j|} = \sum_{j>i} V_{e-e}(r_i - r_j) \tag{2.4d}$$

is the electron–electron interaction.

Considering for the moment only H_e^0, we note that the full many-body electronic wave function solving H_e^0 can be expressed as a product of single-particle wavefunctions (which becomes a Slater determinant when the Fermi statistics of the electrons is taken into account). This leads to a separation of electronic variables and to the complete diagonalization of H_e^0 in terms of single particle molecular orbitals for the system (see, e.g. [2.14]). The computation of these molecular orbitals can be greatly simplified in the case of a periodic arrangement of the nuclear ions by Bloch's theorem. The Slater determinants of Bloch or molecular orbitals $\psi_k(r)$ yield a basis for the many-electron states, and H_e^0 is a diagonal in this basis.

The main assumption of the Hartee–Fock approximation is that the ground state of the full Hamiltonian H_e is still a *single* Slater determinant. However, the single-particle orbitals are not specified in advance but are varied so as to minimize the expectation value of H_e. Thus, even for a periodic arrangement of the nuclei the best wavefunctions are not necessarily of the Bloch–Mulliken type. Instead, Wannier functions $\phi_l(r)$, localized around the lth unit cell and mutually orthogonal, may yield a lower energy. One expects this to be true if either V_{e-n} or V_{e-e} (or both) are large compared to the kinetic energy. V_{e-n} dominates for the tightly bound electrons. The effect of V_{e-e} becomes important for the valence electrons. To see this, we suppose for simplicity that there is one valence electron per unit cell on average. A single Slater determinant in terms of extended Bloch orbitals can be represented as a superposition of determinants of Wannier functions. Among all these terms there are many with an unequal distribution of electrons: for example, some of the cells will be empty, others doubly occupied. The repulsive Coulomb interaction tends to suppress this crowding of electrons and thus enhances the weight of terms with one electron in each cell.

Despite the conceptual simplicity of HF, its realization for moderately large systems is still computationally involved. Therefore, additional approximations are very often made. One obvious approximation involves reducing the number of degrees of freedom, both by focusing on a limited number of electrons, e.g. the valence electrons, and by limiting the number of orbitals which these electrons can occupy. These approximations lead to various truncated Hamiltonians which, while much simpler to study than H_e, capture the essential physical features of conjugated systems. Many such truncated Hamiltonians have been studied, particularly in recent chemical literature. One well-known example is the MNDO (minimal neglect of differential overlap) model, which explicitly ignores the core electrons, uses a minimal basis set for the valence electrons, e.g. 2s, $2p_x$, $2p_y$ and $2p_z$ orbitals for each carbon atom, and provides an explicit (semi-empirical) procedure for treating the Coulomb matrix elements [2.18–20]. It has been used, for example, in studies of the nonlinear excitations in transpolyacetylene [2.21]. Importantly, however, whenever certain interactions are ignored or approximated, the theory is no longer truly microscopic and neglect of certain matrix elements may have to be compensated by (semi-empirical) adjustment of others, introducing a renormalization of the parameters in a model Hamiltonian. We shall return to this critical issue at numerous points in our later discussion.

The HF approach serves as an essentially zeroth-order approximation to the ground state of interacting electrons. By construction, it represents the best effective single-particle theory of the interacting system. It is in this sense that any effects beyond HF are usually referred to as correlation effects. It must be recognized, however, that for a detailed understanding of the theories involving interacting electrons, HF methods are often not quantitatively accurate enough and in some cases can even be qualitatively misleading. This is particularly true in quasi-one-dimensional theories, such as those applied to π-conjugated polymers, where important fluctuation effects are not correctly captured by the HF approximation. This will be illustrated explicitly in our later discussion. It is not hard to convince oneself that any attempt to go beyond HF on an ab initio level represents a very difficult enterprise (see, e.g. [2.22–24]). Thus the alternative approach of focusing on simpler model Hamiltonians as means of clarifying the importance of correlation effects seems more promising.

b) Density-Functional Theory and the Local Density Approximation

In the physics literature, band structure calculations provide an independent-electron approach that has been widely applied to solid state systems. Assuming a periodic structure for the reference geometry, one uses Bloch's theorem to reduce the problem to that of a single unit cell in momentum space, the first Brillouin zone. Since the resulting momentum space band levels are extended Bloch states, the localized regime for dominant Coulomb interactions cannot adequately be described and the band structure approach is thus best applied to the limit of weak correlation effects. Nonetheless, based on the Hohenberg–Kohn–Sham Theorem [2.25, 26] one knows that the total ground-state energy of a many-electron system in the presence of an applied potential is a *functional* of the density of the electrons, $\rho(r)$ (for a current review, see [2.27]). Formally, one can write for the effective single-electron potential,

$$V_{\text{eff}}(r) = V_{\text{n-e}}(r) + e^2 \int d^3 r' \frac{\rho(r')}{|r - r'|} + V_{\text{x-c}}[\rho(r)]. \tag{2.5}$$

Unfortunately, it is as difficult to determine the functional dependence of $V_{\text{x-c}}$ on $\rho(r)$ as it is to solve the full many-body problem from scratch. Hence, the local density approximation (LDA) is typically invoked and the exchange-correlation potential $V_{\text{x-c}}$, a *functional* of $\rho(r)$, is approximated by a spatial integral over a given *function* of the local density, $\rho(r)$, modelled on the behaviour of the free electron gas. One then solves the single-particle problem for an initial form of the potential V_{eff}, calculates the density $\rho(r)$ by filling all single-particle levels up to the Fermi energy and thus determines a new potential V_{eff} using (2.5). This procedure is iterated until self-consistency is achieved. In the area of π-conjugated polymers, the LDA approach has been applied by a number of authors [2.28–37], to the study of both idealized chains and perfect three-dimensional crystals of trans-$(CH)_x$. We shall discuss some of these results for perfect three-dimensional crystals of trans-$(CH)_x$ in Sect. 2.7.

Although computationally very intensive, the LDA method is often simpler than the HF approach. Some correlation effects, which by definition are not included in HF, are also taken into account in LDA, albeit not in a controlled or systematic way. On the other hand, the single-particle levels do not have a simple meaning in LDA, whereas the HF eigenvalues can be interpreted using Koopman's theorem [2.38], which proves that the ith single particle HF level is the ionization energy for removing the ith electron. Nevertheless, within density-functional theory a relationship similar to Koopmans' theorem holds for the single-particle levels associated with the exact energy functional [2.39].

2.2.3 Model Hamiltonians

The primary simplification that model Hamiltonians provide vis-a-vis the ab initio problem is the limitation of the number of degrees of freedom that they treat explicitly. One difficulty that arises immediately when one considers only some of the degrees of freedom is that the parameters describing the interactions among the reduced set of degrees of freedom cannot be readily derived from the underlying microscopic theory. Indeed, although one can derive the *form* of the interactions in the model Hamiltonians fairly reliably, derivation of the *values* of the parameters in these models is nearly as difficult as solving the full ab initio problem. There are two approaches that one can adopt to circumvent this difficulty. Firstly, one can calculate the ground state and low-lying excitations for small systems both for the microscopic Hamiltonian (2.3) and for the parametrized model Hamiltonian, and adjust the parameters of the model so as to reproduce closely the ab initio results. Secondly, a more phenomenological method is to calculate, using a given model Hamiltonian, observable quantities such as the optical absorption spectrum or the magnetic susceptibility and to determine the adjustable parameters in the model by comparing the calculated quantities with experiment. Below we shall discuss various types of model Hamiltonians which are distinguished both by the number of atomic levels considered (all valence electrons or only the π electrons) and by the forms of the interaction terms taken into account (e.g. all Coulomb-type terms or only short-range interaction or no electron–electron interaction at all).

a) Valence Effective Hamiltonians

Quantum chemists have extensively used a scheme in which ab initio HF calculations are performed for a short oligomer, in order to optimize its geometry and to determine the HF bandstructure. Then, in a second step, the HF spectrum is mapped onto an effective single-particle Hamiltonian, a so-called "valence effective Hamiltonian" (VEH), which includes explicitly only the valence electrons [2.8, 40]. The single-particle potential is assumed to act in a reduced function space; typically Gaussian functions are used with appropriate orbital symmetry for the valence electrons of carbon and hydrogen atoms. The relative simplicity of the VEH model permits comparative studies of oligomers composed of many different chemical moieties and thus provides crucial guidance in designing π-conjugated polymers with specific properties (see e.g. [2.3, 8]). Furthermore, it is particularly useful in treating inhomogeneous states (polarons, bipolarons). Thus, this approach can be valuable for discussing high-energy photoemission or optical absorption experiments. However, one must recall that these Hamiltonians contain no (explicit) Coulomb interaction. Putting two electrons into the same level costs twice the energy as adding a single electron to this level. This is in sharp contrast not only to interacting electron theories but also to experiments, as will be shown in Sect. 2.6.

b) Models Involving π-Electrons Only

A more drastic reduction of the degrees of freedom involves focusing solely on the least bound (π) molecular orbitals formed primarily from hybridization of the carbon $2p_z$ atomic orbitals. The remaining three valence electrons in each carbon plus the 1s electron in each hydrogen atom all participate in molecular

σ bonds and, like the carbon core electrons, are not treated explicitly. The effects of all these electrons are incorporated into (1) an effective ion–ion interaction, (2) a renormalized electron–ion interaction and (3) (weakly) screened (π) electron–electron interactions. We will show how this restriction to π electrons leads, starting from a (semi)-microscopic theory, to several of the well-known models that have been used extensively in the study of π-conjugated polymers. In what follows we shall also assume that the π electrons are coupled to the displacements of carbon atoms but not to the displacements of hydrogen atoms. Correspondingly, each CH group will be represented by a *single* ionic coordinate $R_l, l = 1, \ldots, 2n$. For a detailed description of vibrational spectra, the hydrogen coordinates must also be incorporated, as will be discussed in Sect. 2.6.3.

General considerations. Conceptually, we imagine starting with a Hamiltonian of the form (2.4) *except* that (1) the Coulomb interaction between the nuclei and electrons (2.4c) is replaced by a *pseudopotential* $V_P(r_i, \{R\})$ incorporating the (screening and renormalization) effects of core and σ electrons and depending both on the coordinates r_i of the π electrons and on the ionic coordinates R_l, and (2) the electron–electron Coulomb interaction, given by (2.4d), is also replaced by an effective interaction, $V_{e-e}^{eff}\{r\}$, which depends only on the π-electron coordinates and incorporates the screening and renormalization effects of the core and σ electrons. Often, one takes

$$V_{e-e}^{eff}\{r\} \equiv \sum_{j>i} \frac{e^2}{\varepsilon |r_i - r_j|} = \sum_{j>i} V_{e-e}^{eff}(r_i - r_j), \tag{2.6}$$

where ε is a background dielectric constant.

Without specifying either of these effective potentials, we assume that the eigenfunctions of the *single*-particle Hamiltonian, H_e^0 (including V_P) are known. For a periodic configuration of the ions, these are Bloch functions $\psi_k(r)$ from which we can derive, in principle, Wannier functions $\phi_l(r)$ centred around the lth (CH) unit. Thus for this periodic reference configuration, we can write the Hamiltonian entirely in terms of a basis of the Wannier functions. Although it is possible to carry out the ensuing discussion entirely in terms of explicit many-body wave functions constructed from these single particle wave functions, it is notationally *much* simpler—and more in keeping with practice in the literature—to use a second quantized representation. One introduces creation ($c_{l\sigma}^+$) and annihilation ($c_{l\sigma}$) operators acting in occupation number (or Fock) space and satisfying the usual fermion anticommutation relations. This guarantees that the many-body state has the appropriate anti-symmetry under exchange of electrons. In this Fock space $c_{l\sigma}^+(c_{l\sigma})$ creates (annihilates) an electron in the Wannier state $\phi_l(r)$. The full π-electron Hamiltonian can then be written in the form

$$H_{\pi e} = -\sum_{\substack{lm \\ \sigma}} t_{l,m} c_{l\sigma}^+ c_{m\sigma} + \frac{1}{2} \sum_{\substack{ijkl \\ \sigma\sigma'}} V_{ij,kl} c_{i\sigma}^+ c_{j\sigma'}^+ c_{l\sigma'} c_{k\sigma}, \tag{2.7}$$

where the $t_{l,m}$ defined by

$$t_{l,m} \equiv -\int d^3 r \phi_l^*(r) \left[\frac{p^2}{2m_e} + V_P(r, \{R\}) \right] \phi_m(r) \qquad (2.8a)$$

include both the effective site energies and the transfer or resonance integrals, and

$$V_{ij,kl} \equiv \int d^3 r \int d^3 r' \phi_i^*(r) \phi_j^*(r') V_{e-e}^{\text{eff}}(r-r') \phi_k(r) \phi_l(r') \qquad (2.8b)$$

are the Coulomb and direct exchange (as well as three- and four-centred many-electron) matrix elements between π-electron states, modified (as discussed above) to incorporate the effects of the degrees of freedom not explicitly included in the model Hamiltonian.

There are several aspects of (2.7) and (2.8) that deserve comment. Firstly, the matrix elements of $H_{\pi e}$ generally involve Wannier functions localized about different centres, i.e. locations on the backbone (CH) lattice. In chemical terms, the matrix elements are often referred to as two-centre integrals if they involve two distinct sites, three-centre integrals if they involve three distinct sites, and so on. Below we will discuss various approximations which neglect the multi-centre integrals, i.e. those involving more than two different sites. Secondly, both the matrix elements $t_{m,n}$ and $V_{ij,kl}$ depend on the nuclear ion configuration $\{R_\alpha\}$; this is explicit from the dependence of $V_P(r, \{R\})$ but is also implicit in the forms of the Wannier functions, ϕ_l, and of the screening in $V_{e-e}^{\text{eff}}(r-r')$. In general, these dependences are not known, and one therefore starts from a reference ion configuration $\{R_l^0\}$, which is close to the actual configuration, and expands in powers of the (assumed small) displacements, $u_l = R_l - R_l^0$. If the (small) variations in the Wannier functions and in $V_{e-e}^{\text{eff}}(r-r')$ are neglected, then only the dependence of the pseudo-potential, $V_P(r, \{R\})$, remains. The expansion of this term for small deviations from the reference configuration gives rise to couplings between the π electrons and the ionic displacements. These dynamic displacements, when treated collectively and quantized, are known to physicists as *phonons*. Adapting this terminology, we shall refer to the couplings between electrons and ions as electron–phonon (e–p) interactions. Similarly, the interactions resulting from V_{e-e}^{eff} will be referred to as electron–electron (e–e) interactions. Thirdly, the form of the Hamiltonian in (2.7) and (2.8) describes only the π-electronic degrees of freedom dynamically. If one wishes to optimize the nuclear ionic geometry or to calculate the *dynamics* of the nuclear motion, one must add to $H_{\pi e}$ an explicit effective ion Hamiltonian, H_n; typically, this is taken to be

$$H_n \equiv \frac{1}{2M} \sum_l P_l^2 + V_n\{u\}, \qquad (2.9)$$

where P_l is the momentum operator of the *l*th ion (more precisely, the *l*th (CH) unit) and, as noted above, u_l is the (assumed small) deviation of the ions from the reference configuration. Note that the stability of the reference configuration requires that, in the absence of coupling to the π electrons, V_n has a minimum for $u_l = 0$.

Finally, since both V_P and V_{e-e}^{eff} are renormalized or effective interactions, it is not possible a priori to know the relative importance of the e–p and e–e interactions in determining the energetics and dynamics of the π electrons. This uncertainty is the origin of the ongoing debate in the literature concerning e–p versus e–e interactions in π-conjugated polymers. To highlight the central issues of this debate, we shall sketch the explicit derivations and discuss the assumptions of some of the more common π-electron-only models.

Consider the explicit R dependence of the effective one-electron term of $H_{\pi e}$, given by (2.8a). We expand the V_P potential with respect to (small) deviations of the ions from their reference configuration ($u_l \equiv R_l - R_l^0$),

$$V_P(r, \{R\}) = V_P(r, \{R^0\}) + \sum_l \frac{\partial V_P}{\partial R_l}(r, \{R^0\}) \cdot u_l + O(u_l^2).\qquad(2.10)$$

Substituting this expression into (2.8a), we see that the one-electron term in (2.7) actually contains three distinct types of contributions, namely:

$$\sum_{\substack{mn \\ \sigma}} t_{m,n} c_{m\sigma}^+ c_{n\sigma} = \sum_{\substack{m \\ \sigma}} \varepsilon_m c_{m\sigma}^+ c_{m\sigma} + \sum_{\substack{m \neq n \\ \sigma}} t_{m,n}^0 c_{m\sigma}^+ c_{n\sigma} + \sum_{\substack{m,n,l \\ \sigma}} \alpha_{m,n,l} \cdot u_l c_{m\sigma}^+ c_{n\sigma}.\qquad(2.11)$$

The new constants $\{\varepsilon, t^0, \alpha\}$ are matrix elements, i.e. integrals of products of Wannier functions and their derivatives, corresponding to

(1) the site energies

$$\varepsilon_m = \int d^3 r \phi_m^*(r) \left[\frac{p^2}{2m_e} + V_P(r, \{R^0\}) \right] \phi_m(r),\qquad(2.12a)$$

(2) the bare hopping integrals ($m \neq n$)

$$t_{m,n}^0 = \int d^3 r \phi_m^*(r) \left[\frac{p^2}{2m_e} + V_P(r, \{R^0\}) \right] \phi_n(r)\qquad(2.12b)$$

and

(3) the electron–phonon (e–p) interaction terms

$$\alpha_{m,n,l} = \int d^3 r \phi_m^*(r) \frac{\partial V_P}{\partial R_l}(r, \{R\}) \phi_n(r).\qquad(2.12c)$$

The interpretation of the first term, the site energies ε_m, is immediate. Since for trans-$(CH)_x$ all the ionic moieties are the same, and since we are not including any source of disorder, discrete translational invariance implies that $\varepsilon_m = \varepsilon_0$, independent of m. This term then simply amounts to a constant shift of the energy depending only on the total number of π electrons, i.e. a chemical potential term, and is typically ignored.

Despite the rather general forms of the remaining terms in (2.11) and of the e–e interaction term in (2.8b), one feature is clear. The extent to which the Wannier functions are localized is crucial in determining how many parameters must be kept to provide a tractable yet accurate model of π-conjugated polymers.

Normally, one assumes that the overlaps between Wannier functions separated by more than one site are very small[1] and can be neglected. This is a *crucial* assumption: altering it *in a consistent manner* would a priori require keeping longer range overlaps in all the terms described below and would then lead to considerably more complicated (and as yet unstudied) models. For future reference, we introduce the overlap integral S, [2.41,42] and note that

$$S \equiv \int d^3 r \phi_m^*(r) \phi_{m+1}(r) < 1, \tag{2.13}$$

independent of m (by discrete translation invariance along the chain).

In the case of the bare hopping integrals, within a nearest neighbour tight-binding approximation, we ignore all hopping beyond nearest neighbours, i.e. only $t_{l,l\pm1}^0$ is taken to be non-zero. In chemical terms, this corresponds to keeping the resonance integrals only between chemically bonded neighbours. Since for the present case all hoppings take place between identical units, discrete translational invariance shows again that $t_{l,l+1}^0$ is independent of l: $t_{l,l+1}^0 = -t_0$ ($t_0 > 0$, see [2.14]).

For the matrix elements (2.12c) a more detailed analysis is required. Keeping only overlaps involving at most nearest neighbours and assuming the pseudo-potential to be short-ranged, we consider only terms having $|m - n| \le 1$, $|l - m| \le 1, |l - n| \le 1$. For simplicity we choose a reference configuration with a mirror plane through the carbon sites perpendicular to the chain axis, as in undimerized trans-polyacetylene. The gradient of the pseudo-potential is then odd under reflection through this mirror plane. Taking into account also the discrete translational invariance, we are left with the following e–p couplings:

$$\alpha_{l,l,l} = 0, \tag{2.14a}$$

$$\alpha_{l,l+1,l} = \alpha_{l+1,l,l} = -\alpha_{l,l+1,l+1} = -\alpha_{l+1,l,l+1} \equiv \alpha, \tag{2.14b}$$

$$\alpha_{l,l,l+1} = -\alpha_{l,l,l-1} \equiv \beta. \tag{2.14c}$$

Thus the electron–phonon interaction becomes

$$H_{\pi e-p} = \sum_{l\sigma} \{ -\alpha \cdot (u_l - u_{l+1})(c_{l\sigma}^+ c_{l+1\sigma} + c_{l+1\sigma}^+ c_{l\sigma}) + \beta \cdot (u_{l+1} - u_{l-1}) n_{l\sigma} \}, \tag{2.15}$$

where $n_{l\sigma} \equiv c_{l\sigma}^+ c_{l\sigma}$ measures the number of electrons with spin σ at site l. The second term in (2.15) corresponds to the deformation potential interaction, familiar from tight-binding studies of polaron effects.

Turning to the electron–electron e–e interactions, if we again limit ourselves to only nearest neighbour overlaps, we see there are still a variety of terms. Firstly, for $i = k$ and $j = l$, the matrix elements can be written solely in terms of the electron densities in the various Wannier orbitals. Using standard

---

[1] Obviously, if one neglects overlaps entirely, then the problem is immediately diagonal in co-ordinate space and can be solved; but then there is no molecular orbital at all!

notation, these site-diagonal Coulomb repulsion terms are

$$U \equiv V_{ii,ii} = \int d^3r d^3r' |\phi_i(r)|^2 V_{e-e}^{eff}(r-r') |\phi_i(r')|^2 \qquad (2.16a)$$

$$V_1 \equiv V_{ii+1,ii+1} = \int d^3r d^3r' |\phi_i(r)|^2 V_{e-e}^{eff}(r-r') |\phi_{i+1}(r')|^2 \qquad (2.16b)$$

and more generally

$$V_l \equiv V_{i,i+l,i+l} = \int d^3r d^3r' |\phi_i(r)|^2 V_{e-e}^{eff}(r-r') |\phi_{i+l}(r')|^2. \qquad (2.16c)$$

Notice that the relative sizes of these terms do not depend at all on the rate at which the longer range overlaps of Wannier functions fall off, but on how rapidly the (effective) π-electron Coulomb interaction decays. This is an important and incompletely resolved question; furthermore, the relative importance of the V_l is likely to be different in gas phase and dissolved finite polyenes and in condensed phase polymers. The issue of accounting for solid-state screening effects in π-conjugated polymers remains controversial. We shall return to this and related points in our later discussion.

For general $i \neq k, j \neq l$ matrix elements *do* involve, in addition to any screening effects, the overlap of Wannier functions on different sites. Neglecting these terms *entirely* is known as the zero differential overlap (ZDO) approximation and this has been analyzed in both the chemistry [2.14, 43–46] and physics [2.47] literature. If we nonetheless argue that consistency with our earlier approximations requires us to keep at least those terms involving nearest neighbour Wannier orbitals, we see that there are two additional terms:

(1) the *density-dependent hopping* term

$$X \equiv V_{ii,i,i+1} = \int d^3r d^3r' |\phi_i(r)|^2 V_{e-e}^{eff}(r-r') \phi_i^*(r') \phi_{i+1}(r') \qquad (2.17a)$$

and (2) the *bond charge* repulsion term

$$2W \equiv V_{i,i+1,i+1,i} = \int d^3r d^3r' \phi_i^*(r) \phi_{i+1}^*(r') V_{e-e}^{eff}(r-r') \phi_{i+1}(r) \phi_i(r'). \qquad (2.17b)$$

Note that W simply represents the direct exchange between orbitals at sites i and $i+1$. Since neither X nor W can be expressed solely in terms of densities, these terms are often referred to as *off-diagonal*.

Of course, if one relaxes the assumption of nearest-neighbour overlaps, there are longer range versions of these off-diagonal terms. The relative sizes of all these terms depend on *both* the screening in V_{e-e}^{eff} and the fall-off of the Wannier orbitals. Although earlier chemical studies (see [2.14] for a review) suggested that such terms were indeed small in polyenes and thus could be neglected, recently the issue of this approximation has been re-examined in several articles [2.41, 42, 48–52]. We shall discuss specific aspects of this debate later in the article. Here we remark that the bulk of the evidence, based on general overlap considerations [2.41, 42] as given above and on (numerically) exact diagonalizations of small systems [2.52], confirms the essential validity of the ZDO for π-conjugated polymers.

Collecting all the electronic terms discussed and adding to them the purely ionic terms from (2.9), we arrive at the final form for our π-electron Hamiltonian for *trans*-polyacetylene and, *mutatis mutandis*, more general π-conjugated polymers:

$$H_\pi = \sum_l [\varepsilon_0 + \boldsymbol{\beta} \cdot (\boldsymbol{u}_{l+1} - \boldsymbol{u}_{l-1})] n_l - 2 \sum_l [t_0 + \boldsymbol{\alpha} \cdot (\boldsymbol{u}_l - \boldsymbol{u}_{l+1})] p_{l,l+1}$$

$$+ U \sum_l n_{l,\uparrow} n_{l\downarrow} + \sum_{m,l \geq 1} V_l n_m n_{m+l}$$

$$+ X \sum_l (n_l + n_{l+1}) p_{l,l+1} + W \sum_l p_{l,l+1}^2$$

$$+ \sum_l \frac{\boldsymbol{P}_l^2}{2M} + V_n\{\boldsymbol{u}\}. \tag{2.18}$$

Here, in addition to the electron density operators $n_{l\sigma} = c_{l\sigma}^+ c_{l\sigma}$, $n_l = \sum_\sigma n_{l\sigma}$, we have introduced the bond order operators

$$p_{l,l+1} = \frac{1}{2} \sum_\sigma (c_{l\sigma}^+ c_{l+1\sigma} + c_{l+1\sigma}^+ c_{l\sigma}). \tag{2.19}$$

Bilinear terms arising from the transformation of the interaction terms have been incorporated within the (irrelevant) parameter ε_0. Note also that we have reinstated the momentum operator, $\boldsymbol{P}_l$, describing the motion of the (CH) units. In general, calculations are carried out in the Born–Oppenheimer approximation, in which the electronic states are evaluated with respect to a fixed instantaneous lattice configuration. In principle, one can then sum over occupied electronic states to generate a quantum-mechanical (Schrödinger) equation for the nuclear coordinates, but in practice the $\boldsymbol{P}_l$ and $\boldsymbol{u}_l$ are usually treated as classical variables. We shall discuss exceptions to this later.

This form of the π-electron Hamiltonian, motivated on very general grounds, leaves the vital question of the relative importance of the e–p and e–e interactions unanswered because it does not explicitly predict the sizes of the parameters characterizing these interactions. It is thus natural that the two limiting cases, that is, models that neglect one interaction or the other, have been extensively studied. In the next two subsections, we introduce these models.

Electron–Phonon Interaction: the Hückel and Su-Schrieffer-Heeger Models. Historically, the work of Hückel on the molecular orbital approach to conjugated hydrocarbons provided the first independent-electron theory directly relevant to finite polyenes [2.53, 54]. The successes and limitations of this approach have been thoroughly discussed by *Salem* [2.14]. Here we merely note that the neglect of direct electron–electron interactions reduces the problem to that of $2n$ independent electrons moving in $2n$ orbitals associated with the $2n$ CH lattice sites in the reference ionic configuration, and that in the original Hückel approach only the bare resonance integrals $t_{l,l+1}^0$ were included. Despite its simplicity, the Hückel approach played a critical role in the development of the theory of finite polyenes and π-conjugated systems [2.14], with the early

work of *Pople* and *Walmsley* [2.55] containing many of the results (including
the nonlinear, possibly mobile [2.56], soliton excitations!) that have generated
much of the current interest in these systems. More recently, a similar one-
electron theory [2.57, 58] has been proposed as a model specifically for trans-
$(CH)_x$. This celebrated Su-Schrieffer-Heeger (henceforth, SSH) model has become
the theoretical *lingua franca* for interpreting experiments on conjugated polymers.
Besides omitting all two-electron terms in (2.18), the SSH model also neglects
the deformation potential coupling of the phonons to the on-site electronic
charge density. In the case of the half-filled band relevant to π-conjugated
polymers, this coupling is indeed suppressed (see [2.59]). Also, only the
projections $u_{l,x}$ of the displacement vectors $\boldsymbol{u}_l$ on the chain axis are taken into
account; this simplification is motivated by the observation that these are the
relevant variables for bond alternation [2.58]. Furthermore, the lattice potential
$V_n\{\boldsymbol{u}\}$ is modelled in terms of elastic springs between neighbouring sites. Thus
the SSH Hamiltonian is

$$H_{\text{SSH}} = -\sum_{l\sigma}[t_0 + \alpha_x(u_{l,x} - u_{l+1,x})][c_{l\sigma}^+ c_{l+1\sigma} + c_{l+1\sigma}^+ c_{l\sigma}]$$

$$+ \sum_l \frac{P_{l,x}^2}{2M} + \sum_l \frac{K_x}{2}(u_{l,x} - u_{l+1,x})^2. \tag{2.20}$$

In view of its importance and simplicity, we shall discuss this model in detail
in Sect. 2.3.

Electron–Electron Interaction: the Pariser-Parr-Pople Model. In response to
the perceived limitations of Hückel-type theories, chemists in the early 1950s
began to study π-electron models that focused on the e–e interactions in the
ZDO approximation [2.14, 43–46]. Furthermore, in a different context, physicists in the 1960s began to examine related e–e interaction models [2.47]. Retaining only resonance integrals between nearest neighbours, and fixing the zero
of energy such that the diagonal single-particle terms vanish, we obtain the
PPP Hamiltonian (see, e.g. [2.60, 61]), the electronic part of which can be
written as

$$(H_{\text{PPP}})_e = -\sum_{l\sigma} t_{l,l+1}(c_{l\sigma}^+ c_{l+1\sigma} + c_{l+1\sigma}^+ c_{l\sigma})$$

$$+ U\sum_l n_{l\uparrow}n_{l\downarrow} + \sum_{m,l \geq 1} V_l n_m n_{m+l}. \tag{2.21}$$

Although we have indicated the general form $t_{l,l+1}$ in (2.8a), PPP models
typically focus on the bare hopping term $t_{l,l+1}^0$ or on the appropriate generalization for a perfectly bond-alternating system.

Particular forms of the parameters U and V_l determine the specific variants
of the PPP models. In the case $U \neq 0, V_l = 0$, the model becomes the (one-
dimensional version of the) celebrated Hubbard Hamiltonian, which we examine
in Sect. 2.4 as a paradigm of the effects of e–e interactions. For $U \neq 0, V_1 \neq 0$,

but all other $V_l = 0$, the PPP model reduces to the extended Hubbard model. Choosing for the long-range V_l the form

$$V_l = \frac{U}{\left[1 + \left(\frac{|\mathbf{r}_i - \mathbf{r}_j|}{a_0}\right)^2\right]^{1/2}}, \quad i - j \equiv l \tag{2.22a}$$

with U and a_0 fitted to data gives the Ohno [2.62] variant of the PPP model, whereas taking

$$V_l = \frac{U}{\left[1 + \frac{|\mathbf{r}_i - \mathbf{r}_j|}{a_0}\right]}, \quad i - j \equiv l \tag{2.22b}$$

gives the Mataga–Nishimoto [2.63] variant.

Electron–Electron and Electron–Phonon Interaction: Peierls-Hubbard Hamiltonians. The SSH (or Hückel) model can be interpreted in two different ways. One can assert that the effective e–e interactions are negligible and thus that the resonance integrals are precisely given by (2.8a). Or, as in the VEH model, one can adjust the SSH parameters to reproduce approximately an electronic structure calculated ab initio (or extracted from experimental data). The first point of view is simply untenable, as the effective Coulomb repulsion is certainly *not* negligible. Moreover, as we have consistently stressed, the parameters of a model Hamiltonian cannot be viewed as truly microscopic but rather must be taken to be renormalized by the many degrees of freedom that are *not* explicitly included in the models. The second perspective is certainly more flexible and is in part supported by the successes of the related Landau-Fermi liquid theory, in which strongly interacting particles are replaced by weakly interacting quasi-particles. Nonetheless, this perspective is over-simplified and fails to capture even qualitatively some features of the real materials. Thus, for example, the observed occurrence of the 2^1A_g state of finite polyenes *below* the optical gap cannot be explained within the SSH or Hückel Hamiltonians *for any values of* the parameters. To capture this effect, one *must* go beyond any single-particle formalism to include e–e interactions. Of course, including *only* e–e interactions and ignoring totally e–p interactions would contradict the observed alternation of bond lengths in finite polyenes and polyacetylene. Thus one is essentially compelled to consider models incorporating both e–p and e–e interactions. The Peierls–Hubbard Hamiltonians, H_{PH}, are important examples of such models. In the case of *trans*-(CH)$_x$, the electronic part of the Peierls–Hubbard Hamiltonian that includes both on-site U and nearest-neighbour V Coulomb interactions is written as

$$H_{SSH} = -\sum_{l\sigma}[t_0 + \alpha_x(u_{l,x} - u_{l+1,x})][c_{l\sigma}^+ c_{l+1\sigma} + c_{l+1\sigma}^+ c_{l\sigma}]$$
$$+ U\sum_l n_{l,\uparrow}n_{l,\downarrow} + V_1\sum_l n_l n_{l+1}. \tag{2.23}$$

The lattice part is the same as in the SSH Hamiltonian (2.20). In Sect. 2.5 we shall study in detail the Hamiltonian (2.23), for we consider it to be the minimal model capable of capturing the essential features of conjugated polymers. Before confronting the full complexity of this model, however, we focus separately on the roles of e–p and e–e interactions, by examining in Sect. 2.3 the Hückel and SSH Hamiltonians and in Sect. 2.4 the Hubbard model.

2.3 The Hückel and SSH Models: Independent-Electron Theories

In this section we discuss independent-electron models in which the (explicit) Coulomb interaction is neglected. As indicated in Sect. 2.2, these models were first introduced for finite polyenes by *Hückel* [2.53, 54] and were studied extensively by *Pople* and *Walmsley* [2.55] and by *Hanna* et al. [2.64, 65]. In the context of π-conjugated polymers, they are often considered in the form of the SSH Hamiltonian. Although we cannot expect these simplified models to describe adequately the detailed electronic, structural and vibrational properties of conjugated polymers, they do provide a qualitative picture for aspects such as the lattice dimerization and inhomogeneous polaronic states. For other properties, especially those associated with excited electronic states, the predictions of Hückel theory can even be qualitatively wrong, as we will demonstrate in subsequent sections.

2.3.1 From Polyethylene to Polyacetylene

In Sect. 2.2 we have already introduced the second quantized version of the SSH Hamiltonian, which is an explicit form of the Hückel theory applicable to trans-polyacetylene. Although we could proceed directly from (2.20) to analyze the consequences of the Hückel theory approach, it is more instructive to backtrack and to introduce the Hückel model starting from a polyethylene chain, $(CH)_{2n}$, which does not have any π electrons and is held together entirely by the C–C σ bonds. We assume that the binding energy of the chain is simply given by the expression

$$E = \sum_l V(r_{l,l+1}), \tag{2.24}$$

which involves a sum over the energies of the single bonds, since $r_{l,l+1}, l = 1, \ldots, 2n$, are the C–C bond lengths between adjacent atoms. This approach ignores interactions between non-bonded atoms. We do not need to determine the function $V(r)$ but only to note that it has a minimum at the equilibrium bond length of polyethylene, r_{PE}.

We proceed now conceptually through the steps indicated by parts a–c of Fig. 2.7 Removing one hydrogen per carbon atom corresponds to creating one π electron per (CH) unit. As discussed in Sect. 2.2, we describe these π electrons in terms of atomic p_z-wavefunctions (or, more precisely, the orthogonal Wannier functions). Since Hückel theory neglects the explicit Coulomb interactions between the π electrons, we need consider only the terms corresponding to the $t_{l,m}$ in (2.7) and (2.8a). Using chemical terminology, these are "two-centre" integrals (see, e.g., [2.14]) between neighbouring sites; in our present discussion, we shall denote these integrals by $t(r_{l,l+1})$. The dependence of t on the bond length provides a coupling between the π electrons and the lattice, and leads (as we shall shortly show) to a shortening of the average bond length in polyacetylene as compared to polyethylene.

Since there are no direct e–e interactions in the Hückel theory, we can work entirely in terms of single-electron levels, with the many-electron states being described in terms of a single product of single-electron wave functions. Thus for the even polyene $C_{2n}H_{2n+2}$ the Hamiltonian is simply a $2n$ by $2n$ matrix. Furthermore, since the matrix elements of the Hamiltonian, described by the $t(r_{l,l+1})$, are non-zero only between adjacent sites, the resulting π-electron Hamiltonian corresponds to a tridiagonal matrix in the Wannier representation. Introducing the single-particle wavefunctions $\psi_{l,\nu}$ whose modulus squared describes the probability of finding an electron in the Wannier orbital localized at the lth (CH) unit for the νth eigenfunction (the νth molecular orbital in

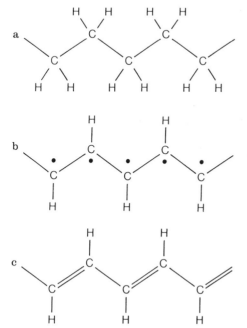

Fig. 2.7 a–c. From polyethylene to polyacetylene. **a** a schematic polyethylene chain, **b** the creation of one π-electron per site by removing one hydrogen atom per monomer, **c** the dimerized or bond-alternated *trans*-(CH)$_x$ chain. Note that in (**a**) the actual bond angles correspond to the tetrahedral carbon sp^3 structure

standard chemical terminology), we obtain the eigenvalue equation

$$\varepsilon_v \psi_{l,v} + t(r_{l,l+1})\psi_{l+1,v} + t(r_{l,l-1})\psi_{l-1,v} = 0. \tag{2.25}$$

Multiplying by $\psi_{l,v}^*$ and summing over all l we find

$$\varepsilon_v \sum_l |\psi_{l,v}|^2 = -\sum_l [t(r_{l,l+1})\psi_{l+1,v} + t(r_{l,l-1})\psi_{l-1,v}]\psi_{l,v}^*. \tag{2.26}$$

Differentiating with respect to $\psi_{l,v}^*$ and using (2.25) we observe that ε_v is stationary with respect to variations in the wavefunctions. Thus, assuming $\psi_{l,v}$ to be normalized, $\sum_l |\psi_{l,v}|^2 = 1$, and differentiating with respect to $r_{l,l+1}$, we obtain

$$\frac{\partial \varepsilon_v}{\partial r_{l,l+1}} = -\frac{dt(r_{l,l+1})}{dr_{l,l+1}}(\psi_{l,v}^*\psi_{l+1,v} + \psi_{l+1,v}^*\psi_{l,v}). \tag{2.27}$$

Approximating the elastic energy of the σ bonds again by (2.24), we write for the total energy of a polyacetylene chain with a given configuration $\{r_{l,l+1}\}$ and given occupation numbers n_v

$$E = \sum_l V(r_{l,l+1}) + \sum_v \varepsilon_v n_v. \tag{2.28}$$

The configuration of lowest energy satisfies the minimization conditions

$$0 = \frac{\partial E}{\partial r_{l,l+1}} = \frac{dV}{dr_{l,l+1}} - 2\langle p_{l,l+1}\rangle \frac{dt}{dr_{l,l+1}}, \tag{2.29}$$

where we have used (2.27) and

$$\langle p_{l,l+1}\rangle = \frac{1}{2}\sum_v (\psi_{l,v}^*\psi_{l+1,v} + \psi_{l+1,v}^*\psi_{l,v})n_v \tag{2.30}$$

is the expectation value of the bond order operator (2.19) in the product state of single-particle wavefunctions. The Schrödinger equation (2.25) and the self-consistency equation (2.29) determine the stationary states and, in particular, the ground state of the system. We consider first the simple case of equal bond lengths, $r_{l,l+1} = r$ and assume periodic boundary conditions (i.e. a ring geometry for $C_{2n}H_{2n}$ to avoid chain-end effects). Equation (2.25) is then easily solved by Fourier transformation, giving a tight-binding spectrum

$$\varepsilon_k = -2t(r)\cos k, \quad k = \frac{\pi v}{n}, \quad -n < v \leq n. \tag{2.31}$$

The ground state is constructed by filling each of the n lowest levels with two electrons (spin up and spin down). This yields a filled Fermi sea with a Fermi wavevector $k_F = \frac{1}{2}\pi$, i.e. a half-filled band. The wavefunctions are

$$\psi_{l,k} = \frac{1}{\sqrt{2n}}e^{ikl}, \tag{2.32}$$

corresponding to a *constant* bond order

$$\langle p_{l,l+1} \rangle = p_0 = \frac{1}{\pi} \int_{-\pi/2}^{\pi/2} dk \cos k = \frac{2}{\pi}. \tag{2.33}$$

From (2.29) the equilibrium bond length r_0 is determined by the solution to

$$\frac{dV}{dr} - 2p_0 \frac{dt}{dr} = 0. \tag{2.34}$$

The length r_0 can be identified with the average bond length of polyacetylene. Note that since dt/dr is less than zero (the hopping is a decreasing function of separation) and since p_0 is positive, dV/dr must be less than zero when evaluated at r_0, the solution to (2.34). Since we have assumed that $V(r)$ has a minimum at r_{PE}, this requires that r_0 be less than r_{PE}; hence, as anticipated, the average bond length in polyacetylene is less than that in PE, due to the interactions of the π electrons with the lattice.

In what follows we will be concerned with small deviations from this average bond length so that $|r - r_0| \ll r_0$. Thus we can expand both $V(r)$ and $t(r)$ around r_0 and write, using (2.34),

$$t(r) = t_0 - \alpha(r - r_0) \tag{2.35a}$$

$$V(r) = V(r_0) - 2\alpha p_0 (r - r_0) + \tfrac{1}{2} K (r - r_0)^2. \tag{2.35b}$$

This provides a very convenient parametrization for an effective independent-electron model of polyacetylene. With this form of $V(r)$, the bond length for polyethylene is predicted to be

$$r_{PE} = r_0 + 2\alpha p_0 / K, \tag{2.36}$$

which, as noted above, is greater than r_0. It is worth reiterating that the SSH model [2.57, 58] uses the projections of the atomic displacements onto the chain axis as coordinates. This implies that SSH parameters in (2.20) have to be multiplied by trigonometric factors in order to be compared to those occurring in (2.35). Assuming that the angles remain constant (otherwise the chain cannot be treated as one-dimensional) and taking for trans-polyacetylene a bond angle of $120°$, gives the relations $\alpha_x = (\sqrt{3}/2)\alpha$, $K_x = (3/4)K$ [2.2].

2.3.2 Bond Alternation

Above we have found a solution of (2.25) and (2.29) for equal bond lengths $r_{l,l+1} = r_0$. This corresponds to a state of high symmetry and amounts to assuming that all bonds are equivalent. As suggested by Fig. 2.7, for this configuration the half-filled π band represents a metallic state. Thus the shortening of the average bond length in $(CH)_x$ as compared to polyethylene simply reflects the metallic binding due to the π electrons. On the other hand, the uniform bond length state does *not* correspond to the usual Kekulé structure, involving a sequence of single and double bonds, as shown in Fig. 2.7c.

The issue of whether the bond lengths in actual finite polyenes and the hypothetical infinite polyene are alternating or not challenged quantum chemists for many years after the introduction of the Hückel model. Indeed, it took more than twenty years until a satisfactory answer was given [2.66]. In the meantime, *Peierls* [2.67] and *Fröhlich* [2.68] had found quite generally that a one-dimensional metal is unstable with respect to a $2k_F$-modulation of the lattice. This modulation opens a gap at the Fermi energy and thus leads to a semiconducting, rather than metallic, ground state. For a half-filled band, a $2k_F$ modulation corresponds to a dimerization, in agreement with the intuitive chemical picture of alternating short and long bonds. Mathematically, this phenomenon is understood by noting that a solution of (2.25) and (2.29) yields an *extremum* but not necessarily a *minimum* of the total energy. In fact, a calculation of the second derivatives of the energy with respect to the bond lengths $r_{l,l+1}$ (about the point $r_{l,l+1} = r_0$) shows that the configuration of equal bond lengths represents a saddle point, rather than a true minimum. Thus the equal bond length state does not correspond to the state of lowest energy. Furthermore, the direction of greatest downward curvature, i.e. strongest instability, has a wavevector $q = \pi$ (e.g. [2.2]), which corresponds to the $2k_F$ phonon mode describing antiphase (optical) motions of adjacent (CH) units. This instability will cause the chain to distort, so that the actual ground state has a broken symmetry, in the form of a *dimerized* or *bond-alternated* lattice where

$$r_{l,l+1} = r_0 + (-1)^l y. \tag{2.37}$$

In view of (2.35a), the resonance integrals are also alternating:

$$t(r_{l,l+1}) = t_0 - \tfrac{1}{2}(-1)^l \Delta_0, \tag{2.38}$$

where

$$\Delta_0 = 2\alpha y. \tag{2.39}$$

Hence this broken symmetry state *does* correspond to the chemical picture of alternating short and long bonds. The parameter Δ_0, which has dimensions of energy, is clearly a measure of the extent of the dimerization of the backbone (CH) lattice; in physics terminology it is known as the dimerization or bond alternation order parameter and the ground state itself is known as a $2k_F$ bond-order wave (BOW). For nonzero y (hence Δ_0) the unit cell contains now two C atoms, and the Brillouin zone is reduced to $-\tfrac{1}{2}\pi < k \leq \tfrac{1}{2}\pi$. Each k value in the reduced zone corresponds to two single-particle levels and we thus seek a solution for the electronic eigenfunction of the form

$$\psi_{l,k} = e^{ikl}(u_k + (-1)^l i v_k). \tag{2.40}$$

Inserting this ansatz into (2.25) gives

$$(\varepsilon + 2t_0 \cos k)u_k - (\Delta_0 \sin k)v_k = 0 \tag{2.41a}$$

$$(\varepsilon - 2t_0 \cos k)v_k - (\Delta_0 \sin k)u_k = 0. \tag{2.41b}$$

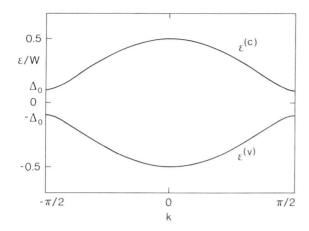

Fig. 2.8. The π-electron spectrum of the dimerized *trans*-(CH)$_x$ chain in units of the bandwidth, $W = 4t_0$. The valence ($\varepsilon^{(v)}$) and conduction ($\varepsilon^{(c)}$) bands are separated by a gap $2\Delta_0$

Solving (2.41) yields a spectrum, as shown in Fig. 2.8, which is split into a valence band $\varepsilon_k^{(v)} = -E_k$ and a conduction band $\varepsilon_k^{(c)} = +E_k$, where

$$E_k = (4t_0^2 \cos^2 k + \Delta_0^2 \sin^2 k)^{1/2}. \tag{2.42}$$

In particular, there is an energy or band gap of $2\Delta_0$ between the top of the valence band, the highest occupied molecular orbital (HOMO), and the bottom of the conduction band, the lowest unoccupied molecular orbital (LUMO). It is straightforward to calculate the bond order using (2.30) and (2.40) to (2.42) together with the normalization condition which, since the total length of the chain is $2n$, reads $2n(|u_k|^2 + |v_k|^2) = 1$. We find

$$P_{l,l+1} = \frac{1}{\pi} \int_0^{\pi/2} dk \frac{1}{E_k} [4t_0 \cos^2 k - 2(-1)^l \Delta_0 \sin^2 k]. \tag{2.43}$$

Inserting this result into the self-consistency condition (2.29) and using (2.35) we find a slightly modified relation for the mean bond length r_0 and, in addition, a relation that determines the gap parameter Δ_0:

$$1 = 4t_0\lambda \int_0^{\pi/2} dk \frac{\sin^2 k}{E_k}. \tag{2.44}$$

Here we have introduced a dimensionless electron–phonon coupling parameter λ, which is given by

$$\lambda = \frac{2\alpha^2}{\pi t_0 K}. \tag{2.45}$$

The ratio of the bandwidth $4t_0$ to the bandgap $2\Delta_0$ is a very important quantity in the theory, since within the single particle theory it plays the role of a coherence length

$$\xi = (2t_0/\Delta_0)a, \tag{2.46}$$

where a is projection of the average C–C spacing onto the chain axis. This coherence length will determine not only the range of influence of defects but also the scale of the nonlinear excitations of the theory. If $\xi \gg a$, details on the scale of the lattice spacing a are not expected to be crucial, and different polymers may exhibit common universal features.

If, consistent with our one-electron model, we interpret the actual experimental gap in polyacetylene as arising *entirely* from the electron–phonon interactions, then from the measured bandgap and bandwidth we find $\xi \sim 7a$. For $\lambda \ll 1$ (2.44) can be solved for the gap parameter in terms of the BCS-like relation

$$\Delta_0 = 8t_0 \exp\left[-\left(1 + \frac{1}{2\lambda}\right) \right]. \tag{2.47}$$

Comparing (2.47) and (2.39), we see that within the single-electron approach the bond alternation and the bandgap, which as we shall later show is precisely the gap for *optical* excitations, are *directly* related and can be described by a *single* parameter Δ_0. Of course, for given Δ_0 (and λ) the actual physical value of the dimerization (y in (2.37)) can be changed by varying α (and also K, to keep λ constant), but the physical origin of the optical gap is the dimerization in the chain. Importantly, as shown in Sects. 2.4 and 2.5, in models with *direct* e–e interactions, this direct relationship is altered, and the e–e interactions themselves also contribute to, or even dominate, the optical gap.

2.3.3 The Strength of the Electron–Phonon Coupling

Unfortunately, there is at present no general agreement on the value of λ appropriate for polyacetylene, but rather there are at least two points of view. In the first approach, λ is simply chosen to reproduce the observed value of $2\Delta_0$ so that, from (2.47), $\lambda \sim 0.2$ in polyacetylene. This amounts to assuming that the single-particle e–p description is sufficiently accurate that the correlation effects due to the neglected (e–e) interactions do not contribute at all to the gap. In the second approach, one tries to determine λ from other quantities, such as vibrational frequencies of related organic molecules or the known relationship between (average) bond order and bond length. In this second approach, it is natural to use the observed bond orders (p_i) and bond lengths (r_i) for different π-electron systems to determine λ. Equations (2.34) and (2.35) imply the following linear relationship

$$K(r_i - r_j) + 2\alpha(p_i - p_j) = 0. \tag{2.48}$$

Such a relationship is indeed suggested by studies of several π-electron systems [2.14]. Inserting the suggested values for graphite ($p = 0.515, r = 1.421$ Å) and benzene ($p = 2/3, r = 1.397$) into (2.47), we find

$$\alpha/K \sim 0.08 \text{ Å}. \tag{2.49}$$

It is interesting to note that using an average bond length $r_0 = 1.4$ Å for polyacetylene together with this value of α/K, we predict from (2.36) a bond length of 1.50 Å for polyethylene. In view of the relatively large change in bond length, this result agrees fairly well with the experimental value of 1.54 Å. In the next section we will argue that effects of electron–electron interactions are not important for the size of the *average* bond length. Therefore, the above determination of Hückel parameters will remain essentially correct even for more realistic models involving e–e interactions. On the other hand, the amplitude of *alternation* of the bond lengths can depend sensitively on electronic correlation; this will be discussed in detail in Sect. 2.5.

To determine λ, we need an additional equation beyond (2.49). We obtain this by noting that the σ-bond force constants for polyacetylene and benzene should be essentially the same. From Raman data on benzene, one knows $K \sim 47.5\,\text{eV}\,\text{Å}^{-2}$ [2.2, 14, 69]; recently, a direct analysis of Raman data for *trans*-$(CH)_x$ has given $K = 46\,\text{eV}\,\text{Å}^{-2}$ [2.70], confirming the theoretical expectation. Thus, using $t_0 \sim 2.5\,\text{eV}$ for the π bandwidth, we obtain a dimensionless coupling constant $\lambda \sim 0.08$, which is considerably less than that required within the single particle theory to reproduce the observed gap in polyacetylene which, as we shall later see, corresponds to the energy difference between the ground state and the lowest optically allowed excited state. This clear disagreement is, from our perspective, a first indication of the failure of the singleparticle model to account correctly for the excited states in polyacetylene. We shall amplify this in later sections. Here we analyze the single-particle model further and show that, despite its limitations, it remains a useful framework for discussing many features of conjugated polymers, including the inhomogeneous structures that result from the coupling between electronic and lattice degrees of freedom.

2.3.4 Stability of the Dimerized State and the Phonon Spectrum

Although we indicated that the uniform ground state was unstable to dimerization and further showed that the dimerized state is a solution of the basic equations (2.25) and (2.29), we have not yet verified that this state corresponds to a true energy *minimum*, that is, that the dimerized state itself has *no* unstable modes. Both analytical calculations in the continuum limit [2.2, 71] and numerical computations for the discrete case [2.72] confirm that the dynamical matrix for fluctuations about the dimerized state is positive definite and therefore that this state is (at least locally) stable. The vibrational spectrum shown in Fig. 2.9 consists of an acoustic branch which is only weakly renormalized by the electron–phonon coupling and an optical branch which is strongly renormalized, in particular for small wavevectors q. The renormalization factor for the optical mode at $q = 0$ can be easily obtained from the second derivative of the total energy with respect to the dimerization parameter y. For weak electron–phonon coupling ($\lambda \ll 1$) one finds for the SSH model (see, e.g. [2.73]).

$$(\Omega/\omega_0)^2 = 2\lambda \tag{2.50}$$

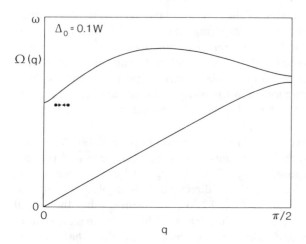

Fig. 2.9. The phonon spectrum of the dimerized *trans*-(CH)$_x$, chain, after [2.72]. Note the strong Kohn anomaly in the optical branch at $q = 0$ (reduced zone scheme)

where ω_0 is the bare phonon frequency at $q = \pi$ in the absence of dimerization and for $\lambda = 0$. From (2.50) we see that for weak e–p coupling, the phonon frequency Ω is considerably reduced from its value for $\lambda = 0$; this is the familiar Kohn anomaly [2.74]. Note that for $\lambda \to 0^+$, $\Omega \to 0$ and the mode becomes soft; this merely reflects the incipient transition to the dimerized system, which by Peierls' theorem occurs for *any* (non-zero) λ.

Of course, a model taking into account only a single lattice degree of freedom per CH group (instead of the actual 6) is certainly much too simple to describe the real lattice dynamics of polyacetylene. On the other hand we are interested in those modes which are most strongly coupled to π electrons. More complete treatments show that the coupling to the bond stretching vibrations is indeed the dominant mode [2.75] and thus the present simple analysis remains qualitatively correct.

The calculation of the phonon spectrum in effect provides a linear stability analysis for fluctuations around the ground state and, since all the eigenvalues are positive, shows that the dimerized state is a true (local) minimum. A rigorous proof that it is the actual ground state, and that states with different lattice configurations have higher energy has been given only recently [2.76].

2.3.5 Spatially Localized Nonlinear Excitations: Solitons, Polarons and Bipolarons

Consider a homogeneously dimerized chain characterized by the order parameter $\Delta = 2\alpha y$. The total energy as a function of Δ is illustrated in Fig. 2.10. It has two minima, $\Delta = \pm \Delta_0$, where Δ_0 is given by (2.47). This double degeneracy of the ground state immediately suggests the existence of excited states that contain domains—regions where Δ has a fixed sign— and domain walls—

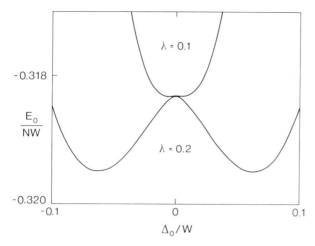

Fig. 2.10. The total energy (per site, in units of the bandwidth, W) as a function of dimerization for two different values of the electron–phonon coupling, λ; note that the undimerized state is an unstable equilibrium

regions over which Δ changes sign rapidly. The existence of these domain walls, in the form of well-localized bond alternation defects (Fig. 2.11), was first proposed in the chemical literature many years ago (see, e.g. [2.14]). Indeed, the possibility of *mobile* spin packets in finite polyenes (in modern parlance, neutral solitons) was explicitly discussed in an early study [2.64, 65], as was the possibility that charge could be transported along the molecule by the motion of the bond-alternation defects and the attendant localized, non-bonding electronic states (in short, mobile *charged* solitons) at about the same time [2.55]. Furthermore, there was an early attempt [2.56] to establish experimentally the presence of the mobile spin defects, by using ESR techniques. As argued in the original articles, the transition from one domain to another in a finite polyene generates a non-bonding orbital, corresponding to a localized defect, and this orbital may be singly occupied, doubly occupied or unoccupied. A neutral defect has a singly occupied non-bonding orbital and therefore carries spin. The two other states are positively or negatively charged and, since the non-bonding orbital is empty or doubly occupied, spinless. In a solid-state language, the non-bonding orbital corresponds to an energy level exactly in the middle of the gap separating the

a

b

Fig. 2.11 a, b. A schematic representation of **a** the neutral bond alternation defect (the neutral soliton, S^0) and **b** the positively charged soliton, S^+, in *trans*-$(CH)_x$

filled valence band and the empty conduction band and having an eigenfunction localized around the bond alternation defect. In the context of conducting polymers these local states were first discussed in terms of *phase kinks* by *Rice* [2.77] and in terms of *solitons* by *Su* et al. [2.57], although an earlier paper by *Brazovskii* [2.78] had already developed a detailed theory of these self-localized excitations applicable to the general Peierls system in one dimension. As we shall now demonstrate, the SSH model provides an essential quantification and embodiment of these ideas.

A general inhomogeneous structure is described by a site-dependent order parameter

$$\Delta_l = 2(-1)^l \alpha (r_{l,l+1} - r_0). \tag{2.51}$$

The homogeneous ground state corresponds to either $\Delta_l = +\Delta_0$ or $\Delta_l = -\Delta_0$. Such a state is 'topologically' incompatible with an odd-numbered ring where Δ_l has to change sign somewhere. The same is true for an infinitely long chain with boundary conditions $\Delta_l \to +\Delta_0$ for $l \to +\infty$ and $\Delta_l \to -\Delta_0$ for $l \to -\infty$. A 'kink' must occur somewhere, to interpolate between the two asymptotic values. This kink solution represents a 'topological' soliton. Mathematically, this means that it belongs to a particular sector of solution space: a finite ring is either even- or odd-membered, whereas an infinite chain in a state with finite energy reaches either the same or different asymptotic limits for $l \to \pm\infty$. Solutions belonging to different sectors cannot be deformed into each other and therefore the kink solutions are stable: it is impossible to produce a *single* kink from the ground state of an even-numbered ring (or of an infinite chain with the same boundary conditions at $x \to \pm\infty$). However, single 'kinks' do exist as the ground state of odd-membered rings (or of an infinite chain with different asymptotic limits). Furthermore, kinks *can* be produced from the uniform ground state in 'kink/anti-kink' pairs, since these configurations interpolate between the *same* ground state as $l \to \pm\infty$.

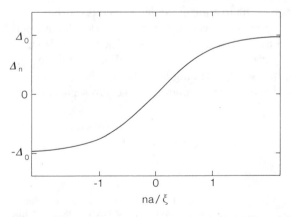

Fig. 2.12. The spatially inhomogeneous bond alternation order parameter for a kink soliton. Note the expected large spatial extent, in contrast to the schematic illustration in Fig. 2.11

The explicit form of the kink can be calculated from (2.25) and (2.29). Instead of reproducing this fairly technical analysis here, we take advantage of our earlier remark that the extent of the inhomogeneity is expected to be of the order of the coherence length. Therefore, we make the ansatz [2.58]

$$\Delta_l = \Delta_0 \tanh(la/\xi) \tag{2.52}$$

which is illustrated in Fig. 2.12. Consider a finite odd-numbered chain with open ends. There exists a zero-energy solution of (2.25) with

$$\psi_{2m+1} = 0 \tag{2.53a}$$

$$\psi_{2m} = (-1)^m \psi_0 \prod_{j=1}^{m} \frac{t(r_{2j-2,2j-1})}{t(r_{2j-1,2j})}. \tag{2.53b}$$

Using the correct generalization of (2.38)

$$t_{l,l+1} = t_0 - \tfrac{1}{2}(-1)^l \Delta_0 \tanh(la/\xi), \tag{2.53c}$$

we find that in the limit $\xi \gg a$ (2.53b) becomes

$$\psi_{2m} \sim (-1)^m \psi_0 \operatorname{sech} \frac{2ma}{\xi}. \tag{2.54}$$

For the case of a neutral kink, this form of ψ_m leads to a spin density which vanishes on odd sites and behaves as shown in Fig. 2.13 on even sites. The charge density is constant for a neutral kink.

These results can readily be understood as simple consequences of electron-hole symmetry. To demonstrate this, we first note that the replacement $\psi_l \rightarrow (-1)^l \psi_l$ in (2.25) transforms any solution with energy ε into a solution with energy $-\varepsilon$ (electron-hole symmetry). We introduce the local density of states

$$\rho_{ll}(\varepsilon) = \sum_v |\psi_{l,v}|^2 \delta(\varepsilon - \varepsilon_v), \tag{2.55}$$

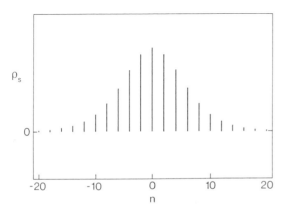

Fig. 2.13. The spin (charge) density distribution in the SSH model for a neutral (charged) soliton. Note that it is exactly zero on alternate sites in this model

which is normalized (in view of the completeness of wavefunctions) as

$$\int_{-\infty}^{\infty} d\varepsilon \rho_{ll}(\varepsilon) = 1. \tag{2.56}$$

For the case of a kink, which produces a midgap state $\psi_{l,0}$ at $\varepsilon = 0$ in addition to valence and conduction band states for $|\varepsilon| > \Delta_0$, the electron-hole symmetry implies

$$\int_{-\infty}^{-\Delta_0} d\varepsilon \rho_{ll}(\varepsilon) + \tfrac{1}{2}|\psi_{l,0}|^2 = \tfrac{1}{2}. \tag{2.57}$$

The electron (charge) density is given by

$$\rho_l = 2 \int_{-\infty}^{-\Delta_0} d\varepsilon \rho_{ll}(\varepsilon) + n_0 |\psi_{l0}|^2 \tag{2.58}$$

$$= \begin{cases} 1 & \text{for a neutral kink} \\ 1 \pm |\psi_{l,0}|^2 & \text{for a charged kink,} \end{cases} \tag{2.58b}$$

where n_0 represents the occupancy of the midgap state. Thus for a charged kink it is the charge density which shows the peculiar alternating pattern of Fig. 2.13, whereas the spin density vanishes everywhere. Much of the excitement surrounding these localized nonlinear excitations originates from their unusual spin and charge quantum numbers which are connected with the topological properties discussed above and are summarized in Table 2.1; for clarity, we indicate both the physicists' and chemists' (in brackets) names for these excitations.

In the above discussion we did not attempt to solve the two basic equations (2.25) and (2.29) *self-consistently*. In general, for the discrete problem, this requires extensive numerical computations. Only a particular 'integrable' variant of the discrete model has been analytically solved so far, using a sophisticated '*inverse scattering transform*' approach [2.79, 80]. However, as indicated above, when the coherence length is large, the relevant fluctuations have long wavelengths

Table 2.1. Quantum numbers for nonlinear excitations

State	Spin	Charge
Charged kink soliton (Mobile charged packet)	0	$\pm e$
Neutral kink soliton (Mobile spin packet)	1/2	0
Polaron (Radical cation/anion)	1/2	$\pm e$
Bipolaron (Dication/Dianion)	0	$\pm 2e$
Singlet exciton	0	0
Triplet exciton	1	0

and the discreteness of the lattice is not important. The continuum limit, in which the difference equations become differential equations, is then appropriate. In this limit, one can obtain analytic solutions for kinks and other local structures [2.81–86].

It is worthwhile to indicate how the continuum limit can be derived from (2.25) and (2.29). The order parameter Δ_l already incorporates the rapid oscillations corresponding to bond alternation, that is, for perfect bond alternation, $\Delta = \Delta_0$, a constant. Thus for long-wavelength fluctuations Δ_l is expected to vary only slowly from site to site. Furthermore, provided that the coupling is weak, electronic states of low energy (close to $\pm k_F = \pm \pi/2$) will be of primary importance. Therefore we generalize (2.40) to the *inhomogeneous* case and make the ansatz

$$\psi_{lv} = i^l [u_v(l) + (-1)^l iv_v(l)] \tag{2.59}$$

where $u_v(k)$ and $v_v(l)$ can be assumed to vary only slowly from site to site. Inserting this form into (2.25) we obtain an expression which contains both a slowly varying and a rapidly oscillating part. Setting these two parts independently equal to zero yields two equations connecting the two functions $u_v(l)$ and $v_v(l)$. We then replace the discrete coordinates la by the continuous variable x through the transformations

$$\Delta_l \to \Delta(x) \tag{2.60a}$$

$$u_v(l) \to \sqrt{a} u_v(x) \tag{2.60b}$$

$$v_l(l) \to \sqrt{a} v_v(x). \tag{2.60c}$$

Expanding

$$u_v(l \pm 1) \to \sqrt{a}\left[u_v(x) \pm a \frac{du_v}{dx} \right] \tag{2.61}$$

we finally obtain the continuum Eqs. [2.78, 81, 87]

$$\varepsilon_v u_v = -i\hbar v_F \frac{du_v}{dx} + \Delta v_v \tag{2.62a}$$

$$\varepsilon_v v_v = i\hbar v_F \frac{dv_v}{dx} + \Delta u_v \tag{2.62b}$$

where $\hbar v_F = 2t_0 a$. The normalization condition becomes

$$\int dx \left[|u_v(x)|^2 + |v_v(x)|^2 \right] = 1. \tag{2.63}$$

Equations (2.62) are often referred to as the Bogoliubov-de Gennes equations, since these authors first derived them in treating the theory of inhomogeneous super-conductors [2.88–90]. These equations represent the continuum limit of the original Schrödinger equation (2.25) and can be viewed as the leading terms in a formal power series expansion in a, the lattice spacing. The self-consistency equation (2.29) similarly leads to a pair of equations, the first connected with slowly varying acoustic-type displacements, the second associated with rapidly oscillating optical-type displacements. Since the optical modes are strongly coupled to π-electrons and the acoustic modes are not, one usually does not consider the acoustic deformations. Indeed, in the formal continuum limit these modes do not couple to leading order in a. Thus we are left with a single self-consistency condition, which takes the form

$$\Delta = -\pi\lambda\hbar v_F \sum_v n_v(u_v^* v_v + v_v^* u_v). \tag{2.64}$$

The derivation of the various localized, nonlinear solutions in the continuum model has been presented in detail several times [2.2, 6]. Since certain technical details of the continuum limit, such as the introduction of a renormalization cut-off procedure, can cause confusion among readers not familiar with field theoretical techniques, it is useful to note that a *semi*-continuum approximation exists [2.91] which retains the finite band width but nonetheless permits analytic solutions. Furthermore, as indicated above, for a modified SSH-like model, one can obtain analytic solutions in the fully discrete case, although the analysis is somewhat challenging [2.79, 80]. In the continuum limit, Eqs. (2.62–2.64), there is an exact solution for the inhomogeneous order parameter corresponding to the kink soliton; this solution has the form

$$\Delta_K(x) = \Delta_0 \tanh(x/\xi), \tag{2.65}$$

with ξ given by (2.46). This is precisely the form of the ansatz used in (2.52), with x replacing la. The electronic states corresponding to this form of $\Delta_K(x)$ include, in addition to the valence band for $\varepsilon < -\Delta_0$ and the conduction band for $\varepsilon > \Delta_0$, a level at *midgap* ($\varepsilon = 0$). This level may be singly occupied (neutral soliton, S^0), or empty (positively charged soliton, S^+), or doubly occupied (negatively charged soliton, S^-) as illustrated in Fig. 2.14. The spatial extent of the soliton is always given by the coherence length ξ, independent of the charge of the soliton. This is an artifact of the independent-electron description, since electron–electron interactions are expected to increase the size of $S_\pm$ relative to that of S^0; this point will be discussed later.

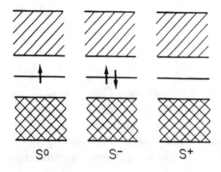

Fig. 2.14. The electronic structure and localized level occupations of the neutral and charged solitons

S^0 S^- S^+

It is important to recognize that the solitons are nonlinear *collective* excitations, involving both the distortion of the bond alternation and the accompanying electronic states. If the lattice is described in terms of classical coordinates, we can define both the position and velocity of the soliton. This will, of course, no longer be possible if both the electron and the lattice are treated quantum mechanically. In this case, the usual representation is in terms of eigenstates of crystal momentum which are (like Bloch orbitals) extended over the whole crystal and lead to a soliton band. We will return to the question of quantum phonon effects later. For the present, we continue to describe the atomic displacements in terms of classical coordinates.

In addition to the rather exotic kink soliton, chains of *trans*-$(CH)_x$ are predicted to support a more conventional localized collective excitation: the *polaron*. It is well-established, both experimentally and theoretically, that polarons can arise when the insertion of an electron into a polarizable semiconductor or insulator can produce a localized distortion of the lattice surrounding the added charge. Qualitatively, the existence of a polaron in *trans*-$(CH)_x$ can be understood in terms of the usual energy balance arguments. Consider adding a single electron to a perfectly dimerized $(CH)_x$ chain. In the absence of lattice relaxation, the lowest empty level would be the electronic state at the bottom of the conduction band, and the energy of the added electron would then be Δ_0. However, through the e–p coupling, the lattice can relax *locally* around the added electron, creating an attractive potential well in which the electron is trapped. Actually, in view of the self-consistency Eq. (2.64), the energies of all the electronic states (both valence and conduction bands) are somewhat changed (see, e.g. [2.84]), but in the end the conclusion that the single added electron can lower its energy by forming a collective localized polaron state remains valid.[2] In the continuum limit, the analytic form for the inhomogeneous order parameter corresponding to a polaron is

$$\Delta_P(x) = \Delta_0[1 - \kappa\xi(\tanh\kappa(x + x_0) - \tanh\kappa(x - x_0))], \tag{2.66a}$$

where x_0 is determined by

$$\tanh 2\kappa x_0 = \kappa\xi. \tag{2.66b}$$

From this analytic form of $\Delta_P(x)$ we see that as κ varies, the shape of the polaron goes from a shallow well for $\kappa \cong 0$ to a widely separated kink anti-kink pair as $\kappa\xi \to 1$. As in the case of the kink soliton, the polaron is a collective excitation consisting of *both* the lattice deformation in (2.66a) and the associated electronic states. For $\Delta_P(x)$, these electronic states include the conduction and valence bands (both modified from their ground-state forms by phase shifts due to the polaron potential, $\Delta_P(x)$), and *two localized states*, with energies symmetrically placed at $\varepsilon_\pm = \pm\omega_0$, where $\omega_0 = \Delta_0\sqrt{1 - (\kappa\xi)^2}$. The actual value of κ for the

---

[2] As discussed above in the case of the soliton, when quantum phonon effects are included, a polaron band will be formed

equilibrium polaron is determined by the self-consistency equation (2.64). For *trans*-(CH)$_x$ one finds $\kappa = \Delta_0/\sqrt{2}$, so that $\omega_0 = \Delta_0/\sqrt{2}$ also. The explicit forms of the corresponding wave functions are in the literature [2.84]. The relation of this two-band polaron to the more conventional single-electron polaron has also been explored [2.92]. The results for the polarons in *trans*-(CH)$_x$ are summarized in Fig. 2.15, which shows that for the negatively charged polaron, the added electron occupies a level below the conduction band edge. Notice also that the spin-charge relationships of the polaron are conventional (Table 2.2).

The energy scale for both kinks and polarons is set by Δ_0, as shown in Table 2.2. Since this is of the order of $1\,\text{eV} \sim 10000\,\text{K}$, the thermal population of these states is negligible. Thus, large external perturbations such as photons in the visible range of the spectrum, strong donor or acceptor molecules or morphological defects (chain ends, cross links, etc.) are likely to be required for producing kinks or polarons in these systems.

Up to now we have considered the degenerate ground state, as in the case of *trans*-polyacetylene. Let us now generalize the model to include the non-degenerate case as, e.g. *cis*-polyacetylene (Fig. 2.2). The stereochemical arrangement of the hydrogen atoms in *cis*-(CH)$_x$ implies that even in the absence of dimerization caused by the electron–phonon coupling two adjacent bonds are a priori inequivalent. We can model this effect in terms of *inequivalent* resonance

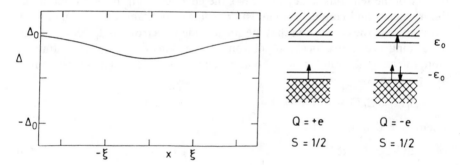

Fig. 2.15. The polaron order parameter in the continuum limit; the characteristic length scale ξ is shown. The intra-gap levels and their occupancies for the electron polaron ($Q = -e$) and hole polaron ($Q = +e$) are shown on the right-hand side of the figure

Table 2.2. Characteristic sizes and energies for the various inhomogeneous states

	Extent	Energy	Gap Levels
Kink soliton ($\gamma = 0$)	ξ	$(2/\pi)\Delta_0$	0
Polaron ($\gamma = 0$)	$\sqrt{2}\xi$	$(2\sqrt{2}/\pi)\Delta_0$	$\pm 2^{-1/2}\Delta_0$
Bipolaron or exciton ($\gamma \ll 1$)	$(\xi/2)\ln(4/\gamma)$	$(4/\pi)\Delta_0[1 + (\gamma/2)\ln(4/\gamma)]$	$\pm \gamma^{1/2}\Delta_0$

integrals $t_0 \pm t_1$ on the two adjacent bonds [2.1, 82, 85]. Thus (2.38) is replaced by

$$t_{l,l+1} = t_0 - (-1)^l(t_1 + \alpha y_l).$$ (2.67)

For a homogeneously dimerized ground state the gap parameter is given by

$$\Delta_0 = \Delta_i + \Delta_e$$ (2.68)

where $\Delta_i = 2\alpha y$ is connected with electron–phonon coupling and dimerization and is thus *intrinsic*, whereas $\Delta_e = 2t_1$ is of *extrinsic* origin. Since the electronic contribution to the energy depends on Δ_0 and the elastic energy on Δ_i, the gap equation (2.44) is modified. One can simply replace λ by $\lambda\Delta_0/\Delta_i$. It follows that the gap equation (2.47) becomes

$$\Delta_0 = 8t_0 e^\gamma \exp\left[-\left(1 + \frac{1}{2\lambda}\right)\right],$$ (2.69)

where $\gamma = \Delta_e/(2\lambda\Delta_0)$ is known, for reasons that will become clear shortly, as the *confinement parameter*.

For small γ, the total energy still has two minima as a function of the dimerization parameter; however they are no longer degenerate. Empirically, the structure of Fig. 2.2a is preferred. Since the two bond alternation sequences in *cis*-(CH)$_x$ are *not* energetically equivalent, there can be no isolated kink soliton solutions in this material. This follows from elementary energetic arguments: namely, the existence of an isolated kink soliton would require an infinitely long segment of the *trans*-cisoid configuration (shown in Fig. 2.2b) and would thus have *infinite* energy relative to the ground state. However a bound kink/anti-kink pair would not have this barrier, since the entire region outside the pair would be in the true *cis-transoid* ground state; this is illustrated in Fig. 2.16. The structure of these confined kink/anti-kink pairs can again be derived analytically on the basis of the continuum equations, with Δ replaced by Δ_i in (2.64). Interestingly, one finds *exactly* the same analytic form as given in (2.66) but with a different value of κ determined by the modified self-consistency equation, which now depends on the confinement parameter γ. The explicit form of this equation is in the literature [2.6]. In Fig. 2.17 we plot the inhomogeneous order parameter corresponding to a confined kink/anti-kink pair and the attendant electronic levels for $\gamma = 0.2$. The first two electronic energy level diagrams illustrate the occupation of the intragap levels for doubly charged excitations; since they have the same topological structure as the polaron but are doubly

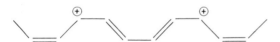

Fig. 2.16. A schematic representation of a confined doubly charged kink/anti-kink pair (a bipolaron) in *cis*-(CH)$_x$

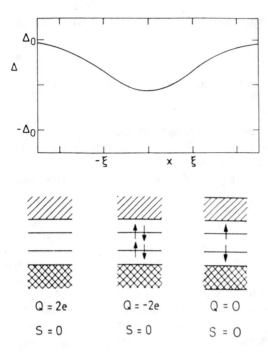

Q = 2e   Q = -2e   Q = 0

S = 0   S = 0   S = 0

Fig. 2.17. Bipolarons in the continuum limit; the sketch corresponds to $\gamma = 0.2$

charged, these excitations are commonly referred to as *bipolarons*. For inter-mediate γ, the bipolaron shape is similar to that of a polaron whereas for very small γ it resembles a loosely bound, and hence widely separated, kink/anti-kink pair. In the latter case, the two electronic levels in the gap can be viewed as the two solitonic midgap levels which are split because of the (small) overlap of the localized wavefunctions around the two distinct solitons. In the limit $\gamma \to 0$, the bipolaron unbinds to form an independent kink and anti-kink. In addition to the charged bipolarons, there exist neutral confined kink/anti-kink structures, as shown in the rightmost inset of Fig. 2.17. Electronically, these correspond to electron–hole pairs bound by the polarization field of the lattice. Therefore they are most appropriately termed *excitons*, even though their binding mechanism is not Coulombic but rather arises from the e–p coupling. Of course, these excitons are unstable against recombination. The quantum numbers for bipolarons and excitons are given in Table 2.1, whereas the charac-teristic length scales and energies are shown in Table 2.2. As expected, in the limit $\gamma \ll 1$, the bipolaron length scale becomes large, corresponding to a widely separated kink/anti-kink pair, and the energy approaches (from below) that of two free kink solitons. One can proceed further in the continuum model of *cis*-$(CH)_x$, analyzing for example the interactions of polarons and bipolarons by studying solutions corresponding to multiple excitations [2.93]; interested readers are referred to the literature.

2.3.6 Predictions of the Model

Although we will discuss experimental results in Sect. 2.6, here we mention, and give selected references to, several of the predictions of the independent-electron models. For the homogeneous ground state, the most immediate predictions are for the dimerization and the band gap and, in particular, for the close relation between them [2.57, 58]. Furthermore, the optical phonon frequency at $q = 0$, (2.50), and the dispersion around $q = 0$ [2.75] are predicted to show strong renormalization effects due to the e–p coupling; these are particularly striking examples of the well-known Kohn anomaly. These predictions can be tested by inelastic neutron scattering experiments, although fairly well-oriented crystalline samples will be necessary to obtain useful results.

Apart from the structure and excitations of the homogeneous state, both the creation and characterization of the localized inhomogeneous states are of particular interest. Numerical simulations [2.94] have shown that kink solitons can be efficiently generated by optical excitation of electron-hole pairs which subsequently relax into kink/anti-kink pairs. In the idealized models, these are free to separate on a *trans*-(CH)$_x$ chain but will be confined for the *cis* isomer.

A natural way to probe the inhomogeneous states is to search for levels in the gap. Many articles have dealt with the changes in $\alpha(\omega)$, the optical absorption as a function of photon frequency, due to the presence of kinks, polarons or bipolarons. In Fig. 2.18 we show the soliton and polaron/bipolaron absorptions within one-electron theory. For kink solitons (see, e.g. [2.95–99]), associated with the single localized electronic level, at midgap there is a strong intra-gap absorption having, in the strictly one-dimensional continuum model, a singularity proportional to $(\omega - \Delta_0)^{-1/2}$ at the absorption edge $\omega = \Delta_0$. By the optical sum rule, the strength in this absorption is removed from the *inter*-band transition, leading to bleaching above $\omega = 2\Delta_0$. Notice that no distinction is made between charged and neutral solitons, since both are expected to absorb precisely at

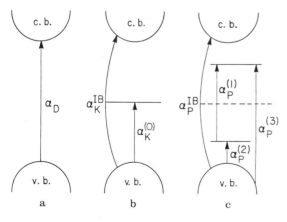

Fig. 2.18 a–c. Subgap excitations within single-particle theory due to (b) a soliton, and (c) a polaron. Part (a) shows the intergap transition of the perfect dimer (after [2.99])

midgap $\sim \Delta_0$. For polarons (see Fig. 2.18 and [2.99]) the *two* localized electronic levels (at $\pm \omega_0$) lead to *three* distinct *intra*-gap absorptions, for which analytic expressions are available [2.99] in the continuum limit: (1) α_1, a δ-function transition at $\omega = 2\omega_0$, (2) α_2, a (strong) low energy feature exhibiting a $(\omega - (\Delta_0 - \omega_0))^{-1/2}$ singularity at the absorption edge $\omega = \Delta_0 - \omega_0$ and (3) α_3, a (weak) high energy feature which vanishes as $(\omega - (\Delta_0 + \omega_0))^{-1/2}$ at the absorption edge $\omega = \Delta_0 + \omega_0$. The relative integrated intensities, as functions of ω_0, the location of the localized electronic levels, are shown in Fig. 2.19. For charged bipolarons [2.99], there are also *two* localized electronic levels (at $\pm \omega_0'$, with $\omega_0' < \omega_0$ for the same bare parameters in the Hamiltonian) but the occupancy of these levels (fully occupied for BP^{--}, empty for BP^{++}) precludes the δ-function transition at $\omega = 2\omega_0'$. Thus the two intra-gap transitions for the bipolaron are the (strong) transition α_2 at $\omega = (\Delta_0 - \omega_0')$ and the (weak) transition α_3 at $\omega = \Delta_0 + \omega_0'$. Within the SSH model, these transitions are symmetrically placed around midgap, so that they satisfy an energy relation that $\hbar\omega_2 + \hbar\omega_3 = 2\Delta_0$. Furthermore, as with the polarons, the ratio of the integrated intensities of these two transitions depends solely on the location of the intra-gap levels, $\pm \omega_0'$, which is in turn a function of the confinement parameter, γ, and the (Peierls) gap, $2\Delta_0$. This is indicated in Fig. 2.19, which illustrates that for most values of ω_0, I_{α_2} is much larger than I_{α_3}. This (typically) large intensity ratio is a consequence of the charge-conjugation [2.98, 100, 101] and mirror-plane symmetries of the SSH theory [2.102].

Much recent work has been devoted to the local vibrational modes around kinks or polarons [2.103–109], especially those which can be detected in infra red absorption. Detailed calculations for the simple SSH model (with one lattice

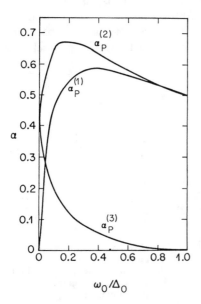

Fig. 2.19. The integrated intensities (arbitrary units) of the transitions involving localized levels for the polaron as a function of the location of the intra-gap electronic level, ω_0. The notation for the transitions is as in Fig. 2.18. Note the order-of-magnitude difference between $\alpha_P^{(2)}$ and $\alpha_P^{(3)}$ for $\omega_P \sim 0.2\Delta_0$ (after [2.99])

degree of freedom) show that, besides the translational mode, an additional local IR-active mode is produced around a soliton [2.105, 110–112]. Even more modes appear around a polaron [2.107, 109]. For the generalized SSH model (several lattice degrees of freedom) three local modes are induced by a soliton for each degree of freedom and correspondingly more around a polaron [2.113–115]. These additional non-universal features may be useful for determining the topology of a defect.

An important and difficult question concerns the possible contributions of solitons (or polarons) to electrical transport and spin diffusion. In the continuum limit these states are translationally invariant and can therefore readily carry charge or spin along a chain. Lattice discreteness gives rise to a periodic potential which produces a Peierls–Nabarro barrier opposing the motion of solitons and polarons, but theoretical estimates suggest that the resulting activation energy is small (of the order of 30 K [2.57, 58]). Detailed calculations for the temperature-dependent soliton diffusion on a single ideal chain have been carried out by several groups [2.116–119].

In summary, Hückel theory as exemplified by the SSH model yields a remarkably rich and intuitively appealing picture of conjugated polymers, and thus forms a valuable framework in which to consider experiments. However, as we shall see, there are many respects in which this simple picture does not reflect the experimental results. A variety of optical and magnetic experiments cannot be understood without including explicit Coulomb interactions. Effects of doping, disorder or interchain coupling can have a major impact on a collective soliton or polaron transport. We shall return to these issues in Sects. 2.6 and 2.7.

2.4 Hubbard Model: A Paradigm for Correlated Electron Theories

In contrast to the relative simplicity of the independent electron theories that incorporate only electron–phonon (e–p) interactions, models that include electron–electron (e–e) interactions representing both short- and long-range Coulomb forces are in general hard to treat adequately. This is particularly true in quasi-one-dimensional systems, where strong fluctuations often cause mean-field methods to fail. If one cannot use any independent-electron approximation, then the full complexity of the many-body problem must be faced; broadly speaking, even considering only the π electrons, for the finite (cyclic) polyene $C_{2n}H_{2n}$ system, one must solve a 4^{2n} by 4^{2n} Hamiltonian matrix. Fortunately, there exists a simplified model, suggested originally by *Hubbard* [2.47], for which many properties are known exactly and which nevertheless captures some of the most important many-body effects of e–e interactions. By studying this model, one can learn a great deal about the striking ways in which strongly correlated electron systems can differ from the independent electron case.

The Hubbard Hamiltonian is

$$H = -t_0 \sum_{l\sigma}(c_{l\sigma}^+ c_{l+1\sigma} + c_{l+1\sigma}^+ c_{l\sigma}) + U\sum_l n_{l\uparrow}n_{l\downarrow}, \tag{2.70}$$

where the operators $c_{l\sigma}^+(c_{l\sigma})$ create (annihilate) an electron with spin σ at site l and as usual $n_{l\sigma} = c_{l\sigma}^+ c_{l\sigma}$. Equation (2.70) is the one-dimensional version of the Hubbard model, which in higher dimensions plays an important role in models of metal-insulator transitions, itinerant magnetism and, recently, high-T_c superconductivity. As suggested by the discussion of Sect. 2.2, the crucial assumption required to derive (2.70) is that the Coulomb interaction is so efficiently screened between neighbouring sites that only an on-site term of strength U survives. This assumption is certainly reasonable in the case of transition metal oxides but is perhaps less justified in the present context. Nevertheless, even for conjugated polymers the on-site Coulomb repulsion is expected to be by far the largest e–e interaction term and thus deserves particular attention.

2.4.1 Ground State and Excitation Spectrum

Among the many eigenstates of the Hubbard model, the ground state and low-lying excited states are of particular interest. We consider first the non-interacting limit ($U = 0$) where the ground state corresponds to a filled Fermi sea, as discussed in Sect. 2.3. Neutral excitations involving the charge degrees of freedom are generated by creating a hole at $k, |k| < k_F$, and an electron at $k + q, |k + q| > k_F$. It is easy to convince oneself that for a half-filled band the excitation energies $\hbar\omega_q = \varepsilon_{k+q} - \varepsilon_k$ for a single electron–hole pair of momentum $\hbar q$ form a continuous spectrum between $2t_0 \sin|q|$ and $4t \sin|q/2|$. Since $q \to 0$ is an allowed state for the infinite system, there is no gap for these excitations. Note that this is the spectral range not only for an electron–hole excitation, but also for a spin excitation, obtained by flipping one of the spins in the ground state.

We next consider the strong coupling limit ($t_0 = 0$); here the ground state for an *average density* of one electron per site corresponds to a real-space configuration with *exactly* one electron at each site. It is highly degenerate ($\sim 2^N$, for a system of N sites) since the energy does not depend on the spin configuration. Therefore, flipping a spin does not cost any energy. On the other hand, it costs an energy U to transfer an electron from one site and put it at another, already (singly) occupied site, i.e. for charge transport. Thus in this limit, the charge excitation spectrum has a gap U, whereas the magnetic excitations are gapless.

The general case $t_0 > 0$, $U > 0$ has been solved by *Lieb* and *Wu* [2.120], using the Bethe ansatz technique [2.121]. They were able to reduce the many-particle Schrödinger equation to four simple integral equations for two types of distribution functions, $\rho(k)$ and $\sigma(\Lambda)$. The ground state energy is simply related to the function $\rho(k)$ which becomes identical to the momentum distribution function $n(k)$ both for $U \to 0$ and for $U \to \infty$ [2.122]. Two types of excited

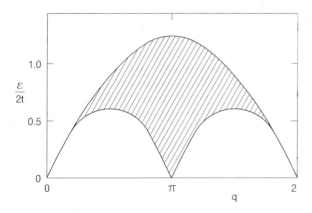

Fig. 2.20. Spin excitation spectrum for the half-filled Hubbard chain ($N = N_a$) for $U = 2t$

states can be distinguished: the charge excitations (mainly determined by $\rho(k)$), which are responsible for electrical transport and optical absorption, and the spin excitations (related to $\sigma(\Lambda)$), which determine the magnetic properties. The excitation spectrum for both charge and spin degrees of freedom was first discussed by *Ovchinnikov* [2.123]. The spectrum for a (triplet) spin wave has no gap and resembles the corresponding dispersion relation for the one-dimensional Heisenberg antiferromagnet [2.124]. This spectrum marks the onset of a multi-spin wave continuum, as illustrated in Fig. 2.20 for the two-spin wave case [2.125]. The charge excitation spectrum, as probed by e.g. optical absorption or electron energy loss spectroscopy, is gapless except for the precisely half-filled band, i.e. N electrons on $N_a = N$ sites [2.126, 127]. In this particular case there is a finite gap E_g for all (positive) values of U with asymptotic behaviour

$$E_g \sim \begin{cases} 8\pi^{-1}(Ut_0)^{1/2}e^{-2\pi t_0/U}, & U \to 0 \\ U - 4t_0 + \dfrac{8t_0^2}{U}\ln 2 + \cdots, & U \to \infty. \end{cases} \qquad (2.71)$$

Therefore the half-filled, one-dimensional system is a Mott–Hubbard insulator for all positive values of U. This contrasts with the expected behaviour for the three-dimensional case which is generally believed to undergo a metal-insulator transition at a finite value of U. The charge excitation spectrum for a single electron–hole pair of momentum $\hbar q$ is shown in Fig. 2.21.

Although not directly accessible experimentally, the value of the ground state energy in the Hubbard model is useful as a theoretical benchmark for approximate solution techniques. Furthermore, its derivatives with respect to t_0 and U provide additional valuable information on ground state properties. The derivative with respect to the resonance integral t_0 yields the bond order p_0. This in turn is proportional to the frequency integral of the conductivity $\sigma(\omega)$ [2.129–131], a relation commonly referred to as the *conductivity* or f sum

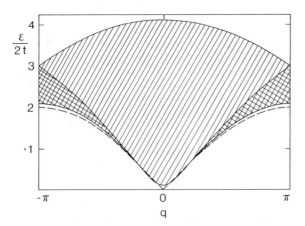

Fig. 2.21. Charge excitation spectrum for the half-filled Hubbard chain ($N = N_a$) for $U = 2t$. The cross-hatched area corresponds to the region where two solutions exist for given q and ε [2.128]. The dashed line indicates the onset of the electron–hole continuum for $U = 0$ (after [2.128])

rule. For $N = N_a$

$$p_0 \sim \begin{cases} (2/\pi)[1 - 7\zeta(3)(U/8\pi t_0)^2], & U \to 0 \\ (4\ln 2)t_0/U, & U \to \infty. \end{cases} \tag{2.72}$$

The derivative with respect to the Hubbard parameter U yields the fraction of doubly occupied sites

$$d \sim \begin{cases} (1/4)[1 + 7\zeta(3)U/(2\pi^3 t)], & U \to 0 \\ \ln 2(4t/U)^2, & U \to \infty. \end{cases} \tag{2.73}$$

For the small U limit in (2.72) and (2.73) we have used the asymptotic expansion for the ground state energy [2.132]. Strictly speaking, as (2.71) suggests, such a small U expansion does not exist [2.133, 134] and should be supplemented by non-analytic "beyond all order" terms, arising from the essential singularity as $U \to 0$. However, these terms are numerically small for small U. For large U the Hubbard model for $N = N_a$ becomes equivalent to the Heisenberg antiferromagnet with exchange constant $J = 4t_0^2/U$ [2.135, 136]. The exact value of the ground state energy of the spin model [2.121, 137] $E_0/N = -J\ln 2$ has been used to derive the constants in the large U limit in (2.72, 2.73). As expected, both the bond order and the double occupancy decrease as functions of U and tend to zero for $U \to \infty$.

The magnetic susceptibility, χ, at $T = 0$ has also been calculated [2.134, 138–140], by solving the Bethe ansatz equations in the presence of a magnetic field. One finds that $\chi(U)$ exhibits the following asymptotic behaviour (at $T = 0$ and for $N = N_a$):

$$\chi(U)/\chi(0) \sim \begin{cases} 1 + (U/4\pi t_0), & U \to 0 \\ U/(\pi t_0), & U \to \infty. \end{cases} \tag{2.74}$$

The enhancement of χ as a function of U can be interpreted in terms of an increase of the effective mass in the small U limit, and of a corresponding increase of the Pauli susceptibility, very much as in Landau's theory of the Fermi liquid. For $U \to \infty$ the system becomes a collection of free spins with a Curie susceptibility, which diverges at zero temperature.

2.4.2 Correlation Functions

Both the kinetic energy (2.72) and the susceptibility (2.74) can be considered as $q = 0$ response functions to, respectively, external electric and magnetic fields. These $q = 0$ response functions are only weakly affected by the interaction as long as U is not too large. Stronger effects are expected to be found for $q = 2k_F$ response functions because of the perfect nesting of the one-dimensional Fermi surface. However, evaluation of these response functions requires the calculation of spatial correlation functions, which are difficult to obtain using Bethe ansatz techniques. Thus one has to rely on approximate analytical methods or numerical simulations. Figure 2.22 shows the spin–spin correlation function as obtained by Monte Carlo calculations [2.141]. The system has short-range antiferromagnetic correlations but no long-range order. This remains true even in the large U limit (i.e. for the Heisenberg model) where the spin–spin correlation function is believed to decrease as $(-1)^l/l$ [2.142, 143]. Figure 2.23 shows the response function for the bond-order operator (recall Sect. 2.2), again from Monte Carlo calculations [2.144]. In sharp contrast to the behaviour of the charge density response function, which is strongly suppressed as a function of U [2.142], the bond-order response first increases, goes through a maximum for $U \cong 4t_0$, and then decreases for larger U. We will show later that this behaviour

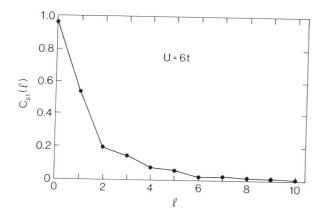

Fig. 2.22. The staggered spin–spin correlation function for the half-filled Hubbard chain ($n = 1$) for $U = 6t_0$, according to [2.141]

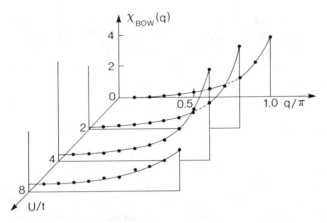

Fig. 2.23. Bond-order response function for the half-filled Hubbard chain ($N = N_a$) for different values of U, according to *Hirsch* and *Scalapino* [2.144]

of the bond-order response function re-emerges in the Peierls–Hubbard model, where coupling to the lattice allows an initial increase of the actual lattice dimerization (a $q = 2k_F$ distortion for the half-filled band) as a function of U.

Dynamic correlation functions, which can provide important information about excited states, are even more difficult to calculate for the Hubbard model. Older results are available for the optical absorption spectrum [2.129, 145] but they depend crucially on the approximations used. Very recently [2.146], the full optical absorption spectrum of the one-dimensional Hubbard model (including the generalization allowing dimerization) as a function of frequency has been studied extensively using exact (numerical) diagonalization of finite size systems and a novel boundary condition averaging technique to control the extrapolation to infinite system size. In Fig. 2.24a, b we show the results for the pure Hubbard model for $U = 8t_0$ and $32t_0$. For comparison, the optical absorption spectrum of the model with $U = 8t_0$ but with dimerization $\delta = 0.07$ (so the alternating hopping integrals are $t_+ = 1.14t_0$ and $t_- = 0.86t_0$) is plotted in Fig. 2.24d.

2.4.3 Relevance for Conjugated Polymers

Although PPP models emphasizing the importance of electron–electron interactions have long been applied by chemists to explain the excited states of finite polyenes [2.14], in the physics literature *Ovchinnikov* and coworkers were the first to propose that the one-dimensional Hubbard model provided a better description of finite polyenes than did the Hückel model [2.147] (see also [2.9]). Their argument was based on the observation that, while both the Hückel and the Hubbard models yield a semiconducting ground state (for $N = N_a$) with

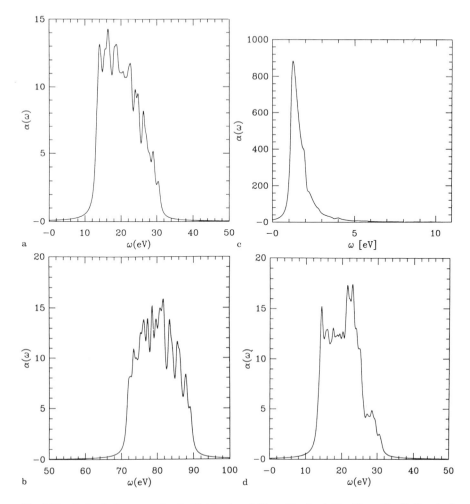

Fig. 2.24 a–d. Optical spectrum for the Hubbard model for **a** $U = 8t_0$, **b** $U = 32t_0$, **c** $U = 0$, dimerized case ($\delta = 0.07$), **d** $U = 8t_0$, dimerized case ($\delta = 0.07$). Results are from Lanczos diagonalizations of 12 site rings

(optical) gaps E_g, given by (2.47) and (2.71) respectively, the anticipated electron–phonon coupling (as deduced from vibrational spectra of polyenes) is too weak to account for the experimental value of $E_g \approx 2$ eV. On the other hand, within the Hubbard model, a very reasonable value of $U \approx 3.5t_0$ can easily reproduce the optical data.

The Hubbard model is also very attractive for describing the effects of doping. In fact, the insulating behavior for all positive values of U is a distinctive feature of the half-filled band case. For all other band fillings the system is always conducting, however large U. This is even true, and particularly easy to visualize, for $U \to \infty$. In this limit (and for $N < N_a$) the ground state configurations are sequences of empty and singly occupied sites, and the holes (empty sites) can move as easily as for $U = 0$. Indeed, in the limit $U \to \infty$ (and for $N < N_a$) it has been shown that the one-dimensional Hubbard Hamiltonian decomposes into two mutually independent terms, a tight-binding Hamiltonian for N spinless fermions on N_a sites and a Heisenberg spin Hamiltonian with a density-dependent exchange integral J [2.148, 149]. For finite values of U, numerical solutions of the Lieb–Wu equations [2.120] confirm that there is no gap for current-carrying excitations for $N_a/N \neq 1$ [2.127]. An expansion of the Lieb–Wu equations in powers of (t_0/U) [2.122] yields the following asymptotic form of the bond order for $U \to \infty$

$$p_0 \sim (1/\pi)\sin[\pi(N_a - N)/N_a] + O(t_0/U) \tag{2.75}$$

as expected for a system of N_a–N non-interacting spinless fermions. In the optical absorption these carriers will produce a Drude-like peak around $\omega = 0$ with intensity increasing proportional to (N_a-N).

Regarding the question of bond alternation, we see from Fig. 2.23 that Monte Carlo calculations [2.144] indicate that the bond order correlation function peaks at $q = \pi$ (units $a = 1$) for $N = N_a$. This correctly suggests that within the Hubbard model for non-zero U the uniform ground state is unstable with respect to the formation of a $2k_F$ BOW (i.e. dimerization), just as the simple Hückel theory is. Similarly, a variational approach à la Gutzwiller finds this result [2.150, 151]. Obviously, in the absence of electron–phonon coupling, no lattice distortion can take place. The result nonetheless indicates that the Hubbard model has an intrinsic tendency towards bond alternation. This is also consistent with results on dimerization in the Heisenberg chain [2.152] which, as noted above, is the $U \to \infty$ limit of the Hubbard model for $N = N_a$.

The standard one-dimensional Hubbard model, without electron–phonon coupling, provides an excellent intuitive framework and analytic benchmark for understanding the role of electron–electron interactions in conducting polymers. However, to obtain a realistic description of these systems, one needs to consider models incorporating both e–p and e–e interactions. Hence, in the next section, we study such Peierls–Hubbard models in detail.

2.5 The One-Dimensional Peierls–Hubbard Model

We turn now to the Peierls–Hubbard Hamiltonians, which incorporate both the Coulomb interaction among π electrons and their coupling to lattice degrees of freedom. In contrast to the models discussed in the previous two sections, these Hamiltonians do not admit exact analytical solutions. Rather, their study requires either extensive numerical computations for relatively small systems or approximations allowing for (partially) analytical treatments.

2.5.1 The Model Hamiltonian and its Parameters

To permit analysis of the full range of π-conjugated polymers, we shall focus on a slight generalization of the Hamiltonian derived in Sect. 2.2. In particular, we shall add firstly an extrinsic dimerization term already discussed in Sect. 2.3 [2.82] to permit us to describe systems with non-degenerate ground states (e.g. cis-(CH)$_x$, polythiophene, polyparaphenylene-vinylene, polypyrrole, and polydiacetylene) and secondly, a site energy term, which can model systems with different electron affinities at alternating sites (e.g. the general diatomic $(A = B)_x$ polymer, polyazacetylene [2.153] and polyaniline [2.154]). The explicit form of the Peierls–Hubbard model we shall consider is

$$H_{PH} = H_{lat} + H_{1-e} + H_{e-e}. \tag{2.76}$$

As discussed in Sect. 2.3, H_{lat} describes the structural energy of the lattice arising from the σ bonds, supplemented by the kinetic energy of the ions. Furthermore,

$$H_{1-e} = -\sum_{l\sigma} \{[t_0 - (-1)^l(\alpha y_l + t_1)](c^+_{l\sigma}c_{l+1\sigma} + c^+_{l+1\sigma}c_{l\sigma}) + \varepsilon_l c^+_{l\sigma}c_{l\sigma}\}$$

$$= H^{(1)}_{1-e} + H^{(2)}_{1-e} + H^{(3)}_{1-e} + H^{(4)}_{1-e} \tag{2.77a}$$

is the one-electron Hamiltonian and

$$H_{e-e} = U\sum_l n_{l\uparrow}n_{l\downarrow} + \sum_{l,j>0} V_j n_l n_{l+j} \tag{2.77b}$$

is the electron–electron interaction, where as usual $n_l = \sum_\sigma n_{l\sigma}$. The notation is the same as in Sects. 2.3 and 2.4, i.e. $y_l = (-1)^l(r_{l,l+1} - r_0)$ is the deviation of the bond length between the lth and $(l+1)$th chemical moieties (CH in polyacetylene, for instance) from the average bond length and $c^+_{l\sigma}(c_{l\sigma})$ creates (annihilates) a π electron at site l with spin $\sigma(= \pm\frac{1}{2})$.

As shown in (2.22a, b), the V_j in (2.77b) depend on the actual distance $r_{l,l+j}$ between the two chemical moieties along the polymer chain. Thus, for a perfectly dimerized polyacetylene chain, for example, there would be *two* values of each V_j, depending on whether l was even or odd. However, for all physically reasonable values of dimerization, this effect is small [2.155] and can be ignored.

The nomenclature in (2.76) is intended to clarify the role of each term in H_{PH}. H_{1-e} includes the tight-binding one-electron terms associated with a bare hopping ($H^{(1)}_{1-e} \propto t_0$) between nearest-neighbour sites and a modification of this hopping due to coupling to the ionic motions of the σ-bonded polymer backbone, ($H^{(2)}_{1-e} \propto \alpha$). This is, of course, just the usual SSH electron–phonon (e–p) coupling term. The additional term ($H^{(2)}_{1-e} \propto t_1$) represents an extrinsic dimerization/bond alternation independent of the coupling to phonons, that breaks the symmetry of the two distinct bond-alternation patterns of the π-conjugated system. Thus, as indicated above, it provides a simple model for those conjugated polymers in which the two possible double-single bond alternation patterns are *not* energetically equivalent. As a consequence, as argued in Sect. 2.3, instead of the free soliton excitations found in *trans*-$(CH)_x$, one has bipolarons, which can be viewed as confined soliton-antisoliton pairs. The site energies ε_l in $H_{1-e}^{(4)}$ may be (a) equal, for systems like $(CH)_x$, (b) alternating, with the form $(-1)^l \varepsilon$, for $(A = B)_x$ polymers, or (c) simply different for heteroatoms, e.g. nitrogen or sulphur, when these are explicitly included [2.156]

H_{e-e} includes direct e-e interactions (in the ZDO approximation, as discussed in Sect. 2.2) between electrons in Wannier orbitals on all pairs of sites and hence can model effects from strongly screened to pure Coulomb interactions. In practice, of course, one does not introduce an infinite number of parameters but works either with an effective short-ranged interaction, typically U and V_1, or with a parametrized form (for example, the Ohno [2.62] or Mataga–Nishimoto [2.63] parametrizations discussed in Sect. 2.2 above) of the long-range interaction. Note that in PPP models the intersite Coulomb term is often written as $\sum_{lj} V_j(n_l - z_l)(n_{l+j} - z_{l+j})$, where z_i is the average number of electrons per site (1 for the half-filled band).

In the infinite chain limit or for rings (periodic boundary conditions), this form merely shifts the chemical potential and adds an overall constant to the energy and hence makes no crucial difference. For finite chains (open boundary conditions), however, this difference is important. In this case the subtraction of z_i is physically reasonable as it mimics the screening effects of the ions.

We have already discussed, in Sects. 2.2 and 2.3, the very important issue of renormalization of the coupling constants and interaction parameters in H_{PH}. Clearly one must consider these parameters as effective, rather than microscopic constants, because they are renormalized by the effects of other degrees of freedom (e.g. the σ electrons) and interactions (e.g. inter-chain screening) that are *not* explicitly included in H_{PH}. However, we believe that one should attempt to understand these renormalizations by making close connections with the molecular parameters measured in related systems, such as the finite polyenes in the case of $(CH)_x$. The renormalizations due to solid-state effects can best be understood after one has determined how results based on including *both* e–p and e–e interactions with the molecular parameters fit the experimental data. For clarity, let us illustrate this approach by discussing the parameter values in the canonical case of *trans*-$(CH)_x$.

Surprisingly, perhaps the most fundamental parameter of all, the π-electron bandwidth in *trans*-$(CH)_x$, $4t_0$, remains uncertain at a level of roughly 20%. The conventional value used in the SSH model, based on one-dimensional band theory calculation [2.28, 29], is $t_0 \cong 2.5$ eV. However, an earlier phenomenological fit to a range of data from finite polyenes [2.69] gave $t_0 \cong 2.9$ eV, and a recent experimental study of the momentum-dependent dielectric function [2.157] suggests $t_0 \cong 3.2$ eV. Of course, estimates based on comparison of materials and experimental data implicitly include effects such as three-dimensional coupling

and Coulomb interactions that ideally one would wish to exclude from determinations of t_0. In this particular case three-dimensionality and Coulomb
repulsion have opposite effects on the apparent bandwidth, the former increasing
it, the latter decreasing it. However, as Fig. 2.21 illustrates, the influence of U
on the onset of the electron-hole continuum in the Hubbard model is fairly
small for weak to intermediate values of U. Hence a value of t_0 in the range
$2.5\,\text{eV} \lesssim t_0 \lesssim 3\,\text{eV}$ seems appropriate for $trans\text{-}(CH)_x$.

A second fundamental parameter is of course the dimensionless e–p coupling
parameter λ, which depends on the bandwidth $4t_0$, the electron–phonon
coupling α and the elastic force constant K through (2.45). In order to estimate
λ it is important to rely on observables which do not depend sensitively on
electron–electron interactions. For instance, it should be clear from the last
two sections that fitting λ to reproduce the optical gap on the basis of (2.47)
represents a strong bias in favor of an effective single-particle theory. To attempt
to determine an unbiased estimate of λ for use in the Peierls–Hubbard Hamiltonian
we first notice that the ratio between bond order and bond length, derived in
Sect. 2.3 for the Hückel model, remains exactly the same for the Peierls–Hubbard
model, *provided* that the bond order is calculated for the full Hamiltonian. The
proof uses the Hellmann–Feynman theorem according to which the derivative
of the (exact) ground state energy with respect to a parameter (e.g. the bond
length $r_{n,n+1}$ for the Peierls–Hubbard Hamiltonians) is identical to the
expectation value of the derivative of the Hamiltonian. Furthermore, in Sect.
2.4 we have seen that for moderate U the bond order of a Hubbard chain is
only slightly decreased due to the e–e repulsion; for instance, for $U \cong 2t_0$, (2.72)
suggests only about a 5% reduction. This is the quantification of our earlier
remark (Sect. 2.3) that e–e interactions do not strongly affect the average bond
order, although they do affect *alternation* of the bond order. Thus even in the
presence of e–e interactions, to determine α/K we can use the linear relation
between bond length and bond order (2.48) with the Hückel values for p_1 and
p_2. Hence we recover exactly the results of Sect. 2.3, namely, $\alpha/K = 0.08\,\text{Å}$ and,
using $K = 47.5\,\text{eV}\,\text{Å}^{-2}$ and, $t_0 = 2.5\,\text{eV}$, $\lambda \cong 0.08$. As noted above, this is in clear
disagreement with the SSH value of λ, determined by fitting the single particle
expression (2.47) for the optical gap. Since, as we shall later demonstrate, even
with this small value for λ one can, as a result of correctly including the effects
of the e–e interactions, obtain the observed value of the dimerization in *trans*-
$(CH)_x$, we feel strongly that this smaller value is the appropriate parameter to
use in the Peierls–Hubbard model.

The third important set of parameters are those characterizing the e–e
interactions. Here one is on somewhat shakier ground, since one must model
both the range (screening) of the Coulomb forces and their magnitudes. An early
estimate in the chemical literature [2.43], based on molecular parameters for
benzene, gave $U = 16.93\,\text{eV}$, $V_1 = 9.027\,\text{eV}$, $V_2 = 5.668\,\text{eV}$, and $V_3 = 4.468\,\text{eV}$.
The Ohno [2.62] and Mataga–Nishimoto [2.63] parametrizations of U and
V_l given in (2.22) represent attempts to limit the number of free Coulomb
parameters, while at the same time fitting excited state data in finite polyenes

[2.158]. An explicit formula for the Coulomb parameters in the Ohno parametrization is [2.158]

$$V_l = \frac{19.39 \, \text{eV}}{\left[\left(\frac{14.397}{U} \right)^2 + r_l^2 \right]^{1/2}}, \tag{2.78}$$

with $U = 11.13 \, \text{eV}$ and r_l being the distance in angstroms between the two atomic sites. For calculating r_l, the true (nominal) geometry of finite polyenes is used, so all c–c bond angles are taken to be 120°, a double bond is taken as 1.35 Å and a single bond is 1.46 Å. In many recent calculations, only the parameters U and V_1 are explicitly taken into account. The effective values of these parameters are probably substantially reduced from the molecular or PPP parameters. We regard reasonable ranges for these parameters to be $1.5 \, t_0 < U < 3.5 t_0$ and $0.5 t_0 < V_1 < 1.5 t_0$. Our reasons will become apparent in the subsequent discussion.

Before proceeding further, we should indicate three approximations which, although typically made, are not essential limitations of the model. Firstly, since the π-conjugated polymers correspond to the case of the half-filled band, we have dropped the deformation potential term mentioned in Sect. 2.2. Secondly, and a related point, the inclusion of only one type of electron–phonon interaction can, at the cost of a modest increase in numerical effort, be relaxed, and additional interactions reflecting more precisely the true structure of the polymer can be incorporated [2.75, 113, 159, 160]. Thirdly, as in the SSH model, the quantum nature of the phonons is usually neglected, and they are treated as classical coordinates. This assumption, which is typically justified by the Born–Oppenheimer approximation, has been tested in recent Monte Carlo simulations involving quantum phonons [2.161–163] and found to be quite adequate for the range of parameters appropriate to the π-conjugated polymers.

2.5.2 Methods

At a general conceptual level, the methods used to study the Peierls–Hubbard model can be separated into two classes: independent-electron approaches and correlated electron methods. Strictly speaking, the independent-electron approaches can be exact only for the one-electron part of H_{PH}, namely the terms in (2.77a) which contain at most bi-linear combinations of electron operators. As stressed in Sect. 2.3, the many-body wave-function in the independent-electron approach can be written as a single Slater determinant of one-particle wave-functions. Accordingly, the problem of diagonalizing the Hamiltonian for a finite (cyclic) polyene of length $2n$ is a $2n \times 2n$ matrix problem, although typically with self-consistency constraints on the parameters occurring in the Hamiltonian. Independent-electron approaches can therefore readily be carried out for the

Peierls–Hubbard model for systems large enough (say, $N \cong 100$) to approximate accurately the infinite polymer limit. In contrast, the correlated electron approaches attempt to treat *exactly* the two-body operators in (2.77b), which involve quadrilinear combinations of electron operators. For a system of N electrons on N sites (including all possible spin configurations), this requires $\binom{2N}{N}$ basis states. For large N, this number goes as 4^N, and thus, as remarked in Sect. 2.4, the complete solution of the many-body Hamiltonian requires the diagonalization of a matrix of size roughly 4^N by 4^N. For any given problem, symmetry and selection rules (total spin, mirror plane or electron-hole symmetry) can be used to reduce the size of the matrix, but the growth with N will still be exponential. Thus these correlated electron methods can in general be applied only to systems of fairly limited size. In view of this limitation, it is natural to ask whether independent-electron methods can provide accurate *approximations* to the results of the full Hamiltonian. In the following subsections, we discuss this question, indicating the ranges of applicability and the limitations of a variety of methods and comparing and contrasting them with the full correlated electron approaches. Readers not interested in the (often subtle) differences among these methods can skip directly to the comparison of the results of the Peierls–Hubbard model with experiments in Sect. 2.6.

a) Hartree–Fock

The Hartree–Fock aproximation, which we have discussed already in Sect. 2.2 for its applications at the ab initio level, can of course also be applied to model Hamiltonians. In the present context, the many-electron interaction term (2.77b) is replaced by an effective single-particle term, involving only bilinear combinations of fermion operators, which is determined self-consistently. The problem is thus immediately reduced to an independent-electron calculation, so that it can be solved by (self-consistent) single-particle states. Since these effective single-particle states include self-consistent interaction effects, strictly speaking one should speak not of electrons but of *quasi-particles*. This observation is important in interpreting the single-particle spectrum of excited states. If these orbitals are assumed to be the same for both up and down spin electrons, one has the restricted Hartree–Fock (RHF) approximation. If the spin-up and spin-down orbitals are allowed to differ, one has the unrestricted Hartree–Fock (UHF) approximation; clearly to obtain a spin-density wave within the HF approach one must use the UHF approximation. In either case, given the single-particle orbitals, HF excited states can be formed by successive promotion of quasi-particles from occupied to unoccupied orbitals to create (multiple) particle-hole pairs. Each of these different many-electron configurations is described by a single Slater determinant. Of course, the relation of these HF excited states to the actual excited states of the full many-body problem must be analyzed in each separate problem; in this regard as noted in Sect. 2.2, Koopmans' Theorem is a useful guide [2.38, 164].

As emphasized in Sect. 2.2, the HF approach serves as an essential zeroth-order approximation to the ground state of theories involving electron–electron interactions. Furthermore, in cases when symmetry arguments can be invoked [2.165], HF results can be essentially exact for certain quantities in these theories. In the area of π-conjugated polymers, HF methods have been extensively applied to the Peierls–Hubbard Hamiltonian [2.59, 97, 166–169]. Early work on Hartree–Fock solutions of the PPP model include those by *Paldus* and *Cizek* [2.170–172]. These authors showed that within the PPP model for large cyclic polyenes, the Hartree Fock single determinant solution is unstable with respect to singlet type monoexcitations if uniform bond lengths are assumed. It was further shown that broken symmetry CDW and BOW solutions can give stable solutions, and for several different PPP parametrizations the BOW was lower in energy than the CDW. This last conclusion would agree with our current results on long-range Coulomb interactions [2.173, 174]. It should, however, be mentioned that such Hartree–Fock approaches are guaranteed to predict equal bond lengths for the Peierls–Hubbard model [2.171]. However, as already noted in Sect. 2.2, these results must be viewed with some caution. In particular, the HF predictions for the ordering of excited states in finite polyenes are simply incorrect [2.158, 175]. Furthermore, the UHF treatment of the simple one-dimensional Hubbard model predicts an SDW ground state with long-range order, in contrast to the exact results quoted in Sect. 2.4. We shall see later in this section that also for the Peierls–Hubbard Hamiltonians, predictions concerning e.g. the lattice dimerization or transition oscillator strengths in the excitation spectra are qualitatively incorrect.

b) Perturbation Theory, Renormalization Group and Configuration Interaction

In response to the limitations of the Hartree–Fock method, systematic ways have been developed for going beyond this approach to incorporate some of the correlation effects arising from e–e interactions. Although it is difficult to characterize succinctly these approximate many-body approaches, in general one can say that they amount to including the contributions of selected subsets of all possible many-electron configurations. Three classes of such approaches are particularly relevant to our discussion: perturbation theory, renormalization group techniques and finite order configuration interaction methods.

Perturbation theory around the independent-electron limit is appropriate for weak electron–electron interactions. If there is a gap between the occupied and unoccupied single particle levels (as in the case of models in which the Peierls instability leads to dimerization and hence to a gap) this perturbation theory is conventional and has a finite radius of convergence [2.176]. In the context of π-conjugated polymers, calculations up to second order have been made for both discrete [2.167] and continuum [2.176] versions of the Peierls–Hubbard model. For the case of the one-dimensional electron gas without phonons and hence with no Peierls gap, the perturbative expansion in the electron–electron interaction contains divergences which must be treated with some care. The well-studied subject of *g-ology* (see, for example, [2.177, 178]) provides a detailed prescription for treating these divergences and for summing to all orders a subset of perturbation theory diagrams. *g*-ology has been widely applied to the full range of quasi-one-dimensional electronic materials and specifically to conducting polymers [2.179, 180]. Recently, the effects of phonons have been incorporated within the *g*-ology approach to π-conjugated polymers [2.181, 182].

In the case of dominant electron–electron interactions, the natural perturbation theory is about the strongly localized or atomic limit. Here the large (2^N-fold) spin degeneracy renders the calculations non-trivial [2.136, 148]. Although calculations performed in this limit are not expected to be immediately relevant to π-conjugated polymers, they are nonetheless very important as benchmarks for numerical methods which are used to probe the difficult but important intermediate coupling regime.

The infinite-order summations in *momentum space* contained in *g*-ology are related to the important concept of scale invariance and the renormalization group techniques [2.177, 178] that originated in field theory and played a major role in clarifying critical phenomena. Conceptually similar ideas have also been applied in *coordinate space* to models of π-conjugated polymers. Firstly, the real space renormalizaton group techniques of iteratively blocking together numbers of lattice sites in real space to form an "effective" single site [2.183, 184] have been used to study the Hubbard model. Secondly, an approach based on creating larger systems by combining two smaller systems—for example, treating an eight-site system as two coupled four-site systems—has been applied to study the combined effects of e–p and e–e interactions in Peierls–Hubbard models [2.185–187].

Within the context of Hartree–Fock approaches, the configuration interaction (CI) method has been used to go beyond the independent-electron level. The electron–electron interaction terms in the many-body Hamiltonian will couple different configurations—ground state and particle-hole excited states—of the HF approximation. By writing the Hamiltonian in the basis set of all HF configurations and then diagonalizing, one can in principle solve the correlated electron problem exactly. Of course, to carry out a full CI calculation requires knowing the matrix elements of the Hamiltonian between all possible HF configurations; simple counting shows that there are $\begin{pmatrix} 2N \\ N \end{pmatrix}$ possible Slater determinants of HF orbitals, each determinant corresponding to an allowed electron configuration. As noted previously, this is precisely the number of basis states needed to specify completely the full many-body problem of N electrons on N sites with one orbital per site. Thus a complete CI calculation is as difficult as any other full many-body calculation. Consequently, substantial efforts have been made to apply partial CI methods to conjugated polymers, that is, calculations in which only a limited number of HF configurations are retained and the Hamiltonian is diagonalized in this subspace (see, e.g. [2.188, 189], for a discussion and earlier references).

c) Variational Methods

Variational approaches form another important general category of approximate methods for incorporating correlation effects. If carried out strictly and without additional approximations, these methods can provide rigorous upper bounds on the energy of the correlated electron system. However, even if they are treated so that their energy estimates remain strictly variational, these methods do not necessarily produce reliable information on the nature of the many-body wave function or physical observables such as the optical gap or the dimerization in the presence of e–e interactions. A relevant example is the Hartree–Fock approximation, which can be formulated variationally and yet produces qualitatively incorrect results for the nature of the ground state of the Hubbard and Peierls–Hubbard models at large U. A separate limitation on these methods is that technical difficulties often compel one to adopt calculational strategies which, in order to render the problem more tractable, spoil its variational character.

Despite these limitations, variational methods can, particularly when coupled with numerical studies of finite clusters, provide essential physical and analytic insight that can be difficult to obtain from more complete but highly numerical schemes. In the area of π-conjugated polymers, variational methods which have been applied include the Gutzwiller *ansatz* [2.190–193], the antisymmetrized product of strongly orthogonal local geminals [2.194–196], and the Jastrow–Feinberg wavefunction methods [2.197].

d) Numerical Many-Body Methods for Finite Clusters

To study the full many-body problem in correlated electron theories requires, as noted above, dealing with matrices that grow in size exponentially with the number of sites. Hence one is in general immediately forced to develop and apply numerical methods to carry out these studies. Here we discuss two important classes of numerical methods which are in principle *exact* in that they make no a priori approximations to the many-body problem.

The first class consists of complete diagonalization of the Hamiltonian matrices for finite systems of up to 14 sites [2.12]. Historically, these studies developed from several classic works on finite polyenes (see, e.g., [2.198–201] and for a review [2.158]). These studies demonstrated the need for going beyond perturbative configuration interaction (CI) approaches, in which only a limited number, typically one or two, excited configurations are kept, to a full CI calculation in order to obtain the true spectrum of excited states within PPP-type models and, more importantly, to fit the observed experimental data on optical properties of these systems. Figure 2.25 shows how the locations and orderings of the excited states of octatetraene vary as functions of the order of the CI calculation.

Since, as noted above, a full CI calculation is simply a complete diagonalization of the many-body Hamiltonian in a self-consistent (Hartree–Fock) basis and since the calculation of this basis is itself quite involved, if one has decided to undertake a complete diagonalization of a given Hamiltonian it is natural to seek a simpler basis set. One such basis is provided by the valence bond diagrams [2.60, 61, 155, 202–204]. With this approach one can obtain the ground state and, importantly, oscillator strengths for optical transitions to several low-lying excited states for PPP models of systems of sizes $N \sim 12$. This permits direct comparison with data on finite polyenes, and, particularly when combined with the quantum Monte Carlo information on larger systems, allows extrapolation to the $N \to \infty$ limit. Crucial to the success of this method has been the application of symmetries to restrict the sizes of the matrices and the use of numerical relaxation techniques to evaluate eigenvalues beyond the ground state (see, e.g., [2.61, 205, 206]). One limitation of this

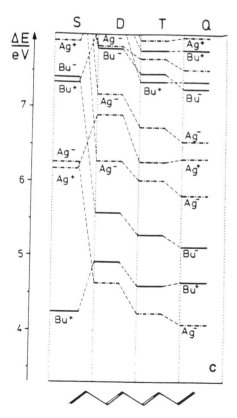

Fig. 2.25. The excited states in the PPP model for octatetraene for four different levels of self-consistent field configuration interaction (CI) S = single, D = double, T = triple, Q = quartic CI (after [2.201])

approach is the difficulty of relaxing the nuclear geometry; each new configuration requires a time-consuming full diagonalization of the Hamiltonian. Furthermore, if spatial symmetries are used to reduce the sizes of the matrices, this will limit the *types* of nuclear geometries that can be considered. This limitation becomes particularly significant for excited states involving localized deformations of the lattice. A second limitation, less recognized, is that extrapolations to the $N \to \infty$ limit often require considerable experience and furthermore, not all physical quantities can be extrapolated for arbitrary values of the parameters. This is particularly so for models with long-range Coulomb interactions (e.g. the PPP model), where finite chains of different lengths actually have different Hamiltonians. Nonetheless, in view of its ability to provide exact results for ground states and electronic excitations (albeit typically in unrelaxed geometries), the diagrammatic valence bond technique provides essential insight into the effect of electron–electron interactions in π-conjugated polymers.

Recently, a related exact diagonalization method using a simple coordinate space electron occupancy basis and relying on the Lanczos algorithm to extract the lowest few eigenstates of the full Hamiltonian has been developed [2.51, 52, 146]. Since in its present form this approach does not rely on the use of symmetries, it should be easier to study nuclear relaxation effects and properties of the phonon spectrum. Furthermore, using a phase randomization technique to reduce finite size effects, this approach appears to yield accurate information on the frequency dependence of the optical absorption for correlated electron theories [2.146].

The second important class of numerical many-body methods is based on Monte Carlo techniques. In these approaches, the problem of the exponential growth in the number of states is attacked by use of stochastic methods and importance sampling. In the area of conducting polymers, several different methods have been used [2.207–212] to investigate variants of the Peierls–Hubbard/PPP models. Without going into too much technical detail one can distinguish three classes of methods:

(1) *Projector methods* [2.209–211, 213] which by projecting with the operator $\exp(-\beta H)$ isolate directly the ground state for β approaching infinity,
(2) *World line* methods [2.141, 214], which by evaluating the trace of the partition function $\mathrm{Tr}(\exp(-\beta H))$ for low temperature also obtain ground state energies, expectation values, and correlation functions and
(3) *Green's function* methods [2.212, 215, 216], which in essence study the operator $\left(\dfrac{1}{H-E}\right)$ and isolate the ground state by performing a (guided) random walk through the space of all possible configurations.

Since quantum Monte Carlo (QMC) methods can provide (numerically) highly accurate results for certain observables in simple models of clusters of fairly large size ($N \sim 30$), they can both establish the true behaviour of many-body models (hence serving as benchmarks for more flexible approximate methods) and come much closer than the complete diagonalization methods to the $N \to \infty$ limit desired for extrapolation of oligomer results to polymers. Further, the world line method can be modified to incorporate quantum phonon effects, so that relaxed (ground state) geometries and finite phonon frequencies have already been investigated [2.161, 162, 208, 217]. In principle, similar effects can be studied within the projector framework. As presently formulated, the projector and world line approaches can only be applied to short range interactions ($V_j = 0, j > 2$), but this is not a limitation in principle. The Green's function approach is immediately suitable for any range of interaction and indeed has been applied to the PPP model [2.212].

In their present incarnations, however, the QMC methods can study only a limited number of physical quantities, including ground state energies, ground state charge and spin densities, and both equal-time (static) and imaginary-time (dynamic) correlation functions. Converting the imaginary time results to the real time domain to study, for example, the optical absorption as a function of frequency, remains one of the outstanding methodological problems for QMC. A second important problem is the development of an approach to excited states. For states with different symmetries from the ground state, e.g. the 1^1B_u state in finite polyenes, the problem should be manageable and, for small molecules, some progress has been made for the general case [2.218]. In view of their ability to produce reliable results for models involving strongly correlated electrons, both QMC and exact diagonalization methods will play very important roles in our discussions of results in Sect. 2.6.

2.6 The Combined Effects of Electron–Phonon and Electron–Electron Interactions: Theory and Experiment

In this section we investigate the Peierls–Hubbard Hamiltonians in detail and illustrate the extent to which these models, properly solved as interacting many-body systems, can in fact describe the experimental situation of π-conjugated polymers. Our approach in describing the comparison between theory and experiment will be to list, and give reference to, most of the relevant work, while at the same time focusing on a few results which we hope will yield lasting insights into the structure of conducting polymers.

2.6.1 Ground State

In the Peierls–Hubbard models, for couplings in the range relevant for conducting polymers ($U > V_j > 0$), the competition for the ground state is among three types of broken symmetries; bond-order wave (BOW), charge-density wave (CDW) and magnetic ordering (spin-density wave (SDW) for small U, antiferromagnetism for large U). Of these, only two (BOW and CDW) can exhibit long-range order in the ground state in these one-dimensional models. (As discussed in Sect. 2.4, even for very large U the antiferromagnetic correlations decay algebraically). In the remainder of this section, we shall be concerned mostly with the BOW and CDW; the magnetic correlations will be emphasized when they lead to dramatic effects as, e.g., in the local spin structure associated with a neutral soliton. Importantly, as we shall discuss in more detail below, both the BOW and the CDW can be driven *either* be electron–phonon *or* by electron–electron interactions.

The manner in which the *inter*-site phonons in the SSH model produce, via the Peierls mechanism, the dimerized BOW ground state has been discussed in detail in Sect. 2.3. Similarly, if one introduces *intra*-site phonons as, e.g., in the molecular crystal model [2.219], a dimerized CDW ground state is produced. Although such *intra*-site phonon effects are not significant in presently available conjugated polymers, they are important in a wide class of other quasi-one-dimensional materials (see, e.g., [2.220–222]). In the half-filled band, the CDW and BOW states compete directly, and it has been shown in models with both intra-site and inter-site electron–phonon coupling (but no Coulomb interaction) that there is essentially no region of coexistence [2.223–225].

All these broken symmetries can be described by order parameters, as discussed for the case of bond alternation in Sect. 2.3. A *homogeneous* ground state is characterized by a site-independent order parameter, such as a constant amplitude of bond alternation. A ring of $2M$ sites containing $2M$ electrons is compatible with a homogeneously broken symmetry. *Inhomogeneous* states occur in finite segments close to chain ends, in rings or chains with odd numbers of sites, and in systems containing slightly more or fewer electrons than sites (nearly half-filled band). As a rule, two different broken symmetries with similar effects on the electronic structure (such as the opening of a gap at the Fermi surface) tend to suppress each other. Therefore they will not in general coexist in the homogeneous ground state. On the other hand, as we will see below, local inhomogeneous structures often exhibit more than one order parameter: a neutral kink in the BOW order parameter introduces a local SDW, whereas a charged kink in the BOW order parameter leads to a local CDW.

In the following we shall focus exclusively on the model for *trans*-$(CH)_x$ with zero site energy differences and only the inter-site electron–phonon coupling. Additional site-energy or electron–phonon interactions can clearly change the nature of the ground state broken symmetry. For instance, if the parameter ε (measuring the energy difference between adjacent sites) is large enough [2.73], or if a strong *intra*-site electron–phonon coupling [2.223–225] is included, a CDW state can occur.

a) Homogeneous Dimerization in the Presence of Electron–Phonon and Electron–Electron Interactions

There is direct experimental evidence that a dimerized BOW state characterized by alternating long and short bonds occurs in both finite polyenes and conjugated polymers. In polyacetylene, for example, from studies of X-ray patterns from polycrystalline, partially ordered trans-$(CH)_x$ films [2.226], C^{13} nutational NMR studies [2.227] and X-ray analysis of highly oriented films [2.228], one finds consistent values of $y_0 \sim 0.05$Å.

In order to facilitate the subsequent discussion, we introduce the dimensionless bond alternation parameter

$$\delta = \alpha y_0/t_0 = 2\alpha_x u_0/t_0, \tag{2.79}$$

which is independent of the trigonometric factors discussed in Sect. 2.3 between the actual change in bond length y_0 and the effective displacement in the chain direction u_0.

The existence of a bond-alternated (dimerized BOW) state clearly establishes the role of inter-site electron–phonon coupling in conjugated polymers: without such coupling, the nuclear positions would not shift even in the presence of an electronic BOW instability. Importantly, however, this result does *not* establish that e–e interaction effects are negligible. The misconception that dimerization implies weak e–e interactions appears to have its origin in the early unrestricted Hartree–Fock (UHF) studies of the Peierls–Hubbard model, which predicted the vanishing of bond alternation beyond a fairly weak critical Hubbard interaction $U_c \cong 1.8t_0$ [2.59, 166, 167, 229–231]. For $U > U_c$, UHF predicts a SDW state with long-range antiferromagnetic order and equal bond lengths. Subsequently, numerically exact approaches have shown that this result is an artifact of the mean-field approximation; proper inclusion of fluctuations modifies this result, and in one dimension the BOW persists for arbitrary U. Since the history of the studies of e–e interaction effects on bond alternation in polyenes and polyacetylene has been thoroughly discussed elsewhere [2.155], we adopt here a primarily methodological perspective, summarizing both the analytical and numerical approaches to the problem.

Exact Numerical and Variational Results for the Simple Peierls–Hubbard Model. For simplicity we consider first the simple Peierls–Hubbard model ($V_j = 0$), in which the electron–electron interaction is described by the single parameter U; as we have previously stressed, in view of the neglect of longer range Coulomb interactions, this U must be viewed as a renormalized or effective interaction. Although the value of the effective U remains the subject of considerable controversy, there is general agreement that it is *not larger* than the bandwidth ($4t_0$) in the π-conjugated polymers. Therefore we concentrate on the regime of weak to intermediate U, that is, U less than or roughly equal to $4t_0$. However, we shall on occasion refer to results from the strong-coupling limit $U > 4t_0$, since despite its quantitative inapplicability to conducting polymers this limit provides considerable insight into certain aspects of their behaviour. For weak to intermediate U there are still two regimes to be distinguished according to the relative importance of electron–electron and electron–phonon interactions. If the electron–phonon interaction dominates, a perturbation expansion about $U = 0$ is reasonable. The ground state can then be characterized as a Peierls semi-

conductor as in the independent-electron models of Sect. 2.3; the e–e interactions provide only small renormalizations of the pre-existing Peierls gap and
the dimerization [2.167, 176]. If, on the other hand, the electron–electron interaction dominates, the system has to be considered a Mott–Hubbard insulator
in which the additional coupling to the lattice stabilizes the BOW; the optical
gap is then essentially a correlation gap, as in the case of the Hubbard model,
whereas the dimerization of the lattice appears as a secondary, independent
effect due to the electron–phonon coupling [2.102, 147, 180, 189, 192, 206].

On the basis of a wide variety of many-body methods, it has been definitively
established within the Hubbard model that for *fixed* values of the electron–
phonon coupling *in the range relevant to conjugated polymers*, even moderate
to strong e–e interaction parameters actually *increase* the bond alternation, in
sharp contrast to the earlier UHF predictions. The effect is *largest* for intermediate values of U, $U \cong 4t_0$. This conclusion has been reached on the basis of
both numerical and analytical studies. The numerical work is naturally divided
into complete diagonalizations for small systems [2.204] and stochastic evalua-
tions (quantum Monte Carlo) for moderately large systems [2.207–209]. The
analytic studies can be subdivided into perturbation expansions about $U = 0$
[2.167, 176], field-theoretical techniques involving the summation of whole classes
of diagrams (in the continuum limit) [2.179, 180] and variational approaches
[2.190, 191, 194–196, 232]. Importantly, in perturbation theory studies of the

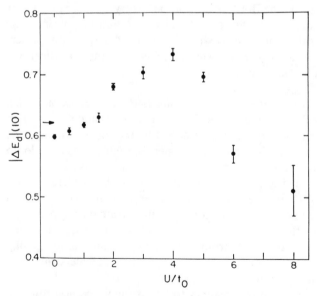

Fig. 2.26. The magnitude of the electronic energy difference between a dimerized $(CH)_x$ chain with
$\delta = 0.1$ (see text) and a uniform chain for an $N = 32$ site ring plotted versus the on-site Hubbard
parameter U. The arrow on the left indicates the exact $U = 0$ result. The enhancement of the energy
gained upon dimerization of intermediate U is clearly shown (after [2.211])

Hubbard model, it is necessary to go to second order in U to see the enhancement of dimerization, since there is no effect of U alone in first order [2.167]. A real space renormalization group approach [2.185] combines both analytical and numerical techniques. In Fig. 2.26 we show for the Hubbard model the electronic energy gained at fixed dimerization—the BOW condensation energy—as a function of U. Note that it increases from its $U = 0$ value and reaches a maximum at $U = 4t_0$; this gain in BOW condensation energy translates directly into an increasing tendency to dimerize. This result is shown directly in Fig. 2.27, which establishes that even for (overly) strong electron–phonon coupling ($\lambda = 0.214$), dimerization is enhanced relative to its $U = 0$ value out to $U/t_0 \sim 6$. In Figs. 2.28 and 2.29 we show plots of the bond alternation parameter as a function of U obtained by application of a real space renormalization group technique and of the Gutzwiller variational wavefunction, respectively. The semi-quantitative agreement with the exact results through the range of intermediate coupling is very encouraging.

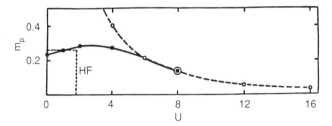

Fig. 2.27. Dimerization amplitude (called m_p in this figure) versus U for the Peierls–Hubbard model with parameters $t_0 = 1, \lambda = 0.25$, $K = 0.24$ and $\omega = 2(K/M)^{1/2} = 0.66$. The short dashes label the Hartree–Fock prediction. The open circles connected by the long dashes are obtained from simulations of the equivalent Heisenberg Hamiltonian (after [2.426])

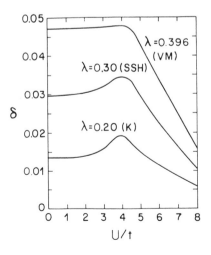

Fig. 2.28. Dimensionless dimerization amplitude (called δ in this figure) of a 16 site chain versus strength of the on-site Coulomb term for three values of λ (after [2.185]). Note that the λ used by these authors is twice that used in our text

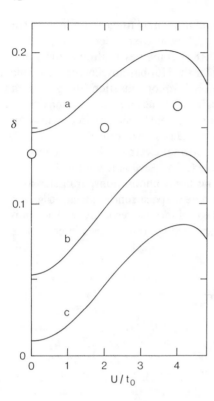

Fig. 2.29. Dimerization amplitude δ as a function of U as obtained using the Gutzwiller ansatz for different values of λ: a) 0.214, b) 0.15, c) 0.10. In contrast to *Baeriswyl* and *Maki* [2.191], no truncation has been used in the evaluation of the various terms. The circles represent Monte Carlo results including quantum fluctuations of the lattice for $\lambda = 0.214$ [2.207]. Note that Hirsch's result was misinterpreted in Fig. 1 of [2.191]

One very important question which combinations of exact and approximate methods are beginning to clarify is the joint dependence of the bond alternation on λ and U. In the limiting case $U = 0$, one obtains the familiar BCS-like result of (2.47), while for $U \to \infty$ at fixed λ, the spin-Peierls results suggest that [2.233, 234],

$$\delta \sim \left(\frac{\lambda t_0}{U}\right)^{3/2}. \tag{2.80a}$$

More detailed recent studies ([2.180]; see also [2.9]) suggest that for small $U(U \ll t)$ and very small $\lambda(\lambda \ll U/t)$, e.g. $U = t_0/4$, $\lambda = 0.1$, the behaviour is

$$\delta \sim (t_0/U)^{1/2}\exp\left[-\frac{2\pi t_0}{U}\right]\lambda^{3/2}. \tag{2.80b}$$

In the intermediate regime, $U \approx 4t_0$, the Gutzwiller approach suggests a slightly super-linear dependence of δ on λ [2.192], as shown in Fig. 2.29, in qualitative agreement with (2.80). All these results deserve further investigations by exact numerical methods. Unfortunately, both results in (2.80) hold only for small λ, a regime where the small finite systems that are presently within reach are *undimerized* because of finite size effects. The same is true for the present renormalization group calculations, while Monte Carlo methods are not accurate enough to check the detailed dependence in (2.80).

Throughout the present discussion we have focused almost exclusively on the BOW. While the UHF method over-emphasizes the SDW, the true result is that the BOW and (a short-range) SDW coexist for all $U > 0$. The only difference is that while the enhancement of the BOW reaches a peak at $U \sim 4t_0$, the SDW is enhanced monotonically until the $U \to \infty$ limit is reached. The reason for this coexistence of the BOW and the SDW is implicit in the valence bond description of broken symmetry [2.155, 173, 204]. Here we discuss it explicitly, since apart from providing physical insight into the BOW and SDW instabilities, it also provides a conceptual starting point for our subsequent discussion of excitations.

Intuitive Valence Bond Interpretation: Barrier to Resonance. Since the discovery of high-temperature superconductivity, there has been a resurgence of interest in *valence bond* theory, originally formulated by *Pauling* [2.235]. In the context of high-temperature superconductivity, several versions of valence bond theory, all usually referred to as *resonating valence bond* (RVB) theory, are being discussed. In the present context we are dealing with spatial broken symmetries only, but it should be realized that the latter, in particular the spin-Peierls dimerization, has also been discussed within valence bond theory. The original work goes back to *Coulson* [2.236], but there has been substantial recent work by *Klein* and collaborators [2.237, 238]. Most recently, however, it has been shown that valence bond ideas can be successfully applied even in the case of *electronic* (as opposed to spin) Peierls transition, provided *all* valence bond diagrams, both covalent and ionic, are retained. Purely covalent diagrams contain no double occupancies, implying, for the half-filled band, no empty sites either and obviously will dominate the ground state wave function in the limit of very strong electron–electron interactions. Purely ionic diagrams contain only double occupancies (and empty sites). Thus a typical valence bond diagram will be of mixed covalent-ionic character. For clarity, several valence bond diagrams for a system of six sites are shown in Fig. 2.30.

The valence bond diagrams correspond directly to electron configurations in coordinate space and thus are eigenstates of H_{e-e} in (2.77b). In this sense, they form a natural basis for analyzing the strong e–e interaction limit, just as the momentum space basis (band theory) is natural for treating the independent-electron (e–p interactions only) limit. However, since they are a complete basis, the valence bond states can be used for any values of e–p and e–e interaction strengths.

Within the valence bond approach (or *any* other configuration space approach) the wave function for an eigenstate of the full Peierls–Hubbard Hamiltonian in (2.77) is a linear combination of all valence bond diagrams that are reached from one another by applications of H_{1-e} in (2.77a). The process of connecting different configurations has been called *resonance* by Pauling. Perfect resonance implies exactly equal contributions to the wavefunction by

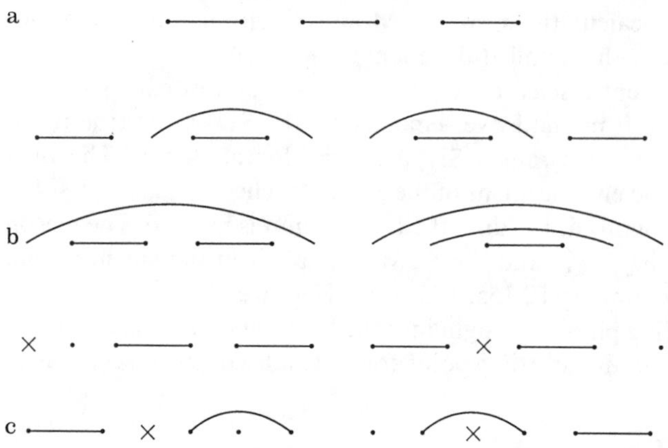

Fig. 2.30 a–c. Several valence bond diagrams for hexatriene: **a** the covalent diagram corresponding to the Kekulé structure, **b** covalent diagrams involving long bonds, **c** some ionic diagrams involving double occupancies (denoted by x) and vacancies (denoted by.)

equivalent coordinate space configurations, where equivalent refers to configurations that are related to one another by a symmetry operation characteristic of the system. Imperfect resonance implies *unequal* contributions of equivalent configurations and results when the symmetry operation is no longer characteristic of the system; thus, in valence bond terminology, broken symmetry is a consequence of imperfect resonance. To cite an example, in the 2-site, 2-electron problem, the configurations $|02\rangle$ and $|20\rangle$, where the numbers denote site occupancies, are equivalent and contribute equally to the wavefunction (which also contains the $|11\rangle$ state) in the symmetric case. When a CDW is present, however, the mirror plane passing through the bond centre is no longer a symmetry operator and the two configurations contribute unequally.

Exactly the same concepts can be used in the infinite system, independent of dimensionality [2.239], band filling [2.240] and correlation parameters [2.173]. For the one-dimensional half-filled band, for any of the three broken symmetries (BOW, SDW or CDW), there are two extreme configurations, related to each other by a lattice translation $l \rightarrow l + 1$. These are the two Néel states in the SDW case, the two alternating sequences of empty and doubly occupied sites in the CDW case, and the two valence bond diagrams known as Kekulé structures

$$(c_{1\uparrow}^{+}c_{2\downarrow}^{+} - c_{1\downarrow}^{+}c_{2\uparrow}^{+})(c_{3\uparrow}^{+}c_{4\downarrow}^{+} - c_{3\downarrow}^{+}c_{4\uparrow}^{+})\cdots|0\rangle$$
$$(c_{2\uparrow}^{+}c_{3\downarrow}^{+} - c_{2\downarrow}^{+}c_{3\uparrow}^{+})(c_{4\uparrow}^{+}c_{5\downarrow}^{+} - c_{4\downarrow}^{+}c_{5\uparrow}^{+})\cdots|0\rangle \tag{2.81}$$

in the BOW case. Clearly the two Kekulé structures in (2.81) correspond to sequences of spin singlet pairs formed from electrons on adjacent sites, with the two different diagrams corresponding to starting from an even or an odd site. Again referring to Fig. 2.30, we see that in valence bond diagrams the singlet pairing between sites is indicated by a line connecting them.

In each case the two extreme configurations can be transformed into each other by a successive application of H_{1-e}, yielding a series of intermediate states with the symmetric configuration half-way between the extremes. One can represent the full Hamiltonian in this space by a tridiagonal matrix with the expectation values of H_{e-e} on the diagonal and the resonance integrals off the diagonal. Clearly, configurations with small diagonal elements have a large weight in the ground state. Thus, for the CDW case the Hubbard U certainly favours the symmetric configuration rather than either extreme, since it always suppresses double occupancy. On the other hand, in the BOW case the expectation value of the Hubbard term *vanishes* for the two extreme configurations and is *positive* for the intermediate states, thus *suppressing* the resonance between the extremes. It follows that U favours one of the two configurations (2.81) more than the other by providing a barrier to resonance between them and thus enhances the tendency towards bond alternation [2.155, 173, 204]. The same argument can be used for the two Néel states to show that antiferromagnetism is also enhanced by U. Since, however, the Néel state is not a good description of the large U limit of the one-dimensional Peierls–Hubbard model (there can be no long-range spin order here, because spin is a continuous symmetry) and since the valence bond diagrams most favourable to the BOW and SDW are not orthogonal, the BOW and SDW coexist in one dimension. For $U > 4t_0$, the antiferromagnetic configurations start to dominate at the expense of the Kekulé structures, implying that the tendency towards bond alternation begins to decrease again.

Long-Range Coulomb Interactions and PPP *Models.* The valence bond arguments presented above for the Peierls–Hubbard model can be extended to the general PPP form (2.77b) of the electron–electron interaction *provided* that the coupling strengths V_j are downward convex, i.e.

$$V_{j+1} + V_{j-1} \geqq 2V_j. \tag{2.82}$$

The crucial quantities are again the expectation values of the Coulomb interaction for the various real space configurations, which can be calculated following an earlier analysis of *Hubbard* [2.241]. One finds that the BOW (SDW) configuration is favoured with respect to both the symmetric and the CDW configuration if

$$\frac{1}{2}U + \sum_{n=1}^{\infty} (V_{2n} - V_{2n-1}) > 0, \tag{2.83}$$

otherwise the CDW configuration is favoured [2.173, 174]. Importantly, both the Ohno and the Mataga–Nishimoto parametrizations of the PPP model satisfy the inequalities (2.82) and (2.83), and therefore one expects a BOW ground state for both parametrizations, in contrast to the UHF results [2.168]. Although the quantitative valence bond analysis summarized by (2.83) is applicable only in the limit of strong coupling, i.e. $t_0 \to 0$, its qualitative validity over the whole range of Coulomb strengths is suggested by the observation that precisely the same criterion (2.83) is found in the weak-coupling limit using g-ology [2.2, 179].

.

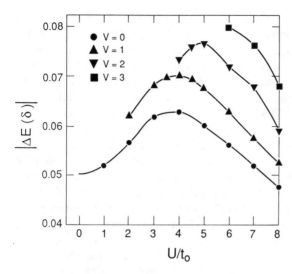

Fig. 2.31. The magnitude of the electronic energy difference between a dimerized and an undimerized periodic ring of $N = 10$ sites as a function of the on-site and nearest neighbour repulsions U and V. For a fixed U, all $V < U/2$ further enhance the dimerization

For weak coupling, where the expression on the left hand side of (2.83) corresponds to the bare vertex for umklapp scattering, the CDW/BOW boundary follows from simple gauge symmetry, as one can easily prove by generalizing earlier work [2.165].

Specific numerical calculations support the qualitative valence bond predictions. Indeed, for the relevant ranges of parameters, bond alternation is found to increase with the strength of the effective Coulomb interactions in both the extended Peierls–Hubbard model [2.155, 210, 211] and the PPP model [2.12, 173, 174, 212, 242]. In the extended Peierls–Hubbard model, the effect of V_1 is to *increase* δ at fixed U in the relevant region of λ [2.155, 173, 174, 191, 208, 211, 243]. As shown in Fig. 2.31, very large enhancements in bond alternation are obtained with nonzero V_1 for all $V_1 \leqq U/2$. Indeed the U at which bond alternation peaks appears to increase monotonically with V_1 until the BOW/CDW instability border at $2V_1 \cong U$. In the PPP models, full many-body calculations also support the qualitative valence bond arguments, showing strongly enhanced dimerization for both the Mataga–Nishimoto and the Ohno parametrizations. Importantly, the large enhancements within PPP models are not due to the distance dependence of the V_j, as suggested recently [2.244], but simply to the fact that $V_j \neq 0$ [2.155].

Off-Diagonal Coulomb Interactions. A recent study [2.48] has questioned the general applicability of extended Hubbard models for describing e–e interactions in conjugated polymers. In essence, some of the exchange and multi-centre interaction terms, which were discussed in Sect. 2.2 and are neglected in the PPP-type Hamiltonians, are reincorporated. Specifically, W and X terms

$$H_W \equiv W \sum_l (p_{l,l+1})^2 \qquad (2.84a)$$

and

$$H_X \equiv X \sum_l p_{l,l+1}(n_l + n_{l+1}), \tag{2.84b}$$

where $p_{l,l+1}$ is the bond-order operator (2.19), are considered. The effects of adding H_W and H_X to the extended Hubbard model (U and V_1 only) have been studied [2.48] in first-order perturbation theory. To this order one finds that W competes with V_1 and tends to *reduce* bond alternation, as long as $3W > V_1$. This analysis has two defects. Firstly, it is based on unreasonably large values of W; secondly, the correlation-induced enhancement due to the U term is completely neglected, since it enters only in second-order perturbation theory. More detailed studies [2.41, 51, 52], employing exact many-body methods in the anticipated ranges of coupling, establish that the effects of W and similar terms do not alter the basic conclusions deduced from the Hubbard and extended Hubbard models. Interestingly, however, the X terms does break charge conjugation/particle–hole symmetry and thus, even if it is numerically small, may have important qualitative effects [2.51, 52, 182], for example, in optical absorption. Furthermore, such density-dependent hopping terms have been studied recently in the contexts of both heavy fermion [2.245] and high-temperature super-conductivity [2.246].

We conclude this subsection by reiterating two crucial observations. Firstly, consistent with experiments on $(CH)_x$, the true ground state of the Peierls–extended Hubbard model Hamiltonian (2.77) in the relevant parameter regimes *is* a bond-alternated state. Thus the existence of a BOW state does not distinguish between independent-electron and correlated-band models. In particular, based on the observed dimerization, one cannot assume that e–e interactions are weak. Secondly, since e–e interactions *enhance* the dimerization at fixed (small) λ, it is possible to fit the observed bond alternation with a much weaker electron–phonon coupling than would be required in the purely single–particle model. This would bring the e–p coupling more into line with the value expected from the molecular parameters while at the same time permitting one to fit observed quantities, for example the spin density, that the independent-electron models cannot explain. We shall return to this point below.

b) Inhomogeneous Ground States: Solitons and Polarons in the Peierls–Hubbard Models

Energetics and Structures. The topological soliton and polaron solutions within the independent-electron theory were introduced explicitly in Sect. 2.3. Although they are most commonly viewed as nonlinear *excitations* of π-conjugated polymers, it is important to recognize that they can in fact be defined entirely as *inhomogeneous ground states* of specific systems. Thus, for example, the negatively-charged polaron is the ground state for $2N + 1$ electrons on $2N$ sites. Furthermore, as long as dimerization persists, the topological kink solitons will also exist in isolated, one-dimensional chains. Indeed, the neutral soliton will be the ground state of $2N + 1$ electrons on $2N + 1$ sites, whereas the positive (negative) soliton will be the lowest energy state of $2N$ $(2N + 2)$ electrons on $2N + 1$ sites.

In the absence of e–e interactions, the inhomogeneous order parameter corresponding to the soliton is given by (2.52). Within the independent-electron theories, neutral and charged solitons are energetically degenerate, have the same spatial extent, and are distinguished only by the spin or charge densities. Furthermore, as shown in Fig. 2.13, the spin or charge densities are zero on alternate sites. This situation changes dramatically for nonzero electron–electron interactions. To go beyond the general remark that topological considerations show that soliton solutions will persist as long as there is bond alternation, one must investigate specific models. Here, again for simplicity, we will treat only the $U \neq 0$, $V_j = 0$, case, as the results for the models with additional Coulomb interactions are at least qualitatively similar [2.209, 210, 212].

Using quantum Monte Carlo methods, one can calculate the ground state energies (E_D and E_S) of an open chain ($N = 21$) both in the rigid dimer (D) and in the (single site) soliton (S) configurations as a function of U [2.210]. The

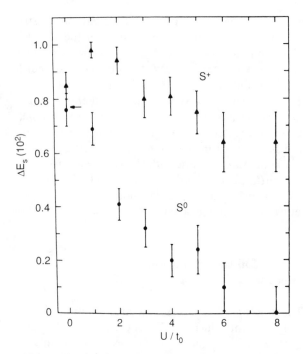

Fig. 2.32. The magnitude of the energy difference per site, $|\Delta E_s|$, between the soliton and the pure dimer configuration on an open $N = 21$ chain versus U. The solitons are always lower in energy. The curve marked by solid triangles and labelled S^+ is for the positively charged soliton, that marked by solid circles and labelled S^0 is for the neutral soliton. Both are in units of t_0. The arrow marks the exact $|\Delta E_s|$ for $U = 0$. Within the pure Hubbard model $\Delta E_{s^+} = \Delta E_{s^-}$ (after [2.209])

soliton stabilization energies $|\Delta E_S| = |E_D - E_S|$ are plotted in Fig. 2.32 as a function of U. Neutral and charged solitons are now strongly non-degenerate. This effect is also observed in Hartree–Fock solutions within the range of U where the ground state is a BOW [2.166–168]. This non-degeneracy of S^0 and $S^{\pm}$ can be understood easily within strong-coupling perturbative approaches that assume $U \to \infty$. In this limit the exactly half-filled band is described by an isotropic Heisenberg spin Hamiltonian with exchange integral $J \sim t^2/U$, while away from the half-filled case the Hamiltonian still has an electronic component $\sim t$. Thus the stabilization energy of the neutral soliton S^0 has only a magnetic contribution, while that of the charged soliton $S^{\pm}$ has contributions from (far larger) electronic terms.

The S^0 and $S^{\pm}$ are surrounded, respectively, by SDW and CDW polarization clouds, whose widths in the SSH limit are the same as the width of the bond alternation domain wall. For nonzero Coulomb interactions, Hartree–Fock methods [2.168] show that in both cases the widths of the kink deformations of the lattice are smaller than those in the SSH limit, with the lattice kink shrinking to a greater extent for the S^0. Furthermore, the widths of the SDW and the CDW profiles are now larger than the bond alternation profile. Based on our unpublished Monte Carlo data, we believe that these specific Hartree–Fock results are valid.

For the polaron and bipolaron states, the subtle energetics have precluded definitive studies in the presence of non-perturbative electron–electron interactions even in the one-dimensional chain limit. Since, however, for weak confinement due to an extrinsic t_1, or to weak interchain coupling, the bipolarons can be viewed as loosely bound soliton/anti-soliton pairs, one anticipates from the stability of the charged solitons that the bipolarons will also survive the inclusion of one-dimensional Coulomb effects [2.209, 210].

Spin Densities. An important observable consequence of electron–electron interactions is the occurrence of negative spin densities on sites which within one–electron theory have zero spin density. Thus let us examine this point in detail. Negative spin densities in finite polyene radicals were demonstrated as early as 1962 [2.64, 65], while the theoretical prediction of negative spin densities (within a Heisenberg antiferromagnetic model) was made even earlier [2.247, 248]. Resonance measurements in polyacetylene itself also have found negative spin densities; early ENDOR (electron nuclear double resonance) and triple-ENDOR results gave $|\rho_-|/|\rho_+| \sim 1/3$, where $\rho_{\pm}$ are the *average* positive and negative spin densities [2.249, 250]. Based on this relatively low value of $|\rho_-|$, a rather small value of the on-site correlation U is required within the Peierls–Hubbard model. In Fig. 2.33 we show the results of a quantum Monte Carlo simulation [2.208] for an open chain of 25 atoms. This simulation required $U = 2t_0$ and $V_1 = t_0$ to obtain a spin density ratio of 1/3. Several Hartree–Fock and perturbative calculations [2.115, 176, 250] within the simple Peierls–Hubbard model ($V_1 = 0$) similarly find that a $U = 3 - 4\,\text{eV}$ ($\cong (1.2 - 1.6)t_0$) can give the above $|\rho_-|/|\rho_+|$. Since these results have often been cited as justifications for

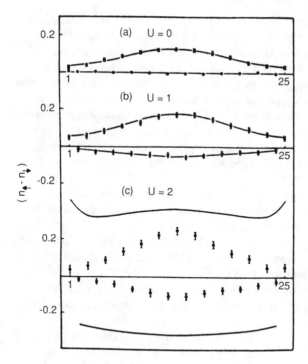

Fig. 2.33. Average spin density for three values of U (in units of t_0) in the presence of a fixed soliton configuration for a 25-site chain. The full lines are the unrestricted Hartree–Fock predictions (after [2.208])

a small U picture, we should examine carefully both the experimental and theoretical studies.

Firstly, it is important to realize that the earlier studies [2.249, 250] did not find the soliton bond alternation profile predicted from theory. Rather, they deduced a highly delocalized (over 49 sites!) defect with *equal* bond lengths and spin densities of *equal magnitudes* on all the carbon atoms. It is therefore not clear at all that comparisons of average spin densities from true soliton profiles with the average densities obtained by *Thomann* et al. is meaningful. Even more importantly, the calculated spin density ratios are strongly chain-length dependent [2.251], and therefore a complete analysis of the chain-length dependence is required before a realistic comparison with the data can be made. Several of the Hartree–Fock studies do not mention which chain lengths were used in the calculations, while the Hirsch and Grabowski calculation is for a chain whose length is half that of the defect size of Thomann. Secondly, other recent ENDOR results [2.252–255] strongly disagree with the earlier work. These authors find a much more localized defect and claim that $|\rho(1)/\rho(0)|$, where $\rho(0)$ is the spin density on the carbon atom at the centre of the soliton and $\rho(1)$, the density on one of the two adjacent carbons, is $\sim 1/2$. Without detailed theoretical modelling, it is not clear how $|\rho(1)/\rho(0)|$ is related to $|\rho_-|/|\rho_+|$; nonetheless, the rather high value for the former suggests a $|\rho_-|/|\rho_+|$ that is larger than that obtained by Thomann.

More recently, the structure of the neutral soliton and the spin density ratio have been reinvestigated [2.256–258]. These studies have used ENDOR, triple ENDOR and multiple quantum spectroscopy techniques on both Feast-type and Shirakawa-type polyacetylene. Pulsed ENDOR techniques were used instead of continuous wave ENDOR. The spin density ratio was found independently from triple resonance studies and was then used as a parameter in the determination of the soliton width from the ENDOR spectrum. Unambiguous proof that the soliton couples to many protons was obtained from multiple quantum ENDOR spectroscopy. The ENDOR spectra for the Feast-type polyacetylene was seen to be temperature-independent up to 200 K and the half width of the soliton was found to correspond to 12 CH units (i.e. total width = 25 units). The authors believe that the soliton in the Feast-type polyacetylene is effectively pinned. The fitted halfwidth was larger in Shirakawa polyacetylene, but motional narrowing was observed here at higher temperature. In both materials a ratio $|\rho_-|/|\rho_+| \cong 0.42$ was found.

This latest figure for the spin density ratio is sufficiently larger than the older value of $1/3$ to lead to interesting consequences. By comparing the Monte Carlo data of *Hirsch* and *Grabowski* [2.208], Mehring (Chap. 5) concludes that their spin density ratio corresponds to an effective $U = 6.25\,\text{eV}$, assuming $t_0 = 2.5\,\text{eV}$; thus $U = 2.5 t_0$. We believe that the comparison is reasonably valid in this case, as the chain length in the calculations and the experimental soliton width are roughly the same (i.e. 25 CH units). Furthermore, a ratio $|\rho_-/\rho_+|$ of ~ 0.41 was also obtained from PPP-Ohno calculations [2.251]. This agreement could be fortuitous, since the value 0.41 was obtained by extrapolation to 49 sites (in order to fit the width suggested by Thomann). Nonetheless, as we point out below, the replacement of the PPP parameters by an effective on-site repulsion requires a value that also lies within the range suggested above. While all these theoretical results seem to agree with the recent experiments, one caveat should be kept in mind: the width of the soliton is a very important parameter and in interacting-electron models the width of the SDW profile and that of the bond alternation profile are different [2.168]. We expect this last factor to change the deduced values of U somewhat. It is interesting that a recent study of the substantially different phenomena of neutral and charged soliton optical absorptions [2.206] finds a range for the effective U (see below) in agreement with Mehring, i.e. $2.25 t_0 < U < 2.75 t_0$.

Finally, we should comment on the use of an effective on-site correlation parameter, U_{eff}, in a Peierls–Hubbard model to reproduce the results of PPP-type models. Firstly, different observables may be affected differently by the V_j terms of the electron–electron interaction. Thus, as discussed above, bond alternation is enhanced by V_1 (for fixed U), whereas the spin exchange integral is independent of V_1, at least for large U [2.259]. Secondly, the often-made assumption that $U_{\text{eff}} = U - V_1$ (see, e.g., [2.12]) is not necessarily valid even if one looks at only a single observable. For example, the spin density ratio that one obtains from exact diagonalizations of PPP models [2.206] *cannot* be fit by a $U_{\text{eff}} = U - V_1$; this is particularly obvious in models of the PPP-Ohno type, where the decay of the potential with distance is very slow. For these models, the U_{eff} is larger than $U - V_1$. Despite these many caveats, since the on-site Coulomb interaction is expected to be by far the largest term, these subtleties will be hard to investigate experimentally and the use of a U_{eff} will be reasonable for many purposes.

2.6.2 Electronic Excitations and Excited States

Most experiments probing a many-body system involve measurements of excitations or excited states. In this section we shall discuss the theoretical expectations and experimental findings for a number of the most prominent excited states of finite polyenes and π-conjugated polymers. Particularly in view of our goal of clarifying the relationships among the chemical concepts typically used to discuss finite polyenes and the physics terminology developed to describe conducting polymers, throughout our discussion we shall attempt to stress the connections among the conventional excited states treated in the chemical literature and the nonlinear soliton and polaron excitations that dominate the physics analyses. Before going into detail, however, we should introduce several very general considerations.

Firstly, in independent-electron theories, e.g. the Hückel or SSH models, of finite polyenes or π-conjugated polymers, at the simplest level excitations from the ground state occur by promoting electrons from occupied molecular orbits (valence band states) to unoccupied molecular orbitals (conduction band states). Of course, when electron–phonon interactions are included, one must include the lattice relaxations in the excited states. This is familiar in the case of solitons and polarons in the SSH model; these excitations are clearly defined in terms of *both* the spatially localized distortions in the bond alternation patterns *and* the associated (momentum space) *intra*-gap, single-particle electronic states (non-bonding orbitals). When electron–electron interactions are included, however, this description in terms of single-particle levels breaks down, and one must view the excitations and excited states in terms of *many*-electron configurations and the associated distortions in the bond-alternation pattern. In particular, for example, for intermediate e–e coupling, it can be misleading to regard the lowest excited state of a negatively charged soliton as a neutral soliton plus an electron: one must look at the full many-electron structure of the excited state. Thus in the correlated band it is often useful to phrase the discussion in terms of a coordinate space description, for example the valence bond diagrams, rather than in terms of the molecular orbital/band theory picture that applies in the non-interacting limit.

Secondly, the separation of the time scales of the electronic interactions and the lattice relaxations which underlies the validity of the Born–Oppenheimer approximation, allows us to simplify our analysis of excited states in many cases. For instance, to describe an experiment involving very short time scales, such as optical absorption in the visible range, a sudden approximation, in which the electronic transition takes place with fixed lattice configuration, is expected to be valid. Of course, since the electronic Coulomb interactions are instantaneous, there may be excitonic effects in these transitions. In contrast, to describe lattice dynamics occurring on a much longer time scale (long compared to the inverse phonon frequency), an adiabatic treatment should be appropriate, in which at each time step the electronic subsystem is in its respective ground state appropriate for the lattice configuration at that instant. A hybrid case is a photogeneration experiment, in which initially the electronic system is excited "suddenly" and subsequently the lattice relaxes on the time scale of the inverse phonon frequency.

Thirdly, independent of the relative strengths of the various interactions, one can generally classify eigenstates of the finite, even, linear polyenes as symmetric (A) or antisymmetric (B) with respect to the two-fold axis perpendicular to the plane of the chain and passing through the centre of the chain (see, e.g., [2.158]). Further classification into gerade (g) and ungerade (u) is possible due to the existence of a centre of inversion. The ground state, for example, is always 1^1A_g, where the first integer is the principal quantum number, indicating the sequence of states of a given symmetry, and the superscript 1 refers to the spin being singlet. Since these symmetries are most readily expressed in coordinate space, the valence bond diagrams form a natural basis for developing intuition concerning the consequences of the symmetries.

Finally, in this subsection we shall focus on the electron excitations and excited states, exploring both excitations from the homogeneous ground state and from inhomogeneous, intrinsic-defect states. In Sect. 2.6.3 we shall deal briefly with the excitations and dynamics associated primarily with the lattice degrees of freedom.

a) The ^1B_u Optical State, the Optical Gap and Excitonic Effects

Since the ground state of finite polyenes has ^1A_g symmetry, the antisymmetric nature of the dipole operator requires that dipole-allowed (i.e., transitions involving single photons) excitations from the ground state lead to ^1B_u states. In the independent-electron limit, the quasi-one-dimensional nature of the polyenes leads to a concentration of the oscillator strength in the transition to the lowest optically allowed state; in physics terminology, this concentration is reflected in the familiar square-root singularity that one-dimensional systems exhibit at the absorption edge. Within *all* independent-electron approximations and mean-field theories, *independent of the level of sophistication*, this optical ^1B_u state is predicted to be the lowest excited state of the finite chains. In fact, as we shall discuss momentarily, experiments indicate a different sequence of excited states, at least in polyenes, and hence provide direct indications for the role of e–e interactions. Accordingly, it is useful to anticipate the nature of the ^1B_u optical state in the presence of (possibly strong) e–e interactions. For this purpose, as noted previously, we can best use a valence bond description. In the limit of stronge e–e coupling, the ground state consists of purely covalent valence bond diagrams with no double occupancies but (since the band is half filled) all sites singly occupied. Since the current operator mediating the optical transition transfers charge between sites, the lowest optically allowed ^1B_u state must have one double occupancy and hence some ionic character. Other states of ^1B_u symmetry made from purely covalent diagrams may lie lower in energy in this limit, but they cannot be reached by (dipole-allowed) optical excitation. This, of course, reflects the result, already noted in Sect. 2.4, that for the Hubbard model $E_g \to U$ at large U (2.71).

Although the existence of a finite excitation energy between the ground state and the lowest optically allowed state is one of the most prominent and obvious features of π-conjugated polymers and their finite analogues, the underlying physical origin of this optical gap remains one of the most contro-versial theoretical issues in the field. From our discussions in Sects. 2.3 and 2.4, it is clear that such a gap can arise *either* from electron–phonon (e–p) coupling (as in the SSH model) *or* from electron–electron (e–e) interactions (as in the Hubbard model). However, attributing the optical gap entirely to either of these interactions seems implausible; recalling our earlier example, fitting the observed optical gap energy $E_g = 1.8\,\text{eV}$ in *trans*-$(CH)_x$ using solely e–p interactions requires a (dimensionless) coupling constant $\lambda \cong 0.20$, which is substantially larger than that suggested by the bond length-bond order relationship observed in related organic molecules. Similarly, although the value of $U(\cong 3.5t_0)$ required to fit this observed gap in the pure Hubbard model is not so unreasonable, the (observed) existence of dimerization requires e–p interactions.

In the presence of both e–p and e–e interactions, one needs a means of determining the *joint* dependence of E_g on these interactions. In the case of simple Peierls–Hubbard models, this amounts to the well-defined question of the dependence of E_g on λ and U. Despite the definiteness of this question, its full answer is as yet unavailable. Thus we must at present be content with the partial answers which we discuss here. We note first that the early Ansatz [2.260, 261]

$$E_g(U, \lambda) = [E_g^2(U) + E_g^2(\lambda)]^{1/2} \tag{2.85}$$

has no basis in theory and is unlikely to be valid. When the electron–phonon coupling dominates, a weak-coupling perturbation expansion in U is appropriate. *Wu* and *Kivelson* [2.176] have deduced the leading behaviour

$$E_g(U, \lambda) = E_g(\lambda)\{1 + (3/4\pi^2)(U/W)^2 \ln^2[4W/E_g(\lambda)]\} \tag{2.86}$$

in the continuum limit, where W is a cut-off energy (not necessarily the band width) and

$$E_g(\lambda) = 4W \exp[-(2\lambda)^{-1}]. \tag{2.87}$$

Identifying (2.87) with (2.47) we find $W = (4/e)t_0$ and (2.86) becomes

$$E_g(U, \lambda) = E_g(\lambda)\{1 + 0.56(U/8\lambda t_0)^2\}. \tag{2.88}$$

On the other hand, on the basis of the Gutzwiller variational wavefunction, the BCS-type formula

$$E_g(U, \lambda) = (16/e)t_0 \exp\{-[2\lambda + 0.52(U/4t_0)^2]^{-1}\} \tag{2.89}$$

has been derived in the limit of weak couplings, $\lambda \ll 1, U \ll 4t_0$ [2.191]. For dominant electron–phonon coupling, the expansion of (2.89) in powers of $(U/4t_0)^2$ yields again (2.88) with the coefficient 0.56 replaced by 0.52. Moreover, for dominant U the Gutzwiller result is quite close to the exact result for the pure Hubbard model [2.192]. This gives additional credibility to (2.89) and yields an estimate for the crossover from the regime of a Mott–Hubbard insulator (e–e interactions dominant) to that of a Peierls semiconductor (e–p interactions dominant): we find

$$\lambda_c \approx 0.26 (U/4t_0)^2. \tag{2.90}$$

Using the molecular value of λ deduced earlier ($\lambda \sim 0.08$) and remembering that it is about a factor of 2 too small to reproduce the observed gap, we observe that for these parameters the electron–phonon coupling and the Hubbard term give about equal contributions to the optical gap. We tentatively conclude that conjugated polymers belong to the crossover regime. From (2.90) and with $\lambda \approx 0.1$ we find $(U/4t_0) \approx 0.7$; recall that this is consistent with the estimate $U \cong 2.5t_0$ obtained from the recent ENDOR experiments [2.256–258]. It seems likely that for such parameter values the perturbative expression (2.86) is no longer valid. Thus it may be worthwhile to mention that a quantitative fit of the optical gap of small polyenes on the basis of PPP parameters gave $\lambda \cong 0.11$ [2.12], in good agreement with the molecular $\lambda \cong 0.08$ proposed in Sect. 2.5.1.

To obtain definitive results for E_g in the intermediate coupling region in which U and λ provide comparable contributions will require further extensions of present many-body results. Exact diagonalization methods can calculate optical gaps [2.146] but these are influenced by finite-size effects. Present quantum Monte Carlo calculations have difficulties generating reliable estimates of E_g, since these estimates require accurate knowledge of either excited states or (equivalently) of the spectrum of the current operator. One method worthy of further investigation is based on the observation that for both the SSH model and the Hubbard model the optical gap energy E_g is the same as the *sum* of the energies required for adding and for removing an electron to the system: that is,

$$E_g = E_0(N + 1) - E_0(N) + E_0(N - 1) - E_0(N), \tag{2.91}$$

where $E_0(N)$ is the ground state energy for a system of N electrons (for a *fixed* lattice configuration). Although (2.91) does not hold if longer range Coulomb interactions are present (because of exciton binding effects), it is tempting to assume that it is valid for the simple Peierls–Hubbard model involving only U and λ; one would then have a way of calculating E_g using only (proven) algorithms for the ground state. Studies using this approach are currently underway.

Obviously, it is pointless to attempt to determine unique values for *several* parameters by fitting to a *single* observable, such as the optical gap. But before discussing how other experimental observables can be described within models incorporating both e–p and e–e interactions, it is instructive to compare the optical properties of polyacetylene to those of polydiacetylenes (PDAs). From Fig. 2.6 we recall that PDAs are π-conjugated polymers in which every alternate double bond of polyacetylene has been replaced by a triple bond, while the carbon atoms joined by the remaining double bond are typically attached to very large side groups (R, R'), whose sizes lead to large separations between PDA chains. Furthermore, many PDAs form high quality crystals in the solid state, with long conjugation lengths in the chain directions. Thus, optically the PDAs, rather than $(CH)_x$, are more likely to resemble an ideal infinite, isolated quasi-one-dimensional π-conjugated chain.

The band structure for polydiacetylenes was solved within an effectively one-dimensional Hückel model by *Wilson* [2.262] and *Cojan* et al. [2.263]. Since there are four "atoms" in the unit cell, there are now four bands, of which the lower two are full and the upper two are empty. The optical gap is given by [2.263]

$$2\Delta = 2t_2\{r^2 + \tfrac{1}{2}(1+s^2) - [\tfrac{1}{4}(s^2-1)^2 + r^2(1+s^2)]^{1/2}\}^{1/2} \qquad (2.92)$$

where $r = t_1/t_2$, $s = t_3/t_2$, and t_1, t_2, t_3 are the hopping integrals for the single, double and triple bonds respectively. Here 2Δ is the gap between the upper occupied band and the lower unoccupied band. While a broad band-like absorption is again expected within the single–particle picture of (2.92), experimentally a sharp excitonic feature at 1.9–2.1 eV is found instead, and the photoconductivity threshold is at 2.5 eV [2.264, 265]. More importantly, if one assumes that the same linear bond length–bond order relationship determines t_3 as well as t_1 and t_2 in (2.48), then a very large t_3 (corresponding to a triple bond length of only 1.2 Å) and consequently large optical gap 2Δ are expected. Indeed, independent of which linear relationship for the e–p coupling is chosen, straightforward comparison of (2.92) and (2.47) shows that Hückel theory predicts for the PDAs optical gaps that are nearly *twice* that of polyacetylene. The experimental gaps, however, are comparable, and one obvious way to reconcile this is to recognize that *bond alternation* (or superalternation in the PDAs) *does not dominate the gaps*. Rather, Coulomb interactions are essential to understand the optical gaps [2.102, 266, 267].

Additional support for this perspective is found on comparing the *finite* analogues of PDAs, the ene-yne molecules, to the finite oligomers of $(CH)_x$, the polyenes. These oligomers and their intermediates have been obtained by direct synthesis [2.268] during the solid state photopolymerization of diacetylenes [2.269, 270]. Comparison of the smallest optical excitation energies in the ene-yne molecules and in the corresponding polyenes with the same number of carbon atoms indicates that the optical gaps are *smaller* in members of the former class for small chain length. Note that the presence of *smaller* gaps in the ene-ynes is in direct contradiction to Hückel theory predictions. It can be shown that neither finite size effects nor sp $-$ sp^2 site energy differences alter this conclusion [2.266]. Thus one is left with the task of explaining simultaneously *both* the *smaller* gaps of the oligomers of PDAs compared to the finite polyenes *and* the *larger* gaps of the PDA polymers compared to *trans*-(CH)$_x$. Two different explanations have been proposed. Firstly, the PDAs have, in addition to the conjugated π-orbitals, localized π'-orbitals and these π and π' are mutually orthogonal. However, Coulomb matrix elements between these orbitals are non-vanishing and can rationalize fully the observed differences between the finite oligomers and the polymers. Secondly, one can assert that due to weaker solid state screening and consequently stronger effective Coulomb interactions in the PDAs than in *trans*-(CH)$_x$, these materials are simply qualitatively different [2.176]. Since the screening affects mostly the long-range part of the interaction it is conceivable that the simple Peierls–Hubbard model is sufficient for describing polyacetylene, but not for treating PDAs, where at least the nearest-neighbour interaction (V_1) appears necessary to explain the strong singlet exciton absorption (see below). In polyacetylene such an exciton absorption is not observed. From our perspective, this second explanation is less appealing, since it offers no explanation of the definitive finite ene-yne data.

In our preceding remarks we have focused primarily on the inter-band optical transition—the optical gap—in polyenes, polyacetylene, and polydiacetylene. In many solid-state materials, however, the onset of optical absorption is dominated by excitonic effects; in conventional physical terminology, an exciton is a neutral excited state consisting of an electron–hole pair in either a spin singlet or spin triplet state, bound by Coulomb forces. Since these forces are instantaneous, i.e. *not* retarded, they can alter the behaviour of the optical absorption at the band edge. Indeed, in polydiacetylene, for example, a singlet exciton is observed as a pronounced peak in optical absorption at 2.0 eV, about 0.5 eV below the inter-band absorption edge [2.271]. For the extended Peierls–Hubbard model the relative values of triplet and singlet excitation energies depend sensitively on the relative values of U and V_1 [2.272].

In all π-conjugated systems, the electron–phonon coupling provides an additional binding mechanism for the exciton; as indicated above, this e–p interaction, however, is considerably retarded and thus for optical processes effectively takes place after the initial electronic transition. However, the subsequent relaxation processes are very important both experimentally and theoretically. Resonant Raman experiments in polydiacetylene [2.273] have demonstrated a whole cascade of phonon emissions associated which e–p couplings. Furthermore, measurements of luminescence from the fully relaxed state provide information about the total exciton binding energy, including both Coulomb and electron–phonon interactions. Unfortunately, the interpretation of these measurements is complicated by the possible role of extrinsic trapping centers.

Although the limit of zero Coulomb interactions is not considered in the traditional excitonic picture, it is nonetheless instructive to consider the relation of these excited states to the evolution and (possible) binding of electron-hole pairs predicted in independent electron models with only e–p interactions. In the SSH model for *trans*-(CH)$_x$, an optically excited electron–hole singlet pair formed on a single chain is predicted [2.274, 275] to evolve (via lattice relaxation arising from the e–p coupling) into an $S^+ - S^-$ pair in a singlet spin state. SSH-like models for systems with non-degenerate ground states (such as the aromatic polymers) predict that analogous lattice relaxations around optically generated electron–hole pairs will produce excitons bound *entirely* by e–p coupling, the exciton being the neutral counterpart of the (charged) bipolarons discussed in Sect. 1.3. In the absence of *inter*-chain coupling, the singlet exciton formed in this manner would presumably be short-lived; as discussed below, when inter-chain coupling is included, it is expected to contribute to the inter-chain charge separation necessary to interpret the subgap optical absorption features.

b) The Triplet Gap and Magnetic Excitations

Although not accessible to the (spin-conserving) optical transitions, triplet excitations from the ground state have long been of interest in studies of models for conducting polymers (see e.g., [2.123]). In the limit of no e–e interactions but in the presence of a (Peierls) gap for the particle–hole excitations, the lowest-lying triplet excitation is a ^3B_u state which, in the absence of lattice relaxation effects and independent of chain length, occurs at the same energy as the optically allowed ^1B_u state. In this limit the triplet state is obtained simply by promoting an electron across the gap from an occupied molecular orbital to an unoccupied orbital and its magnetic character is obtained simply by flipping its spin. Notice that, in terms of valence bond diagrams, this lowest $U = 0$ triplet will contain a mixture of ionic and covalent diagrams, since the band states from which it is made are themselves mixtures. In the strong coupling limit, the triplet state is formed simply by flipping a spin in the singlet ground state; the discussion of the Hubbard model in Sect. 2.4 shows that the energy of this excitation is, in the absence of dimerization, of order $t^2/U \to 0$ as $U \to \infty$. Furthermore, in this limit the lowest triplet will consist of purely covalent valence bond diagrams, as states containing any ionic contributions will be at least $O(U)$ away from the ground state. Thus, the lowest ^3B_u state changes character from mixed ionic-covalent exciton-like for small e–e interactions to a purely covalent spin-like excitation for large e–e interactions. Furthermore, its energy falls from the value of the (Peierls) optical gap, i.e. equal to the energy of the optically allowed ^1B_u state, to zero for $U \to \infty$. Hence the location of the lowest triplet state relative to the optical ^1B_u state provides useful insight into the relative strengths of e–p and e–e interactions.

In both finite polyenes [2.158] and polydiacetylene [2.276] triplet states have been observed. In the polydiacetylene experiments, UV light of energy 3.5 eV was used to create the triplets; since conservation of spin during photoexcitation requires that triplets be produced in pairs, we see that this result implies that

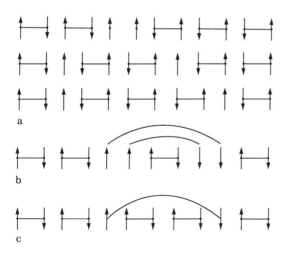

Fig. 2.34 a–c. Valence bond diagrams providing a schematic description of the 1^3B_u and 2^1A_g states: **a** indicates that in general two bond alternation reversals, characteristic of two solitons, occur in the 1^3B_u state, **b** shows a contribution to an A_g state involving *two* long bonds and *four* bond alternation reversals, corresponding to four solitons, **c** the dominant configuration of the 2^1A_g state, involving two bond alternation reversals, corresponding to two solitons. In VB notation, a line linking two spins indicates pairing into a spin singlet; orientations of the individual bonded spins are schematic rather than precise.

the energy of this triplet exciton must be less than 1.75 eV [2.276, 277]. Recalling that in this material the inter-band transition sets in at 2.5 eV, we see that the binding energy of this triplet exciton in polydiacetylene is about 0.75 eV, which is larger than that of the singlet exciton (0.5 eV) in the same material. To date, the location of the lowest-lying triplet state in *trans*-polyacetylene and in aromatic polymers remains unclear (but see [2.278]).

To predict the locations of triplet states in π-conjugated polymers, we must as always take into account the combined e–p and e–e effects, and allow for relaxation of the lattice. Since this has not yet been done systematically, we limit ourselves here to two general remarks. Firstly, in a manner similar to that in which a singlet electron–hole pair in *trans*-CH_x is predicted to evolve into an $S^+ - S^-$ pair, a *triplet* state in that material is expected to evolve into two neutral solitons in a spin triplet configuration [2.279]. Interestingly, based on studies of bond alternation patterns in finite systems with non-zero e–e interactions, this interpretation appears to remain valid even in the limit of strong e–e coupling [2.189]: that is, even in the presence of strong e–e interactions, the two bond reversals characteristic of the two neutral solitons occur in the ^3B_u state. As shown in Fig. 2.34, the ^3B_u state consists of valence bond diagrams where the spins forming the triplet are at neighbour and more distant sites. Secondly, once formed, the lowest triplet state is stable against (dipole-allowed) radiative decay; hence this state may play an important role in, for example, the long-time behaviour of photo-induced photoabsorption.

c) The Optically Forbidden 2^1A_g State

Experimentally, it has been well established for many years [2.158, 280] that the actual lowest singlet excited state in finite polyenes is an optically forbidden 2^1A_g state. Very recently, this has been confirmed in the very long polyene octadecaoctaene, which has eight double bonds. Here the 2^1A_g state occurs [2.175] at 17871 cm^{-1} ($\cong 2.2$ eV), which is 4900 cm^{-1} ($\cong 0.61$ eV) below the 1^1B_u state, which occurs at 22770 cm^{-1} ($\cong 2.82$ eV). Based on this result, plus earlier data, *Kohler* et al. [2.175] suggest that the $N \to \infty$ limits of these two states are $E(1^1B_u) = 1.76$ eV and $E(2^1A_g) = 0.91$ eV. This extrapolation is shown in Fig. 2.35.

Theoretically, there have been extensive studies of the optically forbidden states in finite polyenes and π-conjugated polymers. The location of the 2^1A_g state in finite polyenes *below* the optical gap is well understood within models involving electron–electron interactions [2.12, 60, 61, 186, 188, 189, 198–201]. A very satisfying, intuitive understanding of these theoretical results can be gained by examining the Hubbard model in the case of the half-filled band relevant to π-conjugated polymers; for later purposes, we discuss this using both chemical and physical terminology.

To understand this result in chemical terms we again use the valence bond (VB) approach and consider the probability of double occupancy, $\langle n_{i,\uparrow} n_{i,\downarrow} \rangle$, as a function of U. For $U = 0$, $\langle n_{i,\uparrow} n_{i,\downarrow} \rangle$ is 0.25 in the 1^1A_g ground state. This result follows immediately from the equal probability of all possible occupancies, $0, 1(\uparrow), 1(\downarrow)$, and 2, of any given site in the half-filled band for $U = 0$. As U is increased, $\langle n_{i,\uparrow} n_{i,\downarrow} \rangle$ decreases in the ground state (to obtain the lowest energy, double occupancies must be avoided) and the ground state becomes more covalent, with the ionicity (i.e. number of double occupancies) approaching zero.

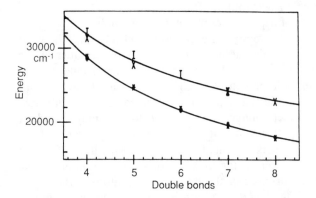

Fig. 2.35. 1^1B_u 0–0 excitation energy (unsubstituted (X) and alkylsubstituted (I)) and 2^1A_g excitation energy (unsubstituted (*) and alkylsubstituted (+)) versus the number of double bonds n for linear polyenes in hydrocarbon solution. The smooth curves are the function $E = A + B/n$ where the parameters A and B were determined by least squares fitting to the data, giving $E(2^1A_g) = 7370$ cm^{-1} + 85,500 cm$^{-1}/n$, $E(1^1B_u) = 14,200$ cm^{-1} + 70,600 cm$^{-1}/n$ (after [2.175])

Similarly, the ionicity of the 1^1B_u also decreases but approaches the value 1, i.e. 1^1B_u at large U is singly ionic. Since in all linear polyenes with more than two π bonds the number of distinct, entirely covalent VB diagrams is greater than one, it is apparent that for large U there are many potential configurations, and hence potential excited states, which should be lower in energy than the optically-allowed 1^1B_u. In Fig. 2.30 we show several covalent diagrams and a few singly ionic diagrams for trans-hexatriene with six carbon atoms. One can see that individual singly ionic diagrams (with one double occupancy and empty site each) cannot have mirror plane symmetry, but basic functions consisting of both plus and minus linear combinations (possessing plus and minus mirror plane symmetry) of these diagrams can be constructed. The former occur in the A subspace, while the latter occur in the B subspace. In contrast, several of the individual covalent diagrams *do* possess mirror plane symmetry and can therefore occur only in the A subspace. While there also exist covalent diagrams without mirror plane symmetry, which thus contribute to both A and B subspaces, minus linear combination of these have the same electron-hole symmetry as the ground state [2.61]. The nature of the dipole operator is such that optical transitions are allowed only between states with different mirror plane *and* electron-hole symmetries, so that the optical B_u subspace can be constructed only from the ionic diagrams. As a result of this basic difference between the two A_g and B_u subspaces with the required electron–hole symmetries, for non-zero U the lowest excited states come from the A_g subspace. Note that the 2^1A_g and higher A_g states are then characterized by long bonds between sites, a point that we will return to later.

In physics terminology, this same argument can be couched in the strong coupling ($U \to \infty$) limit of the Hubbard model [2.126, 135, 136]. Here the 2^1A_g becomes simply a spin-wave excitation, which is gapless in the limit of no dimerization and $U \to \infty$, while the optical ^1B_u state *must* involve a charge excitation, which will be at energy $U \to \infty$ (recall Sect. 2.4).

From the limiting arguments, we see that for the infinite-sized system the 2^1A_g excitation approaches zero energy in the undimerized, large U, whereas it has the same gap as the 1^1B_u state in the dimerized, $U = 0$ system. For finite dimerization, the 2^1A_g state is separated from the ground state by a finite gap, the magnitude of which depends on both U and δ, the dimerization. In the context of the PPP-Ohno model, considerable effort [2.12, 189] has gone into determining the expected 2^1A_g gap in the presence of non-zero dimerization. Assuming the gap to extrapolate linearly as a function of N^{-1} or $(N+1)^{-1}$, where N is the chain length, *Soos* and *Hayden* [2.12] predict a gap of 1.2–1.3 eV from the ground state at $U \to \infty$. In addition, *Tavan* and *Schulten* [2.189] argue that a band involving many covalent states—$3^1A_g, 4^1A_g$, etc.—should be below the optical gap in long polyenes. A somewhat different theoretical approach [2.206] recently investigated within the Peierls–Hubbard model the problem of the minimum U at which the 2^1A_g state goes below the 1^1B_u for dimerized, finite polyenes. With realistic alternation parameters, it is found that in the finite polyenes $4 \le N \le 10$, the *minimum* U at which the states cross is surprisingly independent of N, although in the longer systems the U at which the 3^1A_g, 4^1A_g, etc. come below the 1^1B_u does depend on N. This is shown in Fig. 2.36. The crossing occurs at $U \approx 2t_0$, in excellent agreement with a renormalization group study for $N = 16$ [2.186].

Another aspect of these renormalization group arguments, however, does not fare so well. On the basis of the strong coupling arguments given above, supplemented by energetic considerations [2.126, 186], one can argue that 2^1A_g is actually composed of two triplet states (^3B_u) which are themselves produced by two elementary spin flips (Fig. 2.34). In view of this relation, it might seem that the fully relaxed 2^1A_g should consist of *four* kink solitons. Indeed, the real space renormalization group calculation [2.186] suggested precisely this interpretation. However, a more detailed many-body calculation [2.189] demonstrated that the bond alternation pattern associated with the 2^1A_g state contains just *two* solitons. This situation is depicted in Fig. 2.34 and can be understood qualitatively in terms of valence bond diagrams by noting that the wavefunction of the *lowest* singlet formed from two triplets should be dominated by the configu-

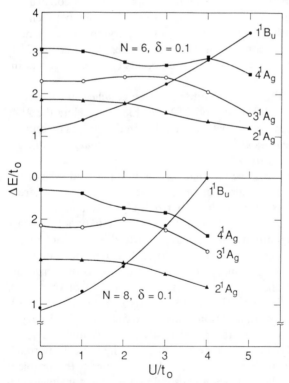

Fig. 2.36. The values of various low-lying excited states for linear hexatriene ($N = 6$) and linear octatetraene ($N = 8$) versus the on-site U for $\delta = 0.1$ in a Peierls–Hubbard model. Note that the values of U at which the 2^1A_g crosses the ^1B_u state is the same in both cases

ration having the fewest possible long bonds. This suggests that the configuration having one long bond, and consequently containing only two bond reversals and hence only two solitons, should in fact dominate.

The experimental situation regarding the location of the 2^1A_g state in *trans*-$(CH)_x$ has for some time been uncertain. An early photoluminescence study [2.281] suggested a two-photon allowed, and hence ^1A_g, state occurred in the range of 1.2–1.3 eV, below the $\cong 1.8$ eV optical gap. More recently, measurement of the frequency-dependent, third-order nonlinear optical susceptibility ($\chi^{(3)}(\omega)$) in *trans*-$(CH)_x$ [2.282] has shown a peak in $\chi^{(3)}$ at 0.91 eV and a second peak at 0.6 eV. The authors of this study associated the peak at 0.91 eV with the two-photon allowed 2^1A_g, thereby suggesting that this state occurs at about 1.8 eV, which is about the same energy as the 1^1B_u. An independent measurement of $\chi^{(3)}$ using a free electron laser [2.283] has confirmed the locations of the two peaks and supports this assignment.

Although the occurrence of the 2^1A_g state at the band edge might seem to imply that U_{eff} must be small, as stressed by *Fann* et al. [2.283], this is not necessarily the case. Firstly, although calculations in the continuum version of the Hückel model reproduce both the features at ~ 0.9 eV and ~ 0.6 eV [2.284], the intensities of these absorptions are considerably different from the experimental results. Secondly, exact diagonalization calculations of finite chains in the PPP-Ohno model [2.205] also show structures similar to the experimental observations, albeit with interpretations which differ from those in the single-particle theory. Thirdly, in the Peierls–Hubbard model, finite chain diagonalizations [2.206] suggest that the minimum value of U for which the 1^1B_u and the 2^1A_g states occur at the same energy is $U = 2t_0$, independent of chain length. Thus at present the overall situation seems quite unresolved, particularly when one recalls that for isolated quasi-one-dimensional finite polyenes, the 2^1A_g is definitely observed *below* the 1^1B_u and, by extrapolation, this result is expected to hold for the isolated quasi-one-dimensional polymer. Clearly further research into intra-chain Coulomb interactions and the effects of inter-chain screening and electron transfer is required to clarify this situation.

In the polydiacetylenes, the experimental situation is, perhaps surprisingly, less clear. From solution studies, it has been suggested [2.285] that the 2^1A_g occurs *above* the 1^1B_u in PDAS. Recent studies of crystalline samples have reported observing the 2^1A_g both above [2.286] and possibly below [2.287] the 1^1B_u state. Although a 2^1A_g above the 1^1B_u in PDA may seem in conflict with the thrust of our discussion of *trans*-$(CH)_x$, there are at least two reasons that the 2^1A_g state here may be higher than the 1^1B_u, *despite* possibly stronger correlation effects in PDA. Firstly, the very strong bond alternation in the PDAs, in which the triple bond is only of length 1.2 Å, raises the energy of the 2^1A_g substantially. Physically, this is understandable from Fig. 2.34: the 2^1A_g is associated with covalent diagrams with long bonds, and the destruction of the very short triple bond required to make such long bonds in the PDAs costs more energy. Secondly, the π–π' Coulomb repulsion between the localized and conjugated π electrons in the PDAs (the repulsion mentioned in our discussion of the optical gap) *lowers* the energy of the 1^1B_u state but raises that of the 2^1A_g [2.266, 267, 288]. Further details may be found in the original references.

Our comparative discussion of the location of the 2^1A_g states in *trans*-$(CH)_x$ and the PDAs demonstrates that the question of the relative strength of electron–electron interactions in the two materials is a subtle one and cannot be answered by the simple statement that these interactions in the PDAs are much stronger than in $(CH)_x$. As we have shown, several additional factors can influence the location of these states, and a fully satisfactory understanding awaits further experimental and theoretical results. Nonetheless, it is clear that both e–p and e–e interactions play substantial roles in each of these materials.[3]

d) Theoretical Perspective on Subgap Excitations in Optical Absorption

In our previous discussion we have already indicated that the striking nonlinear excitations of the single-particle theory are expected to persist even in the presence of relatively strong e–e interactions. Furthermore, within Hückel theory clear signatures of these excitations are additional optical absorptions, having very distinctive shapes, at specific energies below the optical gap. Here we discuss how these sub-gap features are altered in models involving both e–p and e–e interactions. The complexity of the problem compels us to present only an elementary, semi-quantitative analysis. An important challenge to theory over the next few years is to investigate these questions more thoroughly and convincingly.

In systems of interacting electrons, the inherently many-body nature of the problem causes substantial difficulties in the calculations of sub-gap optical absorptions. Thus to date the results are fairly meagre, consisting primarily of perturbative expansions about the weak- and strong-coupling limits and of numerical studies of small, finite systems.

Weak-Coupling Perturbation Theory. To analyze the effects of weak e–e interactions, we consider an arbitrary N-electron eigenstate of H_{1-e}, Ψ_α, as the unperturbed state and the Hubbard term as the perturbation. To first order in U the energy shift is

$$E'_\alpha = U \sum_l \langle \Psi_\alpha | n_{l\uparrow} n_{l\downarrow} | \Psi_\alpha \rangle \tag{2.93a}$$

$$= U \sum_l \langle \Psi_\alpha | n_{l\uparrow} | \Psi_\alpha \rangle \langle \Psi_\alpha | n_{l\downarrow} | \Psi_\alpha \rangle, \tag{2.93b}$$

in view of Wick's theorem. Correspondingly, the interaction effect on an optical transition from an initial state Ψ_i to a final state Ψ_f is $E'_f - E'_i$. In a band-to-band transition the local densities remain unchanged, and there is no linear effect of

---

[3] Since the completion of this review, considerable progress has been made in understanding the nature of the A_g states that produce the two photon resonances in $\chi(3)$ experiments in both *trans*-$(CH)_x$ and PDAs. See, e.g. [2.289, 290]

U on the transition energy, in agreement with (2.88) and (2.89). On the other hand, if localized states are involved, since the local densities differ in general, there is a shift in the optical transition. Consider as an example the transition from the midgap level of a neutral soliton (with spin up) to the conduction band. The local densities are easily derived following the arguments of Sect. 2.3.5: namely,

$$\langle \Psi_i | n_{l\sigma} | \Psi_i \rangle = \tfrac{1}{2}[1 \pm |\psi_{l,0}|^2],$$
(2.94)

where the signs correspond to the spin directions $\sigma = \uparrow, \downarrow$, and

$$\langle \Psi_f | n_{l\uparrow} | \Psi_f \rangle = \tfrac{1}{2}[1 - |\psi_{l,0}|^2] + \frac{1}{N}$$
(2.95a)

$$\langle \Psi_f | n_{l\downarrow} | \Psi_f \rangle = \tfrac{1}{2}[1 - |\psi_{l,0}|^2].$$
(2.95b)

The change in the optical transition is

$$E_f' - E_i' = \tfrac{1}{2}U \sum_l |\psi_{l,0}|^4 = Ua/(3\xi)$$
(2.96)

where (2.54) has been used. Therefore the optical transition associated with the presence of a neutral kink shifts *upward* from mid-gap toward the band edge to $\Delta_0 + U/(3l_K)$ where $l_K = \xi/a$ is the (dimensionless) width of the kink. Similarly, the transition energy for the charged solitons shifts *down* to $\Delta_0 - U/3l_K$ [2.291]. This reduction of the effective Coulomb interaction parameters for an extended nonlinear excitation by the correlation length of that excitation was first noted within perturbation theory by *Kivelson* and *Heim* [2.167]. A similar but less explicit expression of this result was found in early HF calculations of Peierls–Hubbard models [2.166]. The HF results also suggested that, in the region of weak to intermediate Hubbard U where the dimerization persisted, the midgap absorptions should indeed be shifted, the neutral soliton upward and the charged soliton downward.

For a polaron, Hückel theory predicts three subgap transitions at $2\omega_0$, $\Delta_0 - \omega_0$ and $\Delta_0 + \omega_0$ (Sect. 2.3.6); using weak-coupling perturbation theory, we find that the first two transitions remain unshifted, whereas the third is split into *three* transitions, at $\Delta_0 + \omega_0$ and $\Delta_0 + \omega_0 \pm U/(3l_P)$ where l_P is the characteristic polaron length (Table 2.2) [2.292]. For bipolarons the two subgap transitions are both shifted downwards by $U/3l_{BP}$ [2.292]. Thus, in contrast to the single-particle theory, the sum of the two transition energies is less than the gap.

Strong-Coupling Perturbation Theory. To analyze the soliton absorptions in the presence of significant e–e interactions, one can also gain some intuition from the strongly correlated limit (see, e.g. [2.292]). Since, as indicated in Sect. 2.6.1, correlated states associated with solitons do exist even for large U, one can still define (and measure) the associated optical transition energies. For strong coupling, however, one can show that (with U only) the transition for the neutral soliton goes upward to the band edge (U) and the charged soliton absorbs at zero energy. Thus as the strength of the electron–electron interaction parameter U increases, this difference in transition energies crosses over from the weak coupling ($\sim U/l_K$) to the strong coupling $\sim U$ behaviour. With both U and V_1 present the soliton-induced absorptions in the strongly correlated limit are still at zero energy and at the band edge $\sim U - V_1$. Interestingly, the *correlated state* corresponding to the bipolaron is still expected to have *two* intra-gap absorptions, one at V_1 (in the limit of zero bond alternation) the other at $U - 2V_1$. On the other hand, in this strong coupling (undimerized) limit, the charged even chain (the correlated analogue of a polaron) will in fact absorb at the *same* energy, at zero energy in the limit of zero dimerization, as the charged odd chain, which is the correlated analogue of the charged soliton. Hence, in contrast to the behaviour expected in the independent-electron case, in this strong coupling limit the polaron would look optically like a charged soliton. We stress that this statement does not imply that correlations necessarily destabilize the polaron; rather this is just the charged analogue of the result that in strong coupling the neutral soliton absorbs at the same location as the (perfectly dimerized) even chain.

Numerical Results. For intermediate e–e coupling, one must resort to numerical many-body methods. Exact numerical calculations for finite chains described by the PPP-Ohno model [2.293–295] predict that both the charged and neutral soliton absorptions should be considerably shifted from midgap, with the neutral soliton absorbing near the band edge and the charged soliton absorbing at a much lower frequency. More recently, these calculations have been reinvestigated and extended [2.206] with the dual aims of reducing the finite-size effects inherent in all exact diagonalizations and of identifying the range of e–e interaction parameters necessary to fit the experiments. Using a Peierls–Hubbard Hamiltonian with U as the only e–e interaction parameter, one deliberately chooses an artificially large bond alternation, so that even for the short chain lengths, certain essential aspects of the behaviour of the infinite chains can be simulated. In particular, in the limit $U = 0$, the parameters are chosen such that the odd finite chain in the soliton configuration absorbs at *half* the optical gap of the even adjacent chains. As in the case of the 2^1A_g state discussed above, the motivation for this calculation is to determine the *minimum* value of U which reproduces the experimental behaviour. With hopping integrals $t_\pm \equiv t_0(1 \pm \delta)$, *Mazumdar* and *Dixit* [2.206] determined the optical gaps for open chains containing $N = 8, 9$, and 10 atoms. For $N = 9$ a single site soliton configuration, that is, a bond alternation reversal about a single site, was chosen. Then with an artificially large $\delta \cong 0.4 - 0.5$, both the even and the odd chains behave almost like the infinite chain: the even chain optical gap at $U = 0$ is now nearly $4t_0\delta$, while the $N = 9$ gap is close to half of this. The optical gap together with the intra-gap transitions for the neutral and charged solitons are depicted in Fig. 2.37 as functions of U. One deduces parameter range 2.25 to $\leq U_{eff} \leq 2.75t_0$ from a comparison with experimental data. This is consistent with the range of possible U_{eff} values deduced from fits to both spin density and optical gap measurements in *trans*-$(CH)_x$.

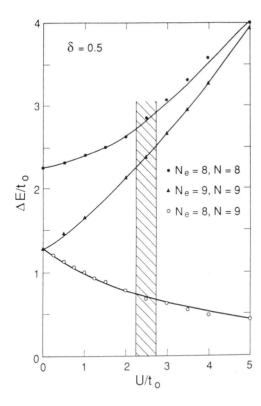

Fig. 2.37. Optical gaps as a function of U/t_0 of the even chain with perfect bond alternation, and of the neutral and charged odd chains in the single site soliton configuration. The striped region indicates the range of U/t_0 suggested by fits to the experimental data for S^0 and S^+ absorption

Experimental Result and Interpretation of Subgap Excitations in Optical Absorption. Subgap optical absorptions have been observed experimentally in all π-conjugated polymers, both in experiments involving doping of the pristine samples and in studies measuring photo-induced photo-absorption. Within the simple theory outlined above, there is no distinction between nonlinear excitations produced by doping and those produced by photoexcitation: both are expected to show similar optical absorption properties. Of course, in doping one generates only like charges, while in photo-induced absorption the excitations must be generated in overall neutral configurations. In addition, in doped samples, the counter ions, through their potentially inhomogeneous distributions and their pinning potentials, can lead to effects beyond those considered in the simple models. One of the successes of the Hückel/SSH model is the unified description it yields for the dopant-induced *and* photo-generated states. In what follows we shall use this framework as a guide for interpreting experimental data, although it will frequently become clear that other effects, in particular (explicit) electron–electron interactions, are by no means small perturbations. Since many of the experimental subtleties of optical absorption have been clearly presented in recent reviews [2.10, 278], we shall be fairly brief, focusing in particular on a limited number of representative experiments for purposes of clarifying our perspective on the present status of theoretical understanding.

Dopant-Induced Transitions. At very weak doping levels, on average less than one dopant ion per uninterrupted conjugation length, in polymers with either degenerate or non-degenerate ground states, one expects to see optical absorptions associated with the polaron excitations. In fact, optical data suggestive of polaron effects are scarce [2.296–298], but there is evidence for the simultaneous occurrence of spin and charge, i.e. of polarons, from magnetic measurements in polypyrrole [2.299], polyaniline [2.300] and polythiophene [2.301] (for a brief review see *Devreux* et al. [2.302]). In contrast, the midgap absorption associated with a *charged soliton* in doped *trans*-$(CH)_x$ has been widely reported [2.303–306]. Even under highly controlled conditions [2.305] the midgap absorption is quite broad and appears somewhat *below* midgap, at an energy $\cong 0.7\,eV$.

Furthermore, although the experiments are not generally interpreted as involving doping, studies of "as-grown" (and hence inadvertently doped) and ammonia-compensated (and hence effectively de-doped) *trans*-$(CH)_x$ [2.307] show that the expected midgap absorption corresponding to *neutral solitons* actually occurs at or near the optical gap (band edge), in sharp contrast to the predictions of the SSH theory, but in good agreement with predictions based on the PPP-Ohno model [2.294, 295].

In polymers such as polypyrrole or polythiophene the non-degeneracy of the two bond alternation patterns suggests that bipolarons should be formed at not too weak doping levels. Spectra with two (broad) sub-gap absorption peaks have indeed been reported [2.297, 298, 308, 309], thus confirming qualitatively the predictions of Hückel theory. However, the two peaks have integrated intensities which differ at most by a factor of two, far from the large ratio expected on the basis of the simplest independent-electron model [2.99]. As we shall discuss below, this intensity anomaly remains an important unresolved theoretical issue. Also it appears to be difficult to determine reliably the relative population of polarons and bipolarons. The optical spectra have been found to depend on the waiting time after the doping [2.308], suggesting that equilibrium statistical mechanics is not applicable. In the case of polythiophene the relative abundance of polarons and bipolarons as deduced from optical spectra appear to differ appreciably from those deduced from ESR data [2.301].

Photogeneration Experiments. Many recent studies of subgap transitions have used photo-excitation to generate the nonlinear excitations (for reviews, see [2.10, 278]). The wealth of data is rather difficult to summarize succinctly, but it is clear that a consistent picture can be obtained only if photo-induced absorption is analyzed together with photo-luminescence, photo-bleaching and photo-conductivity. Furthermore, it has become increasingly evident that photogeneration experiments in π-conjugated polymers give three quite distinct sets of results: (1) those for *trans*- and *cis*-$(CH)_x$, (2) those for polydiacetylenes and (3) those for aromatic polymers, e.g. polythiophene and poly-(para-phenylene vinylene). For clarity, we discuss the three classes of materials separately, in each case presenting both data and minimal interpretations.

In *trans-* and *cis-*$(CH)_x$, a wide variety of pump-probe arrangements have extended our knowledge into the femtosecond regime. Since the original work of *Orenstein* and *Baker* [2.310], many authors have studied different aspects of this problem [2.311–327]. After some initial disagreement, several features in photo-induced absorption are now well established for polyacetylene. In *trans-* $(CH)_x$, a prominent low-energy peak is induced at about 0.45 eV, interpreted as an absorption from a photo-generated soliton, but appreciably lower than the dopant-induced peak around 0.7 eV. This difference is attributed to the effect of dopant ions. A simultaneous photo-bleaching has been reported at 1.40 eV [2.319], and interpreted as the partial suppression of the subgap absorption from neutral kinks in pristine polyacetylene, in agreement with the experiments of *Weinberger* et al. [2.307]. Applying the perturbative results derived above to a $S^{\pm}$ transition at 0.45 eV and a S^0 transition at 1.40 eV, we find $2U/3l_K \cong 1$ eV [2.291, 328]. With $l_K \approx 7$ this implies $U \cong 10$ eV ($\cong 4t_0$), a value well into the intermediate coupling range, where perturbation theory is in fact likely to break down.

Support for associating the low-energy peak with photo-generated S^+/S^- pairs [2.274, 275] comes from the observation of accompanying IR-active modes, showing that a charged species is involved. However, the LE peak survives into the long-time regime [2.278], in contrast to what might be expected if the oppositely charged excitations remained on a single chain so that rapid recombination would be possible. As a consequence, there is now general agreement that a single-chain mechanism is insufficient to explain the LE peak in the PA data and that somehow inter-chain charge separation occurs, leading to the presence of long-lived, like-charged excitations on a single chain. This will be discussed further in Sect. 2.7. Important additional information about the nature of the photo-generated species can be expected from photo-induced spin resonance. Unfortunately, the experimental situation seems to be rather confusing in this respect; even the question of whether the number of spins is increased [2.329] or unchanged upon illumination [2.330, 331] (see also [2.10]) remains controversial.

In addition to photo-induced absorption, photo-conduction in $(CH)_x$ has also been observed at the band edge [2.332–335], consistent with the expectation of the independent-electron theories. However, like induced absorption, photo-conduction also has a fast and a slow component, perhaps again suggesting a strong inter-chain contribution [2.336]. Further details may be found in two recent reviews [2.278, 335].

As a second class of materials where photo-generation experiments have been carried out, we discuss the polydiacetylenes. These materials typically have very large side groups (Fig. 2.6) separating very long conjugated chains. Thus they might well be expected to behave differently from polyacetylene. Indeed, the optical absorption edge is dominated by a singlet exciton at $\cong 2.0$ eV, while photo-conduction sets in only at $\cong 2.5$ eV [2.264, 265] and is considerably more anisotropic than in polyacetylene. Furthermore, the charge recombination is now quite rapid, in the order of picoseconds [2.337–339]. In photo-induced absorption, there is no evidence of a low-energy peak in most polydiacetylenes, but photo-bleaching of the excitonic transition is observed [2.276, 340, 341]. Recently, studies [2.342] on a polydiacetylene with small side groups, $R = (CH_2OH)$, $R' = (CH_3)$, have suggested a low-energy peak at $\cong 0.25\,eV$. With such small side groups, it is plausible that the inter-chain charge separation believed to be essential for a long-lived photo-generated defect can occur. The photo-induced absorption in this system has been interpreted in terms of bipolarons, but as in polyacetylene there is no signature of polaron absorptions even at very short times.

The aromatic π-conjugated polymers provide the third distinct set of results. Within the SSH/independent-electron framework, systems such as polythiophene, polypyrrole and poly-(para-phenylene vinylene) are treated as pseudo-polyenes, in that only the conjugated chain in considered relevant for low-energy excitations. In the case of an aromatic ring involving a heteroatom, the charge-conjugation (electron–hole) symmetry can be broken, but this can readily be modelled along the lines first discussed in the case of the diatomic or $(A = B)_x$ polymer [2.73]. Strictly speaking, the heteroatom makes the carbons *inequivalent* in a way that renders these polymers effectively $(A = B - B = A)_x$ systems with no electron–hole symmetry [2.102]. Independent of this refinement, the inequivalence of the two bond alternation patterns and the resulting non-degeneracy of the ground state again imply, within the simple models, the relevance of bipolarons. Experimentally, many aromatic systems have been investigated, including poly-thiophene [2.309, 343, 344] and its derivatives poly(3-methylthiophene) [2.345] and poly(3-hexylthienylene) [2.346], and poly(para-phenylene vinylene) ([2.278] and references therein). Similar features are seen in all of these systems: two broad, sub-gap peaks and bleaching in the inter-band transition region. The results on polythiophene, which show absorptions at $\cong 0.45\,eV$ and $\cong 1.25\,eV$, are typical [2.344]. Photo-induced IR activity at lower frequencies indicates that the associated defect state is charged. Although ESR measurements indicate the possible formation of spins (associated with polarons), the induced sub-gap absorptions are qualitatively consistent with the formation of bipolarons. How-ever, the experimental data are not in quantitative agreement with predictions of Hückel-type theories (see Sect. 2.3), both in the location of the induced absorption peaks where the measured values of the two energies sum up to less than the gap, and in the relative intensities where the measured peak heights are comparable, in contrast to the large intensity ratio predicted theoretically. The failure of the energy sum rule can be easily understood as an effect of electron–electron interactions, as discussed above. The intensity problem, arising both in dopant- and photo-induced optical absorption, will be discussed in some detail below.

In addition to induced absorption, both photo-conduction and photo-luminescence have been studied in aromatic polymers [2.278]. Photo-conducti-vity is somewhat similar to that in *trans*-$(CH)_x$, occurring at energies slightly higher than the onset of interband absorption. Photo-luminescence appears to result from a polar-exciton, a neutral object bound (presumably) by a combination of e–p and e–e forces. Although the expected radiative decay of this excitation is quite rapid [2.347–350], the very small quantum yield suggests that various additional, non-radiative decay channels [2.278] also play a role.

The absence of a clear signature for polarons in photo-generation experiments remains a puzzle for all quasi-one-dimensional theories. Although (relaxed) polaron energetics are too subtle for easy analysis within correlated models, there is no obvious reason why quasi-one-dimensional Coulomb effects should destroy the polaron but (apparently) leave the (charged) bipolaron intact. A possible explanation for the absence of the polaron absorption is the potential destabilization of the quasi-one-dimensional polarons by transverse coupling effects in the true three-dimensional materials. Since this effect requires transverse coherence, it can at the same time explain that, in contrast to photo-generation, doping may produce stable polarons, as the electrostatic potential of the dopant ions is expected to destroy this coherence. These competing effects of transverse coherence and disorder may also explain the result that in polythiophene the number of photo-induced spins is found to drop below the detectable limit after improvement of structural order [2.344].

The Intensity Anomaly. Let us now turn to the intensity ratio of the high and low energy peaks for polarons and bipolarons. As already mentioned in Sect. 2.3.6 (Figs. 2.18, 19) the intensity ratio I_3/I_2 depends on the parameter ω_0, which is equal to $(\omega_3 - \omega_2)/2$ in the case of the SSH model. Fortunately, in the case of bipolarons this difference in transition energies is not changed to first order in the Hubbard parameter U [2.292], and therefore the relation $2\omega_0 = \omega_3 - \omega_2$ remains approximately valid for the (simple) Peierls–Hubbard model. A typical example for the discrepancy between (simple) theory and experiment is poly-(para-phenylene vinylene) where the measured locations of the gap levels suggest $0.33 < \omega_0/\Delta_0 < 0.4$ [2.278]. The corresponding intensity ratio of 8–12 predicted by the simple independent-electron theory [2.99] strongly disagrees with the observed ratio, which is close to unity [2.278].

As the large intensity ratio predicted by theory is a consequence of symmetries, several different approaches have been pursued, most based on breaking the charge-conjugation and mirror plane symmetries of the SSH model by the effects of additional interactions. Firstly, for $(A = B)_x$ polymers in which these symmetries are not present, it has been explicitly demonstrated that the two intensities can become comparable for certain values of the intrinsic site energy difference between the A and B chemical moieties [2.101]. Secondly, additional couplings of the atomic displacements u_l to the π-electron *density* have been proposed [2.351, 352] which can change this intensity ratio substantially, bringing it closer to the experimental value. A third possibility is an additional next-nearest-neighbour hopping term and, especially, its coupling to phonons [2.353]. Although intriguing, these explanations remain unproved. Firstly, most conjugated polymers do not belong to the AB class. Secondly, whether the additional coupling terms to the density or the next-nearest neighbour hopping are large enough in experimental systems remains to be determined.

A very interesting issue is the role of Coulomb interactions in the intensity anomaly of the SSH model. The Peierls–Hubbard Hamiltonians are invariant under charge conjugation, and the same is true for the homogeneous ground state. However, for an inhomogeneous charge distribution, as for a charged soliton, a polaron or a bipolaron, the electron–electron interactions break locally the charge-conjugation symmetry by means of a local CDW, at least in HF approximation [2.166]. Correspondingly, within the HF approach, the intensity ratios are strongly modified [2.354]. Whether this local symmetry breaking survives more accurate treatments of the e–e interaction is an important problem which deserves further study. A more direct possibility is the addition to Peierls–Hubbard models of the off-diagonal X-term (2.84b) [2.48] which behaves effectively like the additional electron–phonon coupling mentioned above and does break the charge-conjugation symmetry of the Hamiltonian. However, in real materials the X-term is expected to be much smaller than the direct Coulomb interactions [2.49–52] and thus should most probably be considered only after the true effects of the standard Coulomb terms have been accurately assessed.

The High-Energy Peak and Triplet Absorptions. Besides the features described above, in several π-conjugated polymers one observes a high-energy peak in photo-induced absorption, which is *not* accompanied by additional IR modes and therefore corresponds to a neutral defect. In *trans*-$(CH)_x$, this peak appears at about 1.35 eV. It disappears at long times or low temperatures and is also seen in *cis*-rich samples. But even in p-toluenesulphonate polydiacetylene (PTS), where no charged features have been found, there is a high-energy peak at 1.35 eV [2.340, 341]. Optically detected magnetic resonance studies have demonstrated that this peak is associated with a triplet exciton [2.276]. In polydiacetylene with smaller side groups where there is a charged low-energy feature, the high-energy peak has been found at 1.3 eV [2.342]. At present, the spin structure of the high-energy peak in this material remains undetermined; it is conceivable that it is due to the same triplet seen in other polydiacetylenes.

The source of the high-energy peak in *trans*- and *cis*-$(CH)_x$ is less clear. One possibility is that this feature corresponds to a triplet–triplet transition [2.279]. Within the SSH picture, a triplet in *trans*-$(CH)_x$ can be naturally associated with a bound pair of neutral solitons, stabilized by the spin against rapid decay [2.176, 279]. A second interpretation involves possible time-dependent, nonlinear excitations known as *breathers* [2.355–357] which (within first order perturbation theory) appear to survive the inclusion of e–e interactions [2.358]. A third possible interpretation would involve absorption from the 2^1A_g state to a high ^1B_u state, (see, e.g. [2.359]) but the very weak oscillator strengths of such absorptions [2.60, 61] seem to preclude this.

2.6.3 Vibrational Excitation: Raman and Infrared Spectroscopy

In a coupled electron–phonon system the lattice dynamics can provide not only the vibrational spectrum and associated eigenvectors but also important information about the polarizability of the electronic degrees of freedom. Clearly it is possible to represent the (linear) lattice dynamics of a polyacetylene chain in terms of phenomenological force constants for the various types of elastic deformations. Explicit calculations based on such an approach show that nearest-neighbour interactions are not sufficient for reproducing the spectroscopic data of polyacetylene [2.160]. This result follows simply from conjugation, or more precisely from the non-local bond–bond polarizability in π-conjugated systems.

In the context of the SSH model (and also of π-electron models including electron–electron interactions) this bond–bond polarizability produces a softening of the Raman-active bond-stretching mode (cf. (2.50)). It follows that any disorder which induces an inhomogeneous electronic polarizability results in a broadening of Raman lines. The observed spectra are indeed quite broad and furthermore seen to change significantly with varying excitation frequency. There has been considerable discussion whether in polyacetylene the appropriate way of modelling these effects is in terms of a (bimodal) distribution of conjugation lengths [2.360, 361] or in terms of a variation of coupling parameters, λ [2.362, 363]. The disorder in coupling constants looks more attractive since it does not require a bimodal distribution to explain the two-peak structure observed by Raman scattering. On the other hand, there is no doubt that sp^3 defects breaking conjugation do exist and hence that the theorists' infinitely long, perfectly dimerized chain is not very realistic.

A remarkably simple but powerful theory both for the Raman modes of the homogeneous system and for the dopant-induced IR-active vibrational modes, commonly referred to as the amplitude mode theory, has been proposed by *Horovitz* [2.159] and elaborated by a number of groups [2.70, 113, 114, 362–364]. Generalizing the vibrational part of the Hamiltonian to include several normal modes and assuming that all these modes are coupled to the bond order, consistent with (2.35a), Horovitz was able to show that, in the continuum limit, the Raman frequencies are obtained as values of Ω satisfying the equation

$$D_0(\Omega) \equiv \sum_n \frac{\lambda_n}{\lambda} \frac{\omega_n^2}{\Omega^2 - \omega_n^2} = \frac{1}{2\tilde{\lambda} - 1} \tag{2.97}$$

where λ_n and ω_n are, respectively, the coupling parameter and bare frequency of the nth mode and $\lambda = \sum_n \lambda_n$. The parameter $\tilde{\lambda}$ is proportional to the second derivative of the total electronic energy $E_e(\Delta)$ with respect to the (dimerization) order parameter Δ,

$$\tilde{\lambda} = \frac{2\lambda}{\pi h v_F} \frac{\partial^2}{\partial \Delta^2} E_e(\Delta) \tag{2.98}$$

and plays the role of a renormalized electron–phonon coupling parameter, including the effects of electron–electron interactions. For the particular case of the SSH model $\tilde{\lambda} = \lambda$ and (2.97) reduces to (2.50).

In order to describe the IR-active modes introduced by breaking the translational symmetry, for example by creating a (charged) soliton or a polaron in the neighbourhood of a dopant ion, one can add [2.159] the electrostatic potential $V(x)$ between the electronic charge density $\rho(x)$ and the ionic charge and introduce the pinning parameter

$$\alpha = -\frac{e^2}{M_0 \Omega_0^2} \int dx V''(x)\rho(x) \tag{2.99}$$

where $\Omega_0^{-2} = \sum_n (\lambda_n/\lambda)\omega_n^{-2}$ and M_0 is the effective mass of the defect. IR modes are induced at frequencies obtained by solving the equation

$$D_0(\Omega) = (\alpha - 1)^{-1}. \tag{2.100}$$

In the absence of pinning due to dopants (or lattice discreteness) one solution of (2.100) is $\Omega = 0$, representing the Goldstone mode of the broken continuous translational symmetry. It is important to notice that in this case the structure of the defect does not enter at all. Therefore these vibrational modes, when observed [2.325, 364], but their interpretation remains controversial (see also [2.109]).

For a weakly inhomogeneous/disordered system described by a *distribution* of values of the parameter $\tilde{\lambda}$, a systematic study of the resonant Raman spectra calculations are required for predicting the additional non-universal features which may discriminate between the different inhomogeneous states. Up to now this problem has been approached mainly in the framework of the (generalized) SSH model (cf. Sect. 2.3). Electron–electron interactions have been taken into account only on a Hartree–Fock level [2.365, 366], but within the regime where the HF approximation makes sense their effects are very small. On the experimental side, in addition to the well-known T modes, small peaks have been observed [2.325, 364], but their interpretation remains controversial (see also [2.109]).

For a weakly inhomogeneous/disordered system described by *a distribution* of values of the parameter $\tilde{\lambda}$, a systematic study of the resonant Raman spectra as functions of the laser wavelength can provide important information about the underlying mechanism producing the gap and the dispersion of the Raman lines [2.363]. Selective enhancement of the cross section for a laser wavelength matching the electronic gap ($\hbar\omega_L = E_g$) allows the selection of regions in the material where the gap has a particular value. The SSH model variations in $\tilde{\lambda}(= \lambda)$ affect both the gap E_g and the amplitude mode frequency Ω in a specific way, namely so that

$$(\omega_0/\Omega)^2 = \log(E_g/t_0) + \text{const.} \tag{2.101}$$

as is easily seen from (2.47) and (2.50). The experimental analysis [2.70] is indeed consistent with such a relationship. Unfortunately, the modifications of this behaviour due to weak Coulomb interactions are very small, and reliable calculations for intermediate to strong coupling strength are not available. It is interesting to notice that both the gap and the renormalized phonon frequency are independent of λ in the large U limit of the Peierls–Hubbard model, i.e. in the spin-Peierls regime. In fact, the optical gap is then simply given by U, whereas the screening factor $(\Omega/\omega)^2$, as deduced from the theory of *Nakano* and *Fukuyama* [2.233], is 2/3.

2.7 Beyond Simple Models: Discussion and Conclusions

Until now our discussion has focused primarily on models of idealized, one-dimensional, infinite π-conjugated polymers. Our *leitmotif* has been the attempt to understand the relative roles of electron–phonon and electron–electron interactions in these models, to illustrate how various experimental data require the presence of one or the other of these interactions, and to determine at a semi-quantitative level the values of the parameters in the models. In the present section, we shall extend our discussion by examining a number of effects which, while neglected in the simple models, are clearly present in the real materials. In a sense, this section provides a useful caveat against over-interpretation of the results of highly simplified theories.

2.7.1 Effects of Disorder

It is generally accepted that disorder is detrimental to transport and ordering phenomena in solids. The effects are stronger the lower the effective dimensionality of the system. Thus, in materials like polyacetylene, which are both highly dis-ordered and quasi-one-dimensional, these effects are expected to be particularly pronounced. Indeed the experimental results often depend on sample quality.

Thus one must recognize the difficulties that the material aspects of real samples of *trans*-$(CH)_x$ pose for any attempts to extract intrinsic properties of the pure, crystalline polymer. If prepared by the Shirakawa method, using a Ziegler–Natta chemical catalyst, *trans*-$(CH)_x$ films have less than full density $(0.4\,g/cm^3)$ and on a microscopic scale $(200–1000\,\text{Å})$ consist of tangled fibrils which, by stretching, can be partially aligned. Changes in the catalysis conditions can lead to substantial differences in the mesoscale morphology. The degree of microscopic crystallinity is thought to be as high as 75–90% [2.367] but the disordered amorphous regions are not well understood. Furthermore, Shirakawa *trans*-$(CH)_x$ contains about $10^{19}/cm^3$ free spins [2.368], often interpreted as neutral solitons, dangling bonds or radicals, as well as charged impurity states, thought to be related to inadvertent doping by the Ziegler-Natta catalyst, at the level of about $10^{18}/cm^3$ [2.369].

If prepared by the so-called Durham route, using a processible prepolymer and thermal elimination, *trans*-$(CH)_x$ can be made into fully dense $(1.2 = g/cm^3)$ fairly uniform films, either in an oriented form, made by stretching the prepolymer, or a non-oriented form. The former is observed to be highly crystalline and appears to have average conjugation lengths on the order of $n = 36$, whereas the latter seem to be amorphous with an average conjugation length of $n = 11-14$ [2.370]. The levels of intrinsic defects and impurity contamination are not yet fully characterized.

These many different sources and length scales of disorder (the fibrillar morphology, the partial crystallinity, sp^3 defects, chain ends, cross links, *cis* segments within a *trans* chain, and impurities, including catalyst remnants) make it extremely difficult to obtain a quantitative description of the present real material. One should thus use simplified models to describe the effects qualitatively and rely on properties that are not too strongly influenced by disorder. For instance, it appears to be very hard to provide a satisfactory description of electrical conductivity, whereas measurements such as the optical absorption in the visible range or Raman spectra appear easier to describe.

Several studies of the effects of disorder have been based on the Hückel model with a perturbation added, in the form of either a screened Coulomb potential representing a charged impurity [2.371] or a completely localized potential acting on a single site (a site impurity [2.372, 373]) or an extrinsic variation of resonance integrals simulating bond impurities, e.g. *cis* segments within a *trans* chain [2.374]. Typically, the impurities first introduce localized levels of donor and/or acceptor type in the gap. Subsequently, the lattice relaxes around the impurity and the gap levels deepen further. The combined effects of impurities and lattice relaxation leads to a complicated interplay between intrinsic defects (kinks, polarons) and extrinsic disorder [2.375].

An interesting and relevant question concerns the dependence of dimerization on the amount of disorder and the concomitant evolution of the electronic gap. The homogeneous dimerization is found to vanish for large enough disorder [2.376, 377]. Recently, it has been proposed that the electronic gap may vanish even before the dimerization disappears and therefore a gapless Peierls instability may be possible [2.378]. A complete treatment, however, has to allow for inhomogeneous lattice distortions. A local lattice relaxation around an impurity will then always occur, even within the metallic phase [2.379].

2.7.2 Interchain Coupling and Three-Dimensional Effects

In contrast to the extensive efforts devoted to the studies of the one-dimensional models of π-conjugated polymers, relatively little attention has been paid to the effects that arise because the real materials, although highly anisotropic, are in fact inherently three-dimensional. Clearly, the inter-chain couplings that arise in the solid material are the major source of the solid-state effects that can lead to substantial differences between the current forms of conducting polymers and their finite oligomers in the gas phase or in solution. For example, although bond alternation can exist in the ground state of an infinite, idealized one-dimensional polymer, it will not persist at finite temperatures. The familiar Landau argument shows that domain walls (the kink solitons) connecting the two degenerate bond alternation patterns will always be generated by thermal fluctuations for $T \neq 0$ and that any density of these kinks, however small, will destroy the long-range order in one dimension. On the other hand, for polymers in the solid state even a weak inter-chain coupling is sufficient to sustain the long-range order up to a critical temperature which is not much smaller than the mean-field expression $T_c = 0.567 \Delta_0 / k_B \approx 5000\,\mathrm{K}$. One obvious question concerns the nature of this inter-chain coupling. Another question is the nature of the three-dimensional ordering: are the double bonds of neighbouring chains in phase (parallel) or out of phase (antiparallel)? In either case, kink solitons cannot occur individually but are confined within kink–antikink pairs. However, if kink solitons are introduced intentionally, e.g. by doping, the confinement energy may decrease rapidly and vanish at a concentration where the mean size of the pairs is of the order of the mean distance of the pairs [2.380].

Two types of inter-chain coupling have been studied in some detail: (1) coupling through Coulomb and dispersion forces [2.381, 382] and (2) a coupling due to inter-chain electron transfer [2.383, 384]. In the former case, parallel alignment is favoured, although the confinement energy is very small [2.382]. In the latter case, the ordering depends on the variation of transverse resonance integrals, $t_\perp$, along the chains. Purely geometrical considerations show that $t_\perp$ alternates in size and maybe even in sign. Parallel ordering is found if $t_\perp$ does alternate in sign [2.384] and antiparallel ordering if it does not [2.383]. In any case it seems that the confinement energy is larger for this type of coupling than for that due to the Coulomb and dispersion forces.

The actual experimental situation in the (partially) crystalline samples of trans-$(CH)_x$ is at present not definitive. The three-dimensional structure deduced from X-ray experiments on films of Shirakawa trans-$(CH)_x$ [2.226] show a crystallographic structure known as $P_{21/n}$, which would correspond to anti-parallel alignment of double bonds on adjacent chains, but similar studies on films of Durham trans-$(CH)_x$ [2.228, 385] suggest a $P_{21/a}$ structure, with parallel alignment of double bonds on adjacent chains. A recent attempt [2.37] to study this question theoretically by using a first principles density functional calculation of the full three-dimensional structure suggests that the $P_{21/a}$ structure is indeed lower in energy in a perfect crystal, but the $P_{21/n}$ structure is only slightly higher.

The inter-chain electron transfer between midgap states of neutral and charged solitons crossing each other on neighbouring chains has been proposed to be an essential step in the electrical conductivity of lightly doped polyacetylene [2.386, 387], a further step being the subsequent separation of the two solitons. Although this mechanism has received considerable attention in the past [2.388], there is a serious problem. The same quantity $t_\perp$ which is responsible for inter-chain electron transfer in this mechanism leads to a hybridization of the degenerate midgap states and to a strong binding of solitons on neighbouring chains. This binding, which in fact can be viewed as a chemical bond between the two chains, is much stronger than the confinement potential due to mis-aligned chain segments [2.384].

An important question is the extent to which solitons or polarons can move from one chain to another. Given their topological nature, individual solitons cannot hop to adjacent chains, and joint hopping is a process with very small probability. For polarons, the inter-chain hopping amplitude appears to be about an order of magnitude smaller than the bare inter-chain transfer integral $t_\perp$ [2.389]. As already mentioned in Sect. 2.6.2, inter-chain coupling appears to be crucial for understanding photogeneration experiments. The experimental evidence of inter-chain charge separation is quite convincing (see, e.g. [2.323, 324]). Indeed, in view of the finite transverse hopping integral $t_\perp$, there is a non-vanishing probability for direct photo-production of an electron and a hole on adjacent chains [2.336, 384]; a delayed inter-chain charge transfer following an intra-chain optical absorption is of course also possible. Two distinct mechanisms have been proposed for describing the subsequent creation of charged solitons in polyacetylene, as suggested by the appearance of the low-energy peak in photo-induced absorption (see Sect. 2.6.2). The first [2.323, 324, 341] assumes that pre-existing neutral solitons act as traps for the (inter-chain) separated charges generating $S^\pm$ according to $S^0 \pm e \rightarrow S^\pm$. The second [2.327] involves the coalescence of like-charged polarons (themselves generated by the migration of electrons and holes produced in the initial photoabsorption) in the process $P^+ + P^+ \rightarrow S^+ + S^+$. The first mechanism has been supported by photo-induced bleaching at 1.4 eV, possibly corresponding to the loss of S^0 [2.319]. However, other experimental results [2.327] do not show this effect; instead, the expected bleaching at the band edge is seen. This result supports the second mechanism.

That inter-chain coupling can be detrimental to polarons (and bipolarons) is easily shown in the framework of an anisotropic Landau–Ginzburg model [2.390]. This result can be confirmed by explicit calculations for the SSH model supplemented by an inter-chain hopping term [2.391, 392]. One finds that the polaron is destabilized and the charge is spread out over the whole crystal if the energy gained by lattice relaxation (on a single chain) is of the order of the transverse bandwidth $4t_\perp$. Since the polaron binding energy is small ($\sim 0.1\Delta_0$ according to Table 2.2), a small $t_\perp$ is sufficient for destabilizing the polaron. Bipolarons are more stable with respect to transverse delocalization, as long as Coulomb effects are neglected. It is worthwhile mentioning that recent ab initio calculations confirm that in crystalline polyacetylene, polarons and possibly also bipolarons are likely to be unstable [2.37]. Note however that the energy scales determining the polaron stability are small; therefore weak disorder can easily change the delicate energy balance by e.g. destroying the transverse coherence and thus stabilizing the polaron. This may explain the strong dependence of ESR experiments on sample quality.

2.7.3 Lattice Quantum Fluctuations

Despite the explicit appearance of quantum degrees of freedom in the lattice part of the Hamiltonian (2.18), we have assumed in nearly all of our previous discussions that the displacements of the atomic nuclei can be described in terms of classical coordinates. A simple estimate of the zero-point motion of the atoms about their classical equilibrium positions shows that these excursions are of roughly the same order as the classical dimerization amplitude [2.11, 393]. One may thus wonder whether the dimerization survives quantum fluctuations at all. Both numerical and analytical calculations [2.161–163, 394–397] show that the long-range order of the dimerization *does* persist. The effect of quantum fluctuations is to reduce the expectation value of the ground state dimerization amplitude by about 10 to 20% from its value in the absence of fluctuations. This effect can be understood qualitatively. With a decreasing electronic gap the phonons soften due to enhanced screening of force constants; correspondingly, the zero-point energy of the lattice decreases.

An interesting consequence of the lattice quantum fluctuations is the possibility of producing kink–antikink pairs by optical absorption *directly*, below the edge of the optical gap. This effect has been proposed [2.394, 395, 398, 399] as a partial explanation of the band edge tail on the optical absorption and as the possible origin of the nonlinear optical properties of polyacetylene [2.400].

2.7.4 Doping Effects and the Semiconductor-Metal Transition

The discovery of the dramatic increase of electrical conductivity upon doping, observed more than a decade ago in polyacetylene films, marked the beginning of the era of conducting polymers. Optical experiments revealing dopant-induced absorption in the infrared followed quickly. The magnetic susceptibility as a function of dopant concentration y (dopant ions per carbon), a focus of lively discussions initially, has become more clear recently. The Pauli susceptibility remains very small for weak doping until it changes in a step-like fashion at a concentration of about 6% to reach a value characteristic for a π-band [2.401]. The reason for this abrupt (first order?) transition remains unclear. The most puzzling feature is the persistence of dopant-induced IR-active vibrational modes for $y > 6\%$, which suggests the continuation of charge localization into this region, whereas the measured Pauli susceptibility indicates complete delocalization of charge.

The simplest description of doping is based on the assumption that after having removed or added an electron to a chain, the dopant ions have a negligible effect on the electronic properties. Both the high values of the conductivity in the metallic regime and the relatively weak dependence on the dopant species give some credibility to such a hypothesis. With this assumption one can describe the doped materials within the idealized π-electron models for variable electron density, (i.e. away from the half-filled band). For the simple SSH model the system evolves continuously from the dimerized state through the soliton lattice to an incommensurate Peierls distortion [2.402–405]. This picture of a commensurate-incommensurate transition can explain the persistence of the dopant–induced IR modes using the universal amplitude mode theory (recall Sect. 2.6.3) but not the sudden appearance of a Pauli susceptibility. In order to remove this difficulty, *Kivelson* and *Heeger* proposed a first-order transition from a soliton lattice to a polaron lattice [2.406] which although impossible in the one-dimensional SSH theory may be favoured by Coulomb and/or three-dimensional effects [2.407]. Unfortunately, the amplitude of the polaron lattice is found to decrease very quickly for $y > 2\%$, and therefore the dopant-induced IR modes would rapidly lose oscillator strength [2.114, 403], in contrast to the experimental observations [2.408].

An alternative mechanism [2.151] is based on the Peierls–Hubbard model and depends strongly on the relative values of the dimensionless parameters characterizing electron–phonon (λ) and electron–electron ($U/4t_0$) interactions. For a half-filled band, the Hartree–Fock factorization of the Hubbard term does not depend on dimerization since the local electron site density is independent of bond alternation; that is, there is a BOW, but no CDW. On the other hand, for incommensurate band fillings, the $2k_F$ lattice distortion induces *both* a BOW and a CDW. The CDW yields a positive contribution to the (on-site) Coulomb energy and thus tends to destabilize the lattice distortion. One finds [2.151] that the incommensurate distortion is unstable for $U/(4t_0) > \pi\lambda \cos(\pi y/2)$. This inequality can easily be satisfied for reasonable parameter values ($U/(4t_0) = 0.6$, $\lambda = 0.1$). This mechanism, leading to a soliton lattice for weak doping and a transition to a metal for a dopant concentration of a few percent, is able to explain the sudden appearance of a Pauli susceptibility, but unable to explain the persistence of the dopant–induced IR modes.

Neglecting the effect of the impurity potentials caused by the dopant ions is quite a drastic assumption, as arbitrarily weak potentials can produce bound states in one dimension. In fact, an early study by *Mele* and *Rice* [2.409] for the SSH Hamiltonian, including impurity potentials and lattice relaxation, indicates that disorder effectively quenches the (incommensurate) Peierls distortion. The density of states is found to increase smoothly as a function of dopant concentration and to reach the value of the undimerized system for $y \approx 10\%$. More recently, it has become clear that the impurities are not necessarily randomly distributed but can form ordered superlattices [2.367, 410]. These structural studies together with the persistence of the dopant–induced IR modes have been taken as evidence that a soliton lattice persists into the metallic range [2.411]. The electronic gap (between soliton levels and conduction band) is decreased due to the Coulomb potential of the ions and found to vanish at sufficiently high dopant concentration. The aggregation of dopants into ordered structures suggests that finite metallic segments exist which increase in length upon doping. *Conwell* et al. [2.411] propose that the finite level spacings should be visible in the temperature dependence of the magnetic susceptibility, which may be Pauli-like at room temperature but activated at low temperatures.

2.7.5 Transport

Among the wealth of experimental information on conducting polymers, transport data are probably the most complicated to interpret theoretically. The main reason is that transport measurements are not local probes but involve displacements of charge, spin or energy over some distance. Thus, during a dc measurement, macroscopic paths can be involved, so that perforce many different types and length scales of inhomogeneities are also involved [2.412]. Given these complications, we cannot hope to present here a full account of the transport problems. Rather we shall focus on two specific issues: soliton (or polaron) diffusion on the one hand and the ideal conductivity in the metallic regime on the other hand. Considerably more complete discussions of transport can be found in Chap. 3 of this volume and in the review of *Ngai* and *Rendell* emphasizing the low-frequency, complex conductivity [2.413], the review by *Conwell* pointing out virtues and limitations of the soliton model [2.119] and a paper by *Kaiser* making an interesting analogy between conducting polymers and amorphous metals [2.414].

The question whether solitons (and/or polarons) are mobile quasi-particles is not only of fundamental importance, but it may also have practical implications, as illustrated by recent proposals for specific devices to act as building blocks for molecular electronics [2.415, 416]. The diffusion of neutral solitons is commonly believed to be responsible for the narrowing of the electron spin-resonance line with increasing temperature and for the characteristic dependence of the proton spin-lattice relaxation rate on the proton Larmor frequency (see Chap. 5 of this volume).

The effective mass M_s associated with a moving soliton (or polaron) can be readily calculated in the framework of the Born–Oppenheimer approximation. Let

$$u_l = (-1)^l u f\left(\frac{la - vt}{\xi}\right) \tag{2.102}$$

represent a moving displacement pattern, where $f(x)$ has the form of a kink or of a local hump and the length ξ depends in general both on the velocity v and on electron–lattice and electron–electron interactions. The kinetic energy of the lattice (which at this level is responsible for the entire inertia, since the electrons are assumed to move adiabatically), modelled according to the effective Hamiltonian for the (CH) ions (2.9) becomes

$$E_{\text{kin}} = \tfrac{1}{2}M \sum_l \dot{u}_l^2 = \tfrac{1}{2}M_s v^2, \tag{2.103}$$

where

$$M_s = M\frac{u^2}{\xi a} \int_{-\infty}^{\infty} dx [f'(x)]^2 \tag{2.104}$$

is the effective mass in the continuum limit ($\xi \gg a$). M_s is much smaller than the bare mass M (typically, M_s is of the order of a few times the electron mass) since the displacement amplitude u is much smaller than the lattice constant a.

These large polarons are expected to lead to band-like transport along an ideal chain [2.417]. Indeed, explicit calculations indicate that the soliton mobility increases with decreasing temperature due to scattering by thermal phonons. This is easy to understand for optical phonons, as their population dies out exponentially with decreasing temperature [2.116]. Including acoustic phonon scattering yields a slower increase of soliton [2.418] and polaron [2.117] mobilities with decreasing temperature. One-dimensional diffusion of neutral solitons is commonly invoked for explaining magnetic resonance experiments, but trapping at static defects has to be incorporated in the model to qualitatively explain the data [2.118]. The band motion of charged solitons or polarons is less well documented, since they are easily trapped by charged inpurities (such as dopant ions). Such a transport mechanism may be involved in the picosecond photoresponse of trans-polyacetylene, but even this has been questioned recently [2.335].

Recent improvements in sample preparation have made possible the achievement of conductivities (I-doped polyacetylene) of the order of 10^5 S/cm (for details see Chap. 3). Identifying the number of carriers with the number of π-electrons, one finds a mean free path of the order of 100 CH units, which is indicative of band transport [2.419]. However, in contrast to the behaviour of a simple metal, the conductivity still decreases with decreasing temperature. This dependence is well described by Sheng's fluctuation-induced tunnelling between metallic islands [2.420, 421], although recently an alternative model has been proposed where the conductivity is limited primarily by barriers arising from conjugation defects within metallic regions [2.422]. Despite the effects of imhomogeneous doping and of defects of various kinds, calculations of the intrinsic conductivity for an ideal polyacetylene chain (without a conductivity gap) are not purely academic, but can provide estimates of the maximum conductivity which can be reached by further improving the material. An early calculation [2.423] and more recent work [2.424] show that the electron–phonon coupling of the SSH model is inefficient for scattering, especially at low temperatures. The reason is that only backscattering processes contribute to the resistance and these require thermal phonons of large wavevectors and thus of large energies $\omega_0 \sim 2000$ K. The temperature dependence of the conductivity is dominated by an exponential term $\exp(\hbar\omega_0/k_B T)$, yielding estimates of 1.4×10^7 S/cm [2.423] and 2×10^6 S/cm [2.424] at room temperature.

2.7.6 Concluding Remarks

Although there remain many relevant topics which could form the basis for further sections, for example the recent surge of interest in theory and experiments on polyanilines, we must at this point draw our discussion to a close. Throughout this review, we have sought to present a balanced perspective, accessible to both chemists and physicists and stressing both the achievements and limitations of the many competing theoretical models and ideas. On the basis of all the evidence, we feel that one can in the first approximation interpret π-conjugated polymers as quasi-one-dimensional materials in which electron–phonon and electron–electron interactions both play crucial roles. The exotic nonlinear excitations originally associated with purely electron–phonon models are expected to survive, and to remain relevant, in the presence of electron–electron interactions, but detailed calculations of their properties requires the use of many-body techniques substantially more demanding than those useful to solve the independent-electron theories. Accordingly, even within the simple quasi-one-dimensional models there are many theoretical issues yet to be resolved. When one seeks to include the complications that are present in the real materials (three-dimensional couplings, extrinsic defects and impurities, imperfect crystallinity and disorder) the challenge of interpreting experimental data on, and anticipating the properties of, conducting polymers remains. Hence we have every confidence that this field will retain its vitality for many years to come.

In a broader context, π-conjugated, conducting polymers can be viewed as among the first of the new classes of solid state materials in which many competing effects generate experimental interest, technological potential and theoretical difficulty. Together with the charge-transfer solids, conducting polymers have prepared the theoretical community intellectually for the increasing number of challenging novel materials, for instance the organic and high temperature superconductors, in which the subtleties of chemical compositions and (perhaps) many-body effects complicate the process of isolating the underlying physical phenomena. The insights we have gained from studies of the conducting polymers will continue to guide us in these developing areas.

Acknowledgements

During the preparation of this review, we have benefited from interactions with many individuals and hospitality from many institutions. We would like to thank particularly the Institute for Theoretical Physics of the ETH in Zürich, the IBM Research Center in Rüschlikon, the Center for Nonlinear Studies and the Theoretical Division in Los Alamos, the Institute for Scientific Interchange in Torino, the Centre d'Etudes Nucléaires in Grenoble, the National Chemical Laboratory in Pune and the Department of Physics at the University of Arizona. Among the many colleagues who contributed their comments and insights were Alan Bishop, Sergei Brazovskii, Jean-Luc Brédas, José Carmelo, Esther Conwell, Tom DeGrand, Sham Dixit, Art Epstein, Shahab Etemad, Klaus Fesser, Hideo Fukutome, Tinka Gammel, Alan Heeger, Baruch Horovitz, Peter Horsch,

Miklos Kertesz, Steve Kivelson, Bryan Kohler, Valeri Krivnov, Eugene Loh, Kazumi Maki, Eugene Mele, René Monnier, Alexander Ovchinnikov, Michael Rice, Masaki Sasai, Bob Schrieffer, Zoltan Soos, Michael Springborg, Xin Sun, Valy Vardeny, and Peter Vogl. We are most grateful for their interest and assistance.

References

2.1   S.A. Brazovskii, N.N. Kirova: Electron self localization and periodic superstructures in quasi-one-dimensional dielectrics. Sov. Sci. Rev. A, Phys. **5**, 99-264 (1984)

2.2   D. Baeriswyl: "Theroretical Aspects of Conducting Polymers: Electronic Structure and Defect States" in *Theoretical Aspects of Band Structures and Electronic Properties of Pseudo-One-Dimensional Solids*, ed. by R.H. Kamimura (Reidel, Dordrecht, Holland 1985) pp.1-48

2.3   J.L. Brédas, G.B. Street: Polarons, bipolarons and solitons in conducting polymers. Acc. Chem. Res. **18**, 309 (1985)

2.4   R. Silbey: "Electronic Structure of Conducting Polymers" in *Polydiacetylenes* ed. by D. Bloor, R.R. Chance NATO ASI Series E: Applied Sciences, Vol.102 (Nijhoff, Dordrecht 1985) pp.93-104

2.5   S. Kivelson: "Soliton Model of Polyacetylene" in *Solitons*, ed. by S. Trullinger, V. Zakharov, V.L. Pokrovsky, Modern Problems in Condensed Matter Sciences, Vol. 17 (North-Holland, Amsterdam 1986) pp.301-388

2.6   D.K. Campbell, A.R. Bishop, M.J. Rice: "The Field Theory Perspective on Conducting Polymers" in *Handbook of Conducting Polymers*, ed. by T. Skotheim (Dekker, New York 1986) pp.937-966

2.7   R.R. Chance, D.S. Boudreaux, J.L. Brédas, R. Silbey: "Solitons, Polarons, and Bipolarons in Conjugated Polymers" in *Handbook of Conducting Polymers*, ed. by T. Skotheim (Dekker, New York 1986) pp.825-857

2.8   J.L. Brédas: "Electronic Structure of Highly Conducting Polymers" in *Handbook of Conducting Polymers*, ed. by T. Skotheim (Dekker, New York 1986) pp.859-914

2.9   A.A. Ovchinnikov, I.I. Ukrainskii: Electronic processes in quasi-one-dimensional systems: conjugated polymers and donor-acceptor molecular crystals. Sov. Sci. Rev. B, Chem. **9**, 125-207 (1987)

2.10 A.J. Heeger, S. Kivelson, J.R. Schrieffer, W.P. Su: Solitons in conducting polymers. Rev. Mod. Phys. **60**, 781-850 (1988)

2.11 L. Yu: *Solitons and Polarons in Conducting Polymers* (World Scientific, Singapore 1988)

2.12 Z.G. Soos, G.W. Hayden: "Excited States of Conjugated Polymers" in *Electroresponsive Molecular and Polymeric Systems*, ed. by T.A. Skotheim (Dekker, New York 1988) pp.197-266

2.13 D.D.C. Bradley: Precursor-route poly(p-phenylenevinyline): polymer characterization and control of electronic properties. J. Phys. D **20**, 1389-1410 (1987)

2.14 L. Salem: *Molecular Orbital Theory of Conjugated Systems* (Benjamin, London 1966)

2.15 S. Suhai: A priori electronic structure calculations on highly conducting poly-
      mers. I. Hartree-Fock studies on *cis*- and *trans*-polyacetylene (polyenes). J.
      Chem. Phys. **73**, 3843 (1980)

2.16 M. Kertesz: Electronic structure of polymers. Adv. Quantum Chem. **15**, 161-
      214 (1982)

2.17 A Karpfen: "*Ab initio* Studies on Polyynes and Polydiacetylenes: Structure and
      Harmonic Force Fields" in *Polydiacetylenes*, ed. by D. Bloor, R.R. Chance,
      NATO ASI Series E: Applied Sciences, Vol. 102 (Nijhoff, Dordrecht 1985) pp.
      115-124

2.18 M. Kertesz, Y.-S. Lee: Electronic structure of small gap polymers. Synth. Met.
      28, C545-C552 (1989)

2.19 M.J.S. Dewar, W. Thiel: Ground states of molecules 38. The MNDO method
      approximations and parameters: J. Am. Chem. Soc. **99**, 4899-4907 (1977)

2.20 M.J.S. Dewar, W. Thiel: Ground states of molecules 39. MNDO Results for
      molecules containing hydrogen, carbon, nitrogen and oxygen. J. Am. Chem. Soc.
      99, 4907-4917 (1977)

2.21 D.S. Boudreaux, R.R. Chance, J.L. Brédas, R. Silbey: Solitons and polarons in
      polyacetylene: self-consistent-field calculations of the effect of neutral and
      charged defects on molecular geometry. Phys. Rev. **28**, 6927-6936 (1983)

2.22 S. Suhai: Quasiparticle energy band structures in semiconducting polymers:
      correlation effects on the band gap in polyacetylene. Phys. Rev. B **27**,
      3506-3518 (1983)

2.23 S. Suhai: Perturbation theoretical investigation of electron correlation effects in
      infinite metallic and semiconducting polymers. Int. J. Quantum Chem. **23**,
      1239-1256 (1983)

2.24 S. Suhai: Bond alternation in infinite polyene: Peierls distortion reduced by
      electron correlation. Chem. Phys. Lett. **96**, 619-625 (1983)

2.25 P. Hohenberg, W. Kohn: Inhomogeneous electron gas. Phys. Rev. **136**,
      B864-B871 (1964)

2.26 W. Kohn, L.J. Sham: Self-consistent equations including exchange and correla-
      tion effects. Phys. Rev. **140**, A1133-A1138 (1965)

2.27 R.O. Jones, O. Gunnarsson: The density functional formalism, its applications
      and prospects. Rev. Mod. Phys. **61**, 689 (1989)

2.28 P.M. Grant, I.P. Batra: Band structure of polyacetylene, $(CH)_x$. Solid State
      Commun. **29**, 225-229 (1979)

2.29 P.M. Grant, I.P. Batra: Electronic structure of conducting π-electron systems.
      Synth. Met. **1**, 193-212 (1979/80)

2.30 P.M. Grant, I.P. Batra: Self-consistent crystal potential and band structure of
      three dimensional *trans*-polyacetylene. J. Physique **44**, C3, 437-442 (1983)

2.31 J.W. Mintmire, C.T. White: Xα approach for the determination of electronic and
      geometric structure of polyacetylene and other chain polymers. Phys. Rev. Lett.
      50, 101-105 (1983)

2.32 J. Ashkenazi, E. Ehrenfreund, Z. Vardeny, O. Brafman: First principles three-
      dimensional band structure of trans-polyacetylene. Mol. Cryst. Liq. Cryst. **117**,
      193 (1985)

2.33 J. Ashkenazi, W.E. Pickett, B.M. Klein, H. Krakauer, C.S. Wang: First princi-
      ples investigation on the validity of Peierls mechanism for the dimerization
      energy of trans-polyacetylene. Synth. Met **21**, 301 (1987)

2.34 J. Ashkenazi, W.E. Pickett, H. Krakauer, C.S. Wang, B.M. Klein, S.R. Chubb:
      Ground state of *trans*-polyacetylene and the Peierls mechanism. Phys. Rev. Lett.
      62, 2016-2019 (1989)

2.35 M. Springborg: Self-consistent electronic structure of polyacetylene. Phys. Rev.
      B **33**, 8475 (1986)

2.36 P. Vogl, D.K. Campbell, O.F. Sankey: Theory of 3-D structure and intrinsic defects of *trans*-polyacetylene. Synth. Met. **28**, D513-D520 (1989)

2.37 P. Vogl, D.K. Campbell: Three-dimensional structure and intrinsic defects in *trans*-polyacetylene. Phys. Rev. Lett. **62**, 2012-2015 (1989)

2.38 T. von Koopmans: Über die Zuordnung von Wellenfunktionen und Eigenwerten zu den einzelnen Elektronen eines Atoms. Physica 1, 104-113 (1933)

2.39 C.-O. Almbladh, U. von Barth: Exact results for the charge and spin densities, exchange-correlation potentials and density-functional eigenvalues. Phys. Rev. B **31**, 3231 (1985)

2.40 G. Nicolas, Ph. Durand: A new general methodology for deriving transferable atomic potentials in molecules. J. Chem. Phys. **70**, 2020-2021 (1979)

2.41 A. Painelli, A. Girlando: Hubbard models and their applicability in solid state and molecular physics. Solid State Commun. **66**, 273-275 (1988)

2.42 A. Painelli, A. Girlando: Interacting electrons in 1D: applicability of Hubbard models. Synth. Met. **27**, A15-A20 (1988)

2.43 R.G. Parr, D.P. Craig, I.G. Ross: Molecular orbital calculations of the lower excited electronic levels of benzene, configuration interaction included. J. Chem. Phys. **18**, 1561-1563 (1950)

2.44 R.G. Parr: A method for estimating electronic repulsion integrals over LCAO MO's in complex unsaturated molecules. J. Chem. Phys. **20**, 1499 (1952)

2.45 R. Pariser, R.G. Parr: A semi-empirical theory of the electronic spectra and electronic structure of complex unsaturated molecules II. J. Chem. Phys. **21**, 767-776 (1953)

2.46 J.A. Pople: Electron interaction in unsaturated hydrocarbons. Trans. Faraday Soc. **49**, 1375-1385 (1953)

2.47 J. Hubbard: Electron correlations in narrow energy bands. Proc. R. Soc. London, A **276**, 238 (1963)

2.48 S. Kivelson, W.P. Su, J.C. Schrieffer, A.J. Heeger: Missing bond charge repulsion in the extended Hubbard model: effects in polyacetylene. Phys. Rev. Lett. **58**, 1899-1902 (1987)

2.49 D. Baeriswyl, P. Horsch, K. Maki: Missing bond-charge repulsion. Phys. Rev. Lett. **60**, 70 (1988)

2.50 J.T. Gammel, D.K. Campbell: Comment on the missing bond-charge repulsion in the extended Hubbard model. Phys. Rev. Lett. **60**, 71 (1988)

2.51 D.K. Campbell, J.T. Gammel, E.Y. Loh, Jr.: The extended Peierls-Hubbard model: off diagonal terms. Synth. Met. **27**, A9-A14 (1988)

2.52 D.K. Campbell, J.T. Gammel, E.Y. Loh, Jr.: Bond-charge Coulomb repulsion in Peierls-Hubbard models. Phys. Rev. B **38**, 12043-12046 (1988)

2.53 E. Hückel: Quantentheoretische Beiträge zum Benzolproblem. I. Die Elektronenkonfiguration des Benzols und verwandter Verbindungen. Z. Phys. **70**, 204-286 (1931)

2.54 E. Hückel: Quantentheoretische Beiträge zum Problem der aromatischen und ungesättigten Verbindungen III. Z. Phys. **76**, 628-648 (1932)

2.55 J.A. Pople, S.H. Walmsley: Bond alternation defects in long polyene molecules. Mol. Phys. **5**, 15-20 (1962)

2.56 M. Nechtschein: Sur la nature des centres paramagnétiques détectés par RPE dans les polymères conjugués. J. Polym. Sci. C **4**, 1367-1376 (1963)

2.57 W.-P. Su, J.R. Schrieffer, A.J. Heeger: Solitons in polyacetylene. Phys. Rev. Lett. **42**, 1698-1701 (1979)

2.58 W.-P. Su, J.R. Schrieffer, A.J. Heeger: Soliton excitations in polyacetylene. Phys. Rev. B **22**, 2099-2111 (1980); Erratum ibid. **28**, 1138 (1983)

2.59 R. Saito, H. Kamimura: Vibronic states of polyacetylene, $(CH)_x$. J. Phys. Soc. Jpn. **52**, 407-416 (1983)

2.60 S. Ramasesha, Z.G. Soos: Optical excitations of even and odd polyenes with molecular PPP correlations. Synth. Met. **9**, 283-294 (1984)

2.61 S. Ramasesha, Z.G. Soos: Correlated states in linear polyenes, radicals and ions: exact PPP transition moments and spin densities. J. Chem. Phys. **80**, 3278-3287 (1984)

2.62 K. Ohno: Some remarks on the Pariser-Parr-Pople method. Theor. Chim. Acta **2**, 219 (1964)

2.63 N. Mataga, K. Nishimoto: Electronic structure and spectra of nitrogen heterocycles. Z. Physik. Chemie **13**, 140-157 (1957)

2.64 M.W. Hanna, A.D. McLachlan, H.H. Dearman, H.M. McConnell: Radiation damage in organic crystals III. Long polyene radicals. J. Chem. Phys. **37**, 361-367 (1962)

2.65 M.W. Hanna, A.D. McLachlan, H.H. Dearman, H.M. McConnell: Erratum and further comments: radiation damage in organic crystals. III. Long polyene radicals. J. Chem. Phys. **37**, 3008 (1962)

2.66 H.C. Longuet-Higgins, L. Salem: The alternation of bond lengths in long conjugated chain molecules. Proc. R. Soc. London A **251**, 172-185 (1959)

2.67 R.E. Peierls: *Quantum Theory of Solids* (Clarendon, Oxford 1955) p.108

2.68 H. Fröhlich: On the theory of superconductivity: The one-dimensional case. Proc. R. Soc. London A **223**, 296-305 (1954)

2.69 T. Kakitani: Theoretical study of optical absorption curves of molecules III. Prog. Theor. Phys. **51**, 656-673 (1974)

2.70 E. Ehrenfreund, Z. Vardeny, O. Brafman, B. Horovitz: Amplitude and phase modes in *trans*-polyacetylene: resonant Raman scattering and induced infrared activity. Phys. Rev. B **36**, 1535-1553 (1987)

2.71 M. Nakahara, K. Maki: Quantum corrections to solitons in polyacetylene. Phys. Rev. B **25**, 7789-7797 (1982)

2.72 H.J. Schulz: Lattice dynamics and electrical properties of commensurate one-dimensional charge-density-wave systems. Phys. Rev. B **18**, 5756-5767 (1978)

2.73 M.J. Rice, E.J. Mele: Elementary excitations of a linearly conjugated diatomic polymer. Phys. Rev. Lett. **49**, 1455-1459 (1982)

2.74 W. Kohn: Theory of the insulating state. Phys. Rev. **133**, A171-181 (1964)

2.75 E.J. Mele, M.J. Rice: Electron-phonon coupling in intrinsic *trans*-polyacetylene. Solid State Commun. **34**, 339-343 (1980)

2.76 T. Kennedy, E.H. Lieb: Proof of the Peierls instability in one dimension. Phys. Rev. Lett. **59**, 1309-1312 (1987)

2.77 M.J. Rice: Charged π-phase kinks in lightly doped polyacetylene. Phys. Lett. **71A**, 152 (1979)

2.78 S.A. Brazovskii: Electronic excitations in the Peierls-Fröhlich state. Pis'ma Zh. Eksp. Teor. Fiz. **28**, 656 (1978) [JETP Lett. **28**, 606 (1978)]

2.79 S.A. Brazovskii, I.E. Dzyaloshinskii, I.M. Krichever: Exactly solvable discrete Peierls models. Zh. Eksp. Teor. Fiz. **83**, 389 (1982) [Sov. Phys.-JETP **56**, 212-225 (1982)]

2.80 I.E. Dzyaloshinskii, I.M. Krichever: Sound and charge density wave in the discrete Peierls model. Zh. Eksp. Teor. Fiz. **85**, 1771-1789 (1983) [Sov. Phys.-JETP **58**, 1031-1040 (1983)]

2.81 H. Takayama, Y.R. Lin-Liu, K. Maki: Continuum model for solitons in polyacetylene. Phys. Rev. B **21**, 2388-2393 (1980)

2.82 S.A. Brazovskii, N.N. Kirova: Excitons, polarons and bipolarons in conducting polymers. Pis'ma Zh. Eksp. Teor. Fiz **33**, 6-10 (1981) [JETP Lett. **33**, 4 (1981)]

2.83 D.K. Campbell, A.R. Bishop: Solitons in polyacetylene and relativistic field theory models. Phys. Rev. B **24**, 4859-4862 (1981)

118   D. Baeriswyl et al.

2.84 D.K. Campbell, A.R. Bishop: Soliton excitations in polyacetylene and relativistic field theory models. Nucl. Phys. B **200**, [Fs4], 297-328 (1982)
2.85 A.R. Bishop, D.K. Campbell, K. Fesser: Polyacetylene and relativistic field theory models. Mol. Cryst. Liq. Cryst. **77**, 252-264 (1981)
2.86 A.R. Bishop, D.K. Campbell: "Polarons in Polyacetylene" in *Non-Linear Problems: Present and Future*, ed. by A.R. Bishop, D.K. Campbell, B. Nicolaenko (North-Holland, Amsterdam, 1982) pp.195-208
2.87 S.A. Brazovskii: Self-localized excitations in the Peierls-Fröhlich state. Zh. Eksp. Teor. Fiz. **78**, 677-699 (1980) [Sov. Phys.-JETP **51**, 342-353 (1980)]
2.88 N.N. Bogoliubov, V.V. Tolmachev, D.V. Shirkov: *A New Method of Superconductivity* (Consultants Bureau, New York 1959)
2.89 P.G. De Gennes: *Superconductivity of Metals and Alloys* (Benjamin, New York 1966)
2.90 J. Bar-Sagi, C.G. Kuper: Self-consistent pair potential in an inhomogeneous superconductor. Phys. Rev. Lett. **28**, 1556-1559 (1972)
2.91 J.T. Gammel: Finite band continuum model of polyacetylene. Phys. Rev. B **33**, 5974-5975 (1986)
2.92 D.K. Campbell, A.R. Bishop, K. Fesser: Polarons in quasi-one-dimensional systems. Phys. Rev. B **26**, 6862-6874 (1982)
2.93 Y. Onodera: Polarons, bipolarons, and their interactions in *cis*-polyacetylene. Phys. Rev. B **30**, 775-785 (1984)
2.94 W.-P. Su, J.R. Schrieffer: Soliton dynamics in polyacetylene. Proc. Natl. Acad. Sci. **77**, 5626-5629 (1980)
2.95 N. Suzuki, M. Ozaki, S. Etemad, A.J. Heeger, A.G. MacDiarmid: Solitons in polyacetylene: effects of dilute doping on optical absorption spectra. Phys. Rev. Lett. **45**, 1209-1213 (1980); Erratum, ibid. **45**, 1463 (1980)
2.96 J.T. Gammel, J.A. Krumhansl: Theory of optical absorption in *trans*-polyacetylene containing solitons. Phys. Rev. B **24**, 1035-1039 (1981)
2.97 H. Fukutome, M. Sasai: Theory of electronic structures and lattice distortions in polyacetylene and itinerant Peierls system I. Prog. Theor. Phys. **67**, 41-67 (1982)
2.98 S. Kivelson, T.-K. Lee, Y.R. Lin-Liu, I. Peschel, L. Yu: Boundary conditions and optical absorption in the soliton model of polyacetylene. Phys. Rev. B **25**, 4173-4184 (1982)
2.99 K. Fesser, A.R. Bishop, D.K. Campbell: Optical absorption from polarons in a model of polyacetylene. Phys. Rev. B **27**, 4804-4825 (1983)
2.100 B. Horovitz: Optical conductivity of solitons in polyacetylene. Solid State Commun. **41**, 593-596 (1982)
2.101 T. Martin, D.K. Campbell: Optical absorption from polarons in a diatomic polymer. Phys. Rev. B **35**, 7732-7735 (1987)
2.102 S. Mazumdar: "When Kinks are not Elementary Excitations" in *Nonlinearity in Condensed Matter*, ed. by A.R. Bishop, D.K. Campbell, P. Kumar, S.E. Trullinger, Springer Ser. Solid-State Sci., Vol.69 (Springer, Berlin, Heidelberg 1987) pp.94-105
2.103 E.J. Mele, M.J. Rice: Vibrational excitations of charged solitons in polyacetylene. Phys. Rev. Lett. **45**, 926-929 (1980)
2.104 J.C. Hicks, G.A. Blaisdell: Lattice vibrations in polyacetylene. Phys. Rev. B **31**, 919-926 (1985)
2.105 H. Ito, A. Terai, Y. Ono, Y. Wada: Linear mode analysis around a soliton in *trans*-polyacetylene $(CH)_x$. J. Phys. Soc. Jpn. **53**, 3520-3531 (1984)
2.106 E.J. Mele, J.C. Hicks: Continuum theory for defect vibrations in conjugated polymers. Phys. Rev. B **32**, 2703-2706 (1985)
2.107 A. Terai, H. Ito, Y. Ono, Y. Wada: Linear mode analysis around a polaron in the continuum model of *trans*-$(CH)_x$. J. Phys. Soc. Jpn. **54**, 196-202 (1985)

2.108 A. Terai, H. Ito, Y. Ono, Y. Wada: Phonons around a soliton in *trans*-(CH)$_x$: dependence on the electron momentum cut-off. J. Phys. Soc. Jpn. **54**, 4468-4469 (1985)

2.109 A. Terai, Y. Ono: Phonons around a soliton and a polaron in Su-Schrieffer-Heeger's model of *trans*-(CH)$_x$. J. Phys. Soc. Jpn. **55**, 213-221 (1986)

2.110 Y. Ono, A. Terai, Y. Wada: Phonons around a soliton in a continuum model of *trans*-polyacetylene. J. Phys. Soc. Jpn. **55**, 1656-1662 (1986)

2.111 J.C. Hicks, J.T. Gammel: Solitary and infrared-active modes in polyacetylene. Phys. Rev. Lett. **57**, 1320-1323 (1986)

2.112 J.T. Gammel, J.C.Hicks: Electrons and phonons in semi-continuum model of polyacetylene. Synth. Met. **17**, 63-68 (1987)

2.113 E.J. Mele: "Phonons and the Peierls Instability in Polyacetylene" in *Handbook of Conducting Polymers*, ed. by T. Skotheim (Dekker, New York 1986) pp.795-823

2.114 J.C. Hicks, J.T. Gammel, H.-Y. Choi, E.J. Mele: Dynamical conductivity in polyacetylene. Synth. Met. **17**, 57-62 (1987)

2.115 J.C. Hicks, J.T. Gammel: Effects of the Hubbard interaction and electrostatic pinning in polyacetylene. Phys. Rev. B **37**, 6315-6324 (1988)

2.116 K. Maki: Soliton diffusion in polyacetylene. Phys. Rev. B **26**, 2181-2191, 4539-4542 (1982)

2.117 S. Jeyadev, E.M. Conwell: Polaron mobility in *trans*-polyacetylene. Phys. Rev. B **35**, 6253-6259 (1987)

2.118 S. Jeyadev, E.M. Conwell: Soliton mobility in *trans*-polyacetylene. Phys. Rev. B **36**, 3284-3293 (1987)

2.119 E.M. Conwell: Transport in *trans*-polyacetylene. IEEE Trans. EI.22, 591 (1987)

2.120 E.H. Lieb, F.Y. Wu: Absence of Mott transition in an exact solution of the short-range, one-band model in one dimension. Phys. Rev. Lett. **20**, 1445-1448 (1968)

2.121 H. Bethe: Theory of metals. Part I. Eigenvalues and eigenfunctions of the linear atomic chain. Z. Phys. **71**, 205 (1931)

2.122 J. Carmelo, D. Baeriswyl: Solution of the one-dimensional Hubbard model for arbitrary electron density and large U. Phys. Rev. B **37**, 7541-7548 (1988)

2.123 A.A. Ovchinnikov: Excitation spectrum in the one-dimensional Hubbard model. Zh. Eksp. Teor. Fiz. **57**, 2137-2143 (1969) [Sov. Phys.-JETP **30**, 1160-1163 (1970)]

2.124 J. Des Cloiseaux, J.J. Pearson: Spin wave spectrum of the antiferromagnetic linear chain. Phys. Rev. **128**, 2131-2135 (1962)

2.125 T.C. Choy, W. Young: On the continuum spin-wave spectrum of the one-dimensional Hubbard model. J. Phys. C **15**, 521-528 (1982)

2.126 I.A. Misurkin, A.A. Ovchinnikov: Exact spectrum of quasi-ionic excitations in Hubbard's one-dimensional model. Fiz. Tverd. Tela **12**, 2524 (1970) [Sov. Phys.-Solid State **12**, 2031 (1971)]

2.127 C.F. Coll III: Excitation spectrum of the 1-D Hubbard model. Phys. Rev. B **9**, 2150-2158 (1974)

2.128 F. Woynarovich: Excitations with complex wavenumbers in a Hubbard chain I: States with one pair of complex wavenumbers. J. Phys. C **15**, 85-96 (1982)

2.129 P.F. Maldague: Optical spectrum of a Hubbard chain. Phys. Rev. B **16**, 2437-2446 (1977)

2.130 S. Mazumdar, Z.G. Soos: Valence bond analysis of extended Hubbard models: charge transfer excitations of molecular conductors. Phys. Rev. B **23**, 2810-2823 (1981)

2.131 D. Baeriswyl, J. Carmelo, A. Luther: Correlation effects on the oscillator strength of optical absorption: sum rule for the one-dimensional Hubbard model. Phys. Rev. B **33**, 7247-7248 (1986); (E) ibid. B **34**, 8976 (1986)

2.132 W. Metzner, D. Vollhardt: Correlated lattice fermions in d = ∞ dimensions. Phys. Rev. Lett. **62**, 324-327 (1989)

2.133 E.N. Economou, P.N. Poulopoulos: Ground-state energy of the half-filled one-dimensional Hubbard model. Phys. Rev. B **29**, 4756-4758 (1979)

2.134 M. Takahashi: On the exact ground state energy of Lieb and Wu. Prog. Theor. Phys. **45**, 756-760 (1971)

2.135 P.W. Anderson: New approach to theory of superexchange interactions. Phys. Rev. **115**, 2-13 (1959)

2.136 D.J. Klein, W.A. Seitz: Perturbation expansion of the linear Hubbard model. Phys. Rev. B **8**, 2236-2247 (1973)

2.137 L. Hulthen: Über das Austauschproblem eines Kristalls. Arkiv Mat. Astron. Fysik **26A**, A (1938)

2.138 M. Takahashi: Magnetization curve for the half-filled Hubbard model. Prog. Theor. Phys. **42**, 1098-1105 (1969); Erratum, ibid. **43**, 860 (1970)

2.139 M. Takahashi: Magnetic susceptibility for the half-filled Hubbard model. Prog. Theor. Phys. **43**, 1619 (1970)

2.140 H. Shiba: Magnetic susceptibility at zero temperature for the one-dimensional Hubbard model. Phys. Rev. B **6**, 930-938 (1972)

2.141 J.E. Hirsch, R.L. Sugar, D.J. Scalapino, R. Blankenbecler: Monte Carlo simulations of one-dimensional fermion systems. Phys. Rev. B **26**, 5033-5055 (1982)

2.142 A. Luther, I. Peschel: Calculation of critical exponents in two dimensions from quantum field theory in one dimension. Phys. Rev. B **12**, 3908-3917 (1975)

2.143 N.M. Bogoliubov, A.G. Izergin, V.E. Korepin: Critical exponents for integrable models. Nucl. Phys. B **275**, [FS 17], 687 (1986)

2.144 J.E. Hirsch, D.J. Scalapino: 2p$_F$ and 4p$_F$ instabilities in the one-dimensional Hubbard model. Phys. Rev. B **29**, 5554-5561 (1984)

2.145 S.K. Lyo, J.P. Gallinar: Absorption in a strong-coupling half-filled-band Hubbard chain. J. Phys. C **10**, 1693-1702 (1977)

2.146 E.Y. Loh Jr., D.K. Campbell: Optical absorption in the extended Peierls-Hubbard models. Synth. Met. **27**, A499-A508 (1988)

2.147 A.A. Ovchinnikov, I.I. Ukrainskii, G.V. Kventsel: Theory of one-dimensional Mott semiconductors and the electronic structure of long molecules having conjugated bonds. Sov. Phys.-Usp. **15**, 575-591 (1973)

2.148 D.J. Klein, W.A. Seitz: Partially filled linear Hubbard model near the atomic limit. Phys. Rev. B **10**, 3217-3227 (1974)

2.149 J. Bernasconi, M.J. Rice, W.R. Schneider, S. Strässler: Peierls transition in the strong-coupling Hubbard chain. Phys. Rev. B **12**, 1090-1097 (1975)

2.150 J. Carmelo: Exact and variational results for many-electron models of one-dimensional conductors. Ph.D. Thesis, Copenhagen (1986)

2.151 D. Baeriswyl, J. Carmelo, K. Maki: Coulomb correlations in one-dimensional conductors with incommensurate band fillings and the semiconductor-metal transition in polyacetylene. Synth. Met. **21**, 271-278 (1987)

2.152 P.P. Bendt: Nearest-neighbor correlation effects in spinless one-dimensional conductors. Phys. Rev. B **30**, 3951-3962 (1984)

2.153 P. De Loth, J.P. Malrieu, D. Maynau: Polyazacetylene; metallic conductor vs Mott insulator. Phys. Rev. B **36**, 3365-3367 (1987)

2.154 M.C. Dos Santos, J.L. Brédas: Nonlinear excitations in pernigraniline, the oxidized form of polyaniline. Phys. Rev. Lett. **62**, 2499-2502 (1989)

2.155 S.N. Dixit, S. Mazumdar: Electron-electron interaction effects on Peierls dimerization in a half-filled band. Phys. Rev. B **29**, 1824-1839 (1984)

2.156 D. Bertho, A. Laghdir, C. Jouanin: Energetic study of polarons and bipolarons in polythiophene: importance of Coulomb effects. Phys. Rev. B **38**, 12531-12539 (1988)

2.157 J. Fink, G. Leising: Momentum-dependent dielectric functions of oriented *trans*-polyacetylene. Phys. Rev. B **34**, 5320-5328 (1986)

2.158 B.S. Hudson, B.E. Kohler, K. Schulten: "Linear Polyene Electronic Structure and Potential Surfaces" in *Excited States*, ed. by E.C. Lim (Academic, New York 1982) pp.1-95

2.159 B. Horovitz: Infrared activity of Peierls systems and application to polyacetylene. Solid State Commun. **41**, 729-734 (1982)

2.160 G. Zannoni, G. Zerbi: Lattice dynamics of polyacetylene as a one-dimensional system; Part I: The case of the *trans* perfect monomes. J. Mol. Struct. **100**, 485-504 (1983); Part II Doping-induced disorder. ibid **100**, 505-530 (1983)

2.161 J.E. Hirsch, E. Fradkin: Effects of quantum fluctuations on the Peierls instability: A Monte Carlo study. Phys. Rev. Lett. **49**, 402-405 (1982)

2.162 E. Fradkin, J.E. Hirsch: Phase diagram of one-dimensional electron-phonon systems. I. The Su-Schrieffer-Heeger model. Phys. Rev. B **27**, 1680-1697 (1983)

2.163 W.-P. Su: Quantum ground state of a polyacetylene ring. Solid State Commun. **42**, 497-499 (1982)

2.164 C. Kittel: *Quantum Theory of Solids* (Wiley, New York 1963)

2.165 D. Baeriswyl, J.J. Forney: On the instabilities of the 1-d interacting electron gas. J. Phys. C **13**, 3203-3226 (1980)

2.166 K.R. Subbaswamy, M. Grabowski: Bond alternation, on-site Coulomb correlations, and solitons in polyacetylene. Phys. Rev. B **24**, 2168-2173 (1981)

2.167 S. Kivelson, D.E. Heim: Hubbard versus Peierls and the Su-Schrieffer-Heeger model of polyacetylene. Phys. Rev. B **26**, 4278-4292 (1982)

2.168 H. Fukutome, M. Sasai: Theory of electronic structures and lattice distortions of polyacetylene and itinerant Peierls systems III. Prog. Theor. Phys. **69**, 373-395 (1983)

2.169 M. Sasai, H. Fukutome: Continuum version for the PPP model of polyacetylene and other half-filled Peierls systems. Prog. Theor. Phys. **73**, 1-17 (1985)

2.170 J. Cizek, J. Paldus: Stability conditions for the solutions of the Hartree-Fock equations for atomic and molecular systems. Application to the π-electron model of cyclic polyenes. J. Chem. Phys. **47**, 3976 (1967)

2.171 J. Paldus, J. Cizek: Comment on the paper by Harris and Falicov: Self-consistent theory of bond alternation in polyenes: normal states, charge-density waves, and spin-density waves. J. Chem. Phys. **53**, 1619 (1970)

2.172 J. Paldus, J. Cizek: Stability conditions for the solutions of the Hartree-Fock equations for atomic and molecular systems VI. Singlet-type instabilities and charge-density-wave Hartree-Fock solutions for cyclic polyenes. Phys. Rev. A **2**, 2268 (1970)

2.173 S. Mazumdar, D.K. Campbell: Broken symmetries in a one-dimensional half-filled band with arbitrarily long-range Coulomb interactions. Phys. Rev. Lett. **55**, 2067-2070 (1985)

2.174 S. Mazumdar, D.K. Campbell: Bond alternation in the infinite polyene: effect of long-range Coulomb interactions. Synth. Met. **13**, 163-172 (1986)

2.175 B.E. Kohler, C. Spangler, C. Westerfield: The $2^1 A_g$ state in the linear polyene 2,4,6,8,10,12,14,16-octadecaoctaene. J. Chem. Phys. **89**, 5422-5428 (1988)

2.176 W.-K. Wu, S. Kivelson: Theory of conducting polymers with weak electron-electron interaction. Phys. Rev. B **33**, 8546-8557 (1986)

2.177 V.J. Emery: "Theory of the One-Dimensional Electron Gas" in *Highly Conducting One-Dimensional Solids*, ed. by J.T. Devreese, R.P. Evrard, V.E. van Doren (Plenum, New York 1979) pp.247-303

2.178 J. Sólyom: The Fermi gas model of one-dimensional conductors. Adv. Phys. **28**, 201-303 (1979)

2.179 B. Horovitz, J. Sólyom: Charge-density waves with electron-electron interactions. Phys. Rev. B **32**, 2681-2684 (1985)

2.180 V.Ya.Krivnov, A.A. Ovchinnikov: Peierls instability in weakly nonideal one-dimensional systems. Zh. Eksp. Teor. Fiz. **90**, 709-723 (1986) [Sov. Phys.-JETP **63**, 414-421 (1986)]

2.181 J.H. Schulz, J. Voit: Electron-phonon interaction and phonon dynamics in one-dimensional conductors: spinless fermions. Phys. Rev. B **36**, 968-979 (1987)

2.182 J. Voit: Electronic interactions, charge conjugation symmetry breaking, and phonon dynamics in an extended SSH model. Synth. Met. **27**, A33-A40 (1988)

2.183 J.E. Hirsch: Renormalization group study of the Hubbard model. Phys. Rev. B **22**, 5259-5266 (1980)

2.184 C. Dasgupta, P. Pfeuty: Real space renormalization group study of the 1-d Hubbard model. J. Phys. C **14**, 717-735 (1981)

2.185 G.W. Hayden, E.J. Mele: Renormalization-group studies of the Hubbard-Peierls Hamiltonian for finite polyenes. Phys. Rev. B **32**, 6527-6530 (1985)

2.186 G.W. Hayden, E.J. Mele: Correlation effects and excited states in conjugated polymers. Phys. Rev. B **34**, 5484-5497 (1986)

2.187 E.J. Mele, G.W. Hayden: Self-localized excitations in conjugated polymers. Synth. Met. **17**, 107-113 (1987)

2.188 P. Tavan, K. Schulten: The low-lying electron excitations in long polyenes: a PPP-MRD-DI study. J. Chem. Phys. **85**, 6602-6609 (1986)

2.189 P. Tavan, K. Schulten: Electronic excitations in finite and infinite polyenes. Phys. Rev. B **36**, 4337-4358 (1987)

2.190 P. Horsch: Correlation effects on bond alternation in polyacetylene. Phys. Rev. B. **24**, 7351-7360 (1981)

2.191 D. Baeriswyl, K. Maki: Electron correlations in polyacetylene. Phys. Rev. B **31**, 6633-6642 (1985)

2.192 D. Baeriswyl, K. Maki: Coulomb correlations and optical gap in polyacetylene. Synth. Met. **17**, 13-18 (1987)

2.193 D. Baeriswyl: "Variational Schemes for Many Electron Systems" in *Nonlinearity in Condensed Matter*, ed. by A.R. Bishop, D.K. Campbell, P. Kumar, S.E. Trullinger, Springer Ser. Solid State Sci., Vol.69 (Springer, Berlin, Heidelberg, 1987) p.183-194

2.194 I.I. Ukrainskii: New variational function in the theory of quasi-one-dimensional metals. Teor. Mat. Fiz. **32**, 392 (1977) [Sov. Phys-Theor. Math. Phys. **32**, 816 (1978)]

2.195 I.I. Ukrainskii: Effect of electron interaction on the Peierls instability. Sov. Phys.-JETP **49**, 381-386 (1979)

2.196 I.I. Ukrainskii: Effective electron-electron interaction in conjugated polymers. Phys. Status Solidi (b) **106**, 55-62 (1981)

2.197 C.-Q. Wu, X. Sun, K. Nasu: Electron correlation and bond alternation in polymers. Phys. Rev. Lett. **59**, 831-834 (1987)

2.198 K. Schulten, I. Ohmine, M. Karplus: Correlation effects in the spectra of polyenes. J. Chem. Phys. **64**, 4422-4441 (1976)

2.199 I. Ohmine, M. Karplus, K. Schulten: Renormalized configuration interaction method for electron correlation in excited states of polyenes. J. Chem. Phys. **68**, 2298-2318 (1978)

2.200 P. Tavan, K. Schulten: Correlation effects in the spectra of polyacenes. J. Chem. Phys. **70**, 5414-5421 (1979)

2.201 P. Tavan, K. Schulten: The $2^1A_g - 1^1B_u$ energy gap in the polyenes: an extended configuration interaction study. J. Chem. Phys. **70**, 5407-5413 (1979)

2.202 S. Mazumdar, Z.G. Soos: Valence bond theory of narrow-band conductors. Synth. Met. **1**, 77-94 (1979/80)

2.203 S.R. Bondeson, Z.G. Soos: The noncrossing rule and degeneracy in Hubbard models: Cyclobutadiene and benzene. J. Chem. Phys. **71**, 3807 (1979); Erratum, ibid. **73**, 598 (1980)

2.204 S. Mazumdar, S.N. Dixit: Coulomb effécts on one-dimensional Peierls instability: the Peierls-Hubbard model. Phys. Rev. Lett. **51**, 292-295 (1983)

2.205 Z.G. Soos, S. Ramesesha: Valence bond approach to exact nonlinear optical properties of conjugated systems. J. Chem. Phys. **90**, 1067-1076 (1989)

2.206 S. Mazumdar, S.N. Dixit: Electronic excited states in conjugated polymers. Synth. Met. **28**, D463-468 (1989)

2.207 J.E. Hirsch: Effect of Coulomb interactions on the Peierls instability. Phys. Rev. Lett. **51**, 296-299 (1983)

2.208 J.E. Hirsch, M. Grabowski: Solitons in polyacetylene: a Monte Carlo study. Phys. Rev. Lett. **52**, 1713-1716 (1984)

2.209 D.K. Campbell, T.A. DeGrand, S. Mazumdar: Soliton energetics in Peierls-Hubbard models. Phys. Rev. Lett. **52**, 1717-1720 (1984)

2.210 D.K. Campbell, T.A. DeGrand, S. Mazumdar: Soliton energetics in extended Peierls-Hubbard models: a quantum Monte Carlo study. Mol. Cryst. Liq. Cryst. **118**, 41 (1985)

2.211 D.K. Campbell, T.A. DeGrand, S. Mazumdar: Quantum Monte Carlo studies of electron-electron interaction effects in conducting polymers. J. Stat. Phys. **43**, 803-814 (1986)

2.212 S. Klemm, M.A. Lee: Quantum simulations of conjugated carbon chains. Superlattices Microstruct. **1**, 535-542 (1985)

2.213 R. Blankenbecler, R. Sugar: Projector Monte Carlo method. Phys. Rev. D **27**, 1304-1311 (1983)

2.214 M. Suzuki: Relationship between d-dimensional quantal spin systems and (d+1)-dimensional Ising systems. Equivalence, critical exponents and systematic approximants of the partition function and spin correlations. Prog. Theor. Phys. **56**, 1454-1469 (1976)

2.215 M. Kalos, D. Levesque, L. Verlet: Helium at zero temperature with hard-sphere and other forces. Phys. Rev. A **9**, 2178-2195 (1974)

2.216 M.A. Lee, K.A. Motakabbir, K.E. Schmidt: Ground state of the extended one-dimensional Hubbard model: a Green's function Monte Carlo algorithm. Phys. Rev. Lett. **53**, 1191-1194 (1984)

2.217 J.E. Hirsch, E. Fradkin: Phase diagram of one-dimensional electron-phonon systems II. The molecular-crystal model. Phys. Rev. B **27**, 4302-4316 (1983)

2.218 R.M. Grimes, B.L. Hammond, P.J. Reynolds, W.A. Lester: Quantum Monte Carlo approach to electronically excited molecules. J. Chem. Phys. **85**, 4749-4750 (1986)

2.219 T. Holstein: Studies of polaron motion. Ann. Phys. **8**, 325-389 (1959)

2.220 M.J. Rice: Organic linear conductors as systems for the study of electron-phonon interactions in the organic solid state. Phys. Rev. Lett. **37**, 36-39 (1976)

2.221 R. Bozio, M. Meneghetti, C. Pecile: Optical properties of molecular conductors: one-dimensional systems with twofold-commensurate charge-density waves. Phys. Rev. B **36**, 7795-7804 (1987)

2.222 R. Bozio, M. Meneghetti, D. Pedron, C. Pecile: Molecular cluster models for the analysis of the optical spectra of organic charge transfer crystals: properties and applications. Synth. Met. **27**, B109 (1988); Optical studies of the interplay between electron-lattice and electron-electron interactions in organic molecular conductors. Synth. Met. **27**, B129 (1988)

2.223 W.-P. Su: Soliton excitations in a commensurability 2 mixed Peierls system. Solid State Commun. **48**, 479-481 (1983)

2.224 S. Kivelson: Solitons with adjustable change in a commensurable Peierls insulator. Phys. Rev. B **28**, 2653-2658 (1983)

2.225 D. Baeriswyl, A.R. Bishop: Localized polaronic states in mixed-valence linear chain complexes. J. Phys. C **21**, 339-356 (1988)

2.226 C.R. Fincher, C.-E. Chen, A.J. Heeger, A.C. MacDiarmid, J.B. Hastings: Structural determination of the symmetry breaking parameter in *trans*-$(CH)_x$. Phys. Rev. Lett. **48**, 100-104 (1982)

2.227 C.S. Yannoni, T.C. Clarke: Molecular geometry of *cis*- and *trans*-polyacetylene by nutation NMR spectroscopy. Phys. Rev. Lett. **51**, 1191-1193 (1983)

2.228 H. Kahlert, O. Leitner, G. Leising: Structural properties of *trans*- and *cis*-$(CH)_x$. Synth. Met. **17**, 467-472 (1987)

2.229 R.A. Harris, L.M. Falicov: Self-consistent theory of bond alternation in polyenes: normal state, charge-density waves and spin-density waves. J. Chem. Phys. **51**, 5034-5041 (1969)

2.230 I. Egri: Theory of 1-D Peierls-Hubbard model. Z. Phys. B **23**, 381-387 (1976)

2.231 I. Egri: Antiferromagnetism and lattice distortion in the Peierls-Hubbard model. Solid State Commun. **22**, 281-284 (1977)

2.232 Y.J. I'Haya, M. Suzuki, S. Narita: Separated pair approximation and electron correlation in linear polyenes. J. Chem. Phys. **77**, 391-395 (1982)

2.233 T. Nakano, H. Fukuyama: Solitons in spin-Peierls systems and applications to polyacetylene. J. Phys. Soc. Jpn. **49**, 1679-1691 (1980)

2.234 J. Kondo: Coulomb enhancement of the Peierls distortion. Physica **9B**, 176-184 (1980)

2.235 L. Pauling: The metallic orbital and the nature of metals. J. Solid State Chem. **54**, 297-307 (1984) and references therein

2.236 C. Coulson, W.T. Dixon: Bond lengths in cyclic polyenes $C_{2n}H_{2n}$. A reexamination from the valence-bond point of view. Tetrahedron **17**, 215-228 (1961)

2.237 D.J. Klein, M.A. Garcia-Bach: Variational localized site cluster expansions. X. Dimerization in linear Heisenberg chains. Phys. Rev. B **19**, 877-886 (1979)
      D.J. Klein: Long-range order for spin pairing in valence bond theory. Int. J. Quantum Chem. (Quantum Chem. Symp.) **13**, 293-303 (1979)

2.238 D.J. Klein, T.G. Schmalz, W.A. Seitz, G.E. Hite: Overview of Hückel- and resonance-theoretical approaches to π-network polymers. Int. J. Quantum Chem. (Quantum Chem. Symp.) **19**, 707-718 (1986)

2.239 S. Mazumdar: Valence bond approach to two-dimensional broken symmetries: application to La_2CuO_4. Phys. Rev. B **36**, 7190-7193 (1987)

2.240 S. Mazumdar, S.N. Dixit, A.N. Bloch: Correlation effects on charge density waves in narrow band one-dimensional conductors. Phys. Rev. B **30**, 4842-4845 (1984)

2.241 J. Hubbard: Generalized Wigner lattices in one dimension and some applications to TCNQ salts. Phys. Rev. B **17**, 494-503 (1978)

2.242 Z.G. Soos, S. Ramasesha: Valence-bond theory of linear Hubbard and Pariser-Parr-Pople models. Phys. Rev. B **29**, 5410-4095 (1984)

2.243 J. Paldus, M. Takahashi: Bond length alternation in cyclic polyenes. IV. Finite order perturbation theory approach. Int. J. Quantum Chem. **25**, 423 (1984)

2.244 G.W. Hayden, Z.G. Soos: Dimerization enhancement in one-dimensional Hubbard and Pariser-Parr-Pople models. Phys. Rev. B **38**, 6075-6083 (1988)

2.245 A. Zawadowski: Two-band model with off-diagonal occupation dependent hopping rate: enhancement of coupling and superconductivity. Phys. Scr. **T27**, 66-73 (1989)

2.246 J.E. Hirsch: Coulomb attraction between Bloch electrons. Phys. Lett. A **138**, 83 (1989)

2.247 H.M. McConnell, D.B. Chesnut: Negative spin densities in aromatic radicals. J. Chem. Phys. 27, 984-985 (1957)

2.248 A.D. McLachlan: The signs of the valence bond wave functions in an alternant hydrocarbon. Mol. Phys. 2, 223-224 (1959)

2.249 H. Thomann, L.R. Dalton, Y. Tomkiewicz, N.S. Shiren, T.C. Clarke: Electron-nuclear double-resonance determination of the ^{13}C and ^{1}H hyperfine tensors for polyacetylene. Phys. Rev. Lett. 50, 533-536 (1983)

2.250 H. Thomann, L.R. Dalton, M. Grabowski, T.C. Clarke: Direct observation of Coulomb correlation effects in polyacetylene. Phys. Rev. B 31, 3141-3143 (1985)

2.251 Z.G. Soos, S. Ramasesha: Spin densities and correlations in regular polyene radicals. Phys. Rev. Lett. 51, 2374-2377 (1983)

2.252 S. Kuroda, M. Tokumoto, N. Kinoshita, T. Ishiquro, H. Shirakawa: Distribution of spin density in undoped $(CH)_x$. J. de Phys. 44, C3,303-306 (1983)

2.253 S. Kuroda, H. Bando, H. Shirakawa: Direct observation of "soliton-like" spin density distribution in undoped cis-rich $(CH)_x$ by ENDOR at 4K. Solid State Commun. 52, 893-897 (1984)

2.254 S. Kuroda, H. Bando, H. Shirakawa: Direct observation of "soliton-like" spin density distribution in undoped cis-rich $(CH)_x$ using ENDOR. J. Phys. Soc. Jpn. 54, 3956-3965 (1984)

2.255 S. Kuroda, H. Shirakawa: Soliton spin density in polyacetylene studied by ENDOR in pristine and ^{13}C-enriched cis-rich sample. Synth. Met. 17, 423-428 (1987)

2.256 A. Grupp, P. Höfer, H. Käss, M. Mehring, R. Weizenhöfer, G. Wegner: "Pulsed ENDOR and TRIPLE Resonance on trans-Polyacetylene a la Durham Route", in Electronic Properties of Conjugated Polymers, ed. by H. Kuzmany, M. Mehring, S. Roth, Springer Ser. Solid-State Sci., Vol. 76 (Springer, Berlin, Heidelberg 1987) pp.156-159

2.257 H. Käss, P. Höfer, A. Grupp, P.K. Kahol, R. Weizenhöfer, G. Wegner, M. Mehring: Electron spin delocalization in Feast-type (Durham route) polyacetylene: pulsed ENDOR investigations. Europhys. Lett. 4, 947-951 (1987)

2.258 M. Mehring, A. Grupp, P. Höfer, H. Käss: The structure of the soliton in trans-polyacetylene: a pulsed ENDOR analysis. Synth. Met. 28, D399-D406 (1989)

2.259 V.J. Emery: Theory of the quasi-one-dimensional electron gas with strong on-site interactions. Phys. Rev. B 14, 2989-2994 (1976)

2.260 N.A. Popov: Use of the alternate-molecular orbit method for calculating the electron spectra of alternate systems. J. Struct. Chem. 9, 766-771 (1968) [Russian original: Zh. Strukt. Khim. 9, 875-880 (1968)]

2.261 N.A. Popov: The alternation of bonds and the nature of the energy gap in the π-electronic spectrum of long polyenes. J. Struct. Chem. 10, 442-448 (1969) [Russian original: Zh. Strukt. Khim. 10, 533 (1969)]

2.262 E.G. Wilson: Electronic excitations of a conjugated polymer crystal. J. Phys. C 8, 727-742 (1975)

2.263 C. Cojan, G.P. Agrawal, C. Flytzanis: Optical properties of one-dimensional semiconductors and conjugated polymers. Phys. Rev. B 15, 909-925 (1977)

2.264 K. Lochner, H. Bässler, B. Tieke, G. Wegner: Photoconduction in polyacetylene multilayer structures in single crystals. Evidence for band-to-band excitation. Phys. Status Solidi (b) 88, 653-661 (1978)

2.265 K.J. Donovan, E.G. Wilson: Photocarrier creation in one dimension. Philos. Mag. B 44, 31-45 (1981)

2.266 Z.G. Soos, S. Mazumdar, S. Kuwajima: Extended PPP model: 2^1A and 1^3B states of polydiacetylene fragments. Physica 143B, 538-541 (1986)

2.267 Z.G. Soos, S. Mazumdar, S. Kuwajima: "Extended PPP Model for Polydiacetylene Excitations" in *Crystallographically Ordered Polymers*, ed. by D.J. Sandman, ACS Symp. Ser. **337**, 190-201 (1987)

2.268 F. Wudl, S.P. Bitler: Synthesis and some properties of poly(diacetylene) (polyenyne) oligomers. J. Am. Chem. Soc. **108**, 4685-4687 (1986)

2.269 H. Gross, H. Sixl: Identification and correlation of the short-chain intermediates and final photoproducts in diacetylene crystals. Chem. Phys. Lett. **91**, 262-267 (1982)

2.270 H. Gross, H. Sixl: Spectroscopy of the intermediates of the low temperature polymerization reaction in diacetylene crystals. Mol. Cryst. Liq. Cryst. **93**, 261-277 (1983)

2.271 L. Sebastian, G. Weiser: One-dimensional wide energy bands in a polydiacetylene revealed by electroreflectance. Phys. Rev. Lett. **46**, 1156-1159 (1981)

2.272 H. Hayashi, K. Nasu: Effect of electron correlation on the ground state, the singlet-exciton states, and the triplet-exciton states of *trans*-polyacetylene. Phys. Rev. B **32**, 5295-5302 (1985)

2.273 D.N. Batchelder, D. Bloor: An investigation of the electronic excited state of polyacetylene by resonance Raman spectroscopy. J. Phys. C **15**, 3005 (1982)

2.274 R. Ball, W.-P. Su, J.R. Schrieffer: Photoproduction of neutral versus charged solitons in *trans*-$(CH)_x$. J. de Phys. Colloq. **44**, C3-429-436 (1983)

2.275 M.J. Rice, I.A. Howard: Photogenerated charged solitons in *trans*- polyacetylene. Phys. Rev. B **28**, 6089-6090 (1983)

2.276 L. Robins, J. Orenstein, R. Superfine: Observation of the triplet excited state of a conjugated-polymer crystal. Phys. Rev. Lett. **56**, 1850-1853 (1986)

2.277 M. Winter, A. Grupp, M. Mehring, H. Sixl: Transient ESR observation of triplet-soliton pairs in a conjugated polymer single crystal. Chem. Phys. Lett. **133**, 482-484 (1987)

2.278 R.H. Friend, D.D.C. Bradley, P. Townsend: Photo-excitation in conjugated polymers. J. Phys. D **20**, 1367-1384 (1987)

2.279 W.-P. Su: Triplet solitonic excitations in *trans*-polyacetylene. Phys. Rev. B **34**, 2988-2990 (1986)

2.280 B.S. Hudson, B.E. Kohler: Linear polyene electronic structure and spectroscopy. Annu. Rev. Phys. Chem. **25**, 437-460 (1974)

2.281 E.A. Imhoff, D.B. Fitchen, R.E. Stahlbush: Infrared photoluminescence in polyacetylene. Solid State Commun. **44**, 329-332 (1982)

2.282 F. Kajzar, S. Etemad, G.L. Baker, J. Messier: $\chi^{(3)}$ of *trans*-$(CH)_x$: experimental observation of $2A_g$ excited state. Synth. Met. **17**, 563-568 (1987)

2.283 W.S. Fann, S. Benson, J.M.J. Madey, S. Etemad, G.L. Baker, F. Kajzar: Spectrum of $\chi^{(3)}(-3\omega;\omega,\omega,\omega)$ in polyacetylene: an application of the free-electron laser in nonlinear optical spectroscopy. Phys. Rev. Lett. **62**, 1492-1495 (1989)

2.284 W.-K. Wu: Nonlinear optical susceptibilities of a one-dimensional semiconductor. Phys. Rev. Lett. **61**, 1119-1122 (1988)

2.285 R.R. Chance, M.L. Shand, C. Hogg, R. Silbey: Three-wave mixing in conjugated polymer solutions: two-photon absorption in polydiacetylenes. Phys. Rev. B **22**, 3540-3550 (1980)

2.286 Y. Tokura, Y. Oowaki, T. Koda, R.H. Baughman: Electro-reflectance spectra of one-dimensional excitons in polyacetylene crystals. Chem. Phys. **88**, 437-442 (1984)

2.287 P.A. Chollet, F. Kajzar, J. Messier: Nonlinear spectroscopy in polydiacetylenes. Synth. Met. **18**, 459 (1987)

2.288 U. Dinur, M. Karplus: Correlation effects in the excited states of polydiacetylene models. Chem. Phys. Lett. **88**, 171-176 (1982)

2.289 Z.G. Soos, P.C.M. McWilliams, G.W. Hayden: Coulomb correlations and two-photon spectra of conjugated polymers. Chem. Phys. Lett. **171**, 14 (1990)

2.290 S.N. Dixit, D. Guo, S. Mazumdar: Essential states mechanism of optical nonlinearity in π-conjugated polymers. Phys. Rev. B **43**, 6781-6784 (1991)

2.291 D. Baeriswyl, D.K. Campbell, S. Mazumdar: Correlations and defect energies. Phys. Rev. Lett. **56**, 1509-1510 (1986)

2.292 D.K. Campbell, D. Baeriswyl, S. Mazumdar: Electron-electron interaction effects in quasi-one-dimensional conducting polymers and related systems. Synth. Met. **17**, 197-202 (1987)

2.293 L.R. Ducasse, T.E. Miller, Z.G. Soos: Correlated states in finite polyenes: exact PPP results. J. Chem. Phys. **76**, 4094-4104 (1982)

2.294 Z.G. Soos, L.R. Ducasse: Electronic correlations and midgap absorption in polyacetylene. J. Physique **44**, C3,467-470 (1983)

2.295 Z.G. Soos, L.R. Ducasse: Electronic correlations and midgap absorption in polyacetylene. J. Chem. Phys. **78**, 4092-4095 (1983)

2.296 S. Etemad, A. Feldblum, A.G. MacDiarmid, T.-C. Chung, A.J. Heeger: Polarons and solitons in *trans*-(CH)$_x$: an optical study. J. Physique **44**, C3, 413-422 (1983)

2.297 G. Harbeke, E. Meier, W. Kobel, M. Egli, H. Kiess, E. Tosatti: Spectroscopic evidence for polarons in poly(3-methylthiophene). Solid State Commun. **55**, 419-422 (1985)

2.298 G. Harbeke, D. Baeriswyl, H. Kiess, W. Kobel: Polarons and bipolarons in doped polythiophenes. Phys. Scr. **T13**, 302-305 (1986)

2.299 F. Genoud, M. Guglielmi, M. Nechtschein, E. Genies, M. Salmon: ESR study of electrochemical doping in the conducting polymer polypyrrole. Phys. Rev. Lett. **55**, 118-121 (1985)

2.300 M. Nechtschein, F. Devreux, F. Genoud, E. Vieil, J.M. Pernaut, E. Genies: Polarons, bipolarons and charge interaction in polypyrrole: physical and electrochemical approaches. Synth. Met. **15**, 59-78 (1986)

2.301 M. Schärli, H. Kiess, G. Harbeke, W. Berlinger, K.W. Blazey, K.A. Müller: E.S.R. of BF$_4^-$-doped polythiophene. Synth. Met. **22**, 317-336 (1988)

2.302 F. Devreux, F. Genoud, M. Nechtschein, B. Villeret: "On Polaron and Bipolaron Formation in Conducting Polymers" in *Electronic Properties of Conjugated Polymers*, ed. by H. Kuzmany, M. Mehring, S. Roth, Springer Ser. Solid State Sci., Vol. 76 (Springer, Berlin, Heidelberg 1987) pp.270-276

2.303 M. Tanaka, A. Watanabe, J. Tanaka: Absorption spectra of the polyacetylene films doped with BF$_3$. Bull. Chem. Soc. Jpn. **53**, 645-647 (1980)

2.304 M. Tanaka, A. Watanabe, J. Tanaka: Absorption and reflection spectra of pure and ClSO$_3$H doped polyacetylene films. Bull. Chem. Soc. Jpn. **53**, 3430-3435 (1980)

2.305 A. Feldblum, J.H. Kaufman, S. Etemad, A.J. Heeger, T.-C. Chung, A.G. MacDiarmid: Opto-electrochemical spectroscopy of *trans*-(CH)$_x$. Phys. Rev. B **26**, 815-826 (1982)

2.306 T.-C. Chung, F. Moraes, J.D. Flood, A.J. Heeger: Solitons at high density in *trans*-(CH)$_x$: collective transport by mobile, spinless, charged solitons. Phys. Rev. B **29**, 2341-2343 (1984)

2.307 B.R. Weinberger, C.B. Roxlo, S. Etemad, G.L. Baker, J. Orenstein: Optical absorption in polyacetylene: a direct measurement using photothermal deflection spectroscopy. Phys. Rev. Lett. **53**, 86-89 (1984)

2.308 J.H. Kaufman, N. Colaneri, J.C. Scott, G.B. Street: Evolution of polaron states into bipolarons in polypyrrole. Phys. Rev. Lett. **53**, 1005-1008 (1984)

2.309 K. Kaneto, S. Hayashi, S. Ura, K. Yoshino: ESR and transport studies in electrochemically doped polythiophene film. J. Phys. Soc. Jpn. **54**, 1146-1153 (1985)

2.310 J. Orenstein, G.L. Baker: Photogenerated gap states in polyacetylene. Phys. Rev. Lett. **49**, 1043-1046 (1982)

2.311 C.V. Shank, R. Yen, R.L. Fork, J. Orenstein, G.L. Baker: Picosecond dynamics of photoexcited gap states in polyacetylene. Phys. Rev. Lett. **49**, 1660-1663 (1982)

2.312 C.V. Shank, R. Yen, J. Orenstein, G.L. Baker: Femtosecond excited-state relaxation in polyacetylene. Phys. Rev. B **28**, 6095-6096 (1983)

2.313 J. Orenstein, Z. Vardeny, G.L. Baker, G. Eagle, S. Etemad: Mechanism for photogeneration of charge carriers in polyacetylene. Phys. Rev. B **30**, 786-794 (1984)

2.314 J. Orenstein: "Photoexcitations of Conjugated Polymers" in *Handbook of Conducting Polymers*, ed. by T. Skotheim (Dekker, New York 1986) pp.1297-1335

2.315 Z. Vardeny, J. Strait, D. Moses, T.-C. Chung, A.J. Heeger: Picosecond photoinduced dichroism in *trans*-$(CH)_x$; direct measurement of soliton diffusion. Phys. Rev. Lett. **49**, 1657-1660 (1982)

2.316 Z. Vardeny, J. Orenstein, G.L. Baker: Photoinduced infrared activity in polyacetylene. Phys. Rev. Lett. **50**, 2032-2035 (1983)

2.317 Z. Vardeny, E. Ehrenfreund, O. Brafman: Photomodulation of soliton defects in polyacetylene, in *Electronic Properties of Polymers and Related Compounds*, ed. by H. Kuzmany, M. Mehring, S. Roth, Springer Ser. Solid State Sci., Vol.63 (Springer, Berlin, Heidelberg 1985) pp.91-95

2.318 Z. Vardeny, M.T. Grahn, L. Chen, G. Leising: Photoexcitation dynamics in oriented *trans*-$(CH)_x$. Synth. Met. **28**, D167-D174 (1989)

2.319 Z. Vardeny, J. Tauc: Method for direct determination of the effective correlation energy of defects in semiconductors: optical modulation spectroscopy of dangling bonds. Phys. Rev. Lett. **54**, 1844-1847 (1985)

2.320 Z. Vardeny, J. Tauc: Response to comment by Baeriswyl, Campbell and Mazumdar. Phys. Rev. Lett. **56**, 1510 (1986)

2.321 G.B. Blanchet, C.R. Fincher, T.-C. Chung, A.J. Heeger: Photo-excitations in *trans*-$(CH)_x$: a Fourier transform infrared study. Phys. Rev. Lett. **50**, 1938-1941 (1983)

2.322 G.B. Blanchet, C.R. Fincher, A.J. Heeger: Excitation profile for photogeneration of solitons in *trans*-$(CH)_x$. Phys. Rev. Lett. **51**, 2132-2135 (1983)

2.323 L. Rothberg, T.M. Jedju, S. Etemad, G.L. Baker: Charged-soliton dynamics in *trans*-polyacetylene. Phys. Rev. Lett. **57**, 3229-3232 (1986)

2.324 L. Rothberg, T.M. Jedju, S. Etemad, G.L. Baker: Picosecond dynamics of photogenerated charged solitons in *trans*-polyacetylene. Phys. Rev. B **36**, 7529-7536 (1987)

2.325 H.E. Schaffer, R.H. Friend, A.J. Heeger: Localized phonons associated with solitons in polyacetylene: coupling to the non-uniform mode. Phys. Rev. B **36**, 7537-7541 (1987)

2.326 P.D. Townsend, R.H. Friend: Photoexcitation in oriented polyacetylene. Synth. Met. **17**, 361-366 (1987)

2.327 N.F. Colaneri, R.H. Friend, H.E. Schaffer, A.J. Heeger: Mechanism for photogeneration of metastable charged solitons in polyacetylene. Phys. Rev. B **38**, 3960-3965 (1988)

2.328 D.K. Campbell, D. Baeriswyl, S. Mazumdar: Coulomb correlation effects in quasi-one-dimensional conductors. Physica **143B**, 533-537 (1986)

2.329 C.G. Levey, D.V. Lang, S. Etemad, G.L. Baker, J. Orenstein: Photo-generation of spins in *trans*-polyacetylene. Synth. Met. **17**, 569 (1987)

2.330 J.D. Flood, A.J. Heeger: Photogeneration of solitons in *trans*-polyacetylene: the reversed spin-charge relation of the photoexcitations. Phys. Rev. B **28**, 2356-2360 (1983)

2.331 F. Moraes, Y.W. Park, A.J. Heeger: Soliton photogeneration in *trans*-polyacetylene: light-induced electron spin resonance. Synth. Met. **13**, 113 (1986)

2.332 Y. Yacoby, S. Roth, K. Menke, F. Keilmann, J. Kuhl: Carrier drift time from pulsed photoconductivity in as-grown *trans*-polyacetylene. Solid State Commun. **47**, 869-871 (1983)

2.333 M. Sinclair, D. Moses, A.J. Heeger: Picosecond photoconductivity in *trans*-polyacetylene. Solid State Commun. **59**, 343-347 (1986)

2.334 H. Bleier, S. Roth, H. Lobentanzer, G. Leising: Anisotropic kinetics of optically excited charge carriers in *trans*-polyacetylene. Europhys. Lett. **4**, 1397-1402 (1987)

2.335 H. Bleier, K. Donovan, R.H. Friend, S. Roth, L. Rothberg, R. Tubino, Z. Vardeny, G. Wilson: Non-solitonic nature of picosecond photoconductivity in *trans*-polyacetylene. Synth. Met. **28**, D189-D195 (1989)

2.336 P.L. Danielsen: Inter-chain versus intra-chain electron-hole photogeneration in *trans*-polyacetylene. J. Phys. C **19**, L741-L745 (1986)

2.337 G.M. Carter, J.V. Hryniewicz, M.K. Thakur, Y.J. Chen, S.E. Meyler: Nonlinear optical processes in a polydiacetylene measured with femtosecond duration pulses. Appl. Phys. Lett. **49**, 998-1000 (1986)

2.338 B.I. Greene, J. Orenstein, R.R. Millard, L.R. Williams: Nonlinear optical response of excitons confined to one dimension. Phys. Rev. Lett. **58**, 2750-2753 (1987)

2.339 B.I. Greene, J.F. Mueller, J. Orenstein, D.H. Rapkine, S. Schmitt-Rink, M. Thakur: Phonon-mediated optical nonlinearity in polydiacetylene. Phys. Rev. Lett. **61**, 325-328 (1988)

2.340 T. Hattori, W. Hayes, D. Bloor: Photoinduced absorption and luminescence in polydiacetylenes. J. Phys. C **17**, L881-L888 (1984)

2.341 J. Orenstein, S. Etemad, G.L. Baker: Photoinduced absorption in a polydiacetylene. J. Phys. C **17**, L297-L300 (1984)

2.342 F.L. Pratt, K.S. Wong, W. Hayes, D. Bloor: Infrared photo-induced absorption in polydiacetylene. J. Phys. C **20**, L41-L46 (1987)

2.343 T. Hattori, W. Hayes, K. Wong, K. Kaneto, K. Yoshino: Optical properties of photoexcited and chemically doped polythiophene. J. Phys. C **17**, L803-L807 (1984)

2.344 Z. Vardeny, E. Ehrenfreund, O. Brafman, M. Nowak, H. Schaffer, A.J. Heeger, F. Wudl: Photogeneration of confined soliton pairs (bipolarons) in polythiophene. Phys. Rev. Lett. **56**, 671-674 (1986)

2.345 Y.H. Kim., S. Hotta, A.J. Heeger: Infrared photoexcitation and doping studies on poly(3-methylthiophene). Phys. Rev. B **36**, 7486-7490 (1987)

2.346 Y.H. Kim., D. Spiegel, S. Hotta, A.J. Heeger: Photoexcitation and doping studies of poly(3-hexylthienylene). Phys. Rev. B **38**, 5490-5495 (1988)

2.347 W. Hayes, C.N. Ironside, J.F. Ryan, R.P. Steele, R.A. Taylor: Picosecond study of luminescence of *cis*-polyacetylene. J. Phys. C **16**, L729-L732 (1983)

2.348 S. Etemad, G.L. Baker, C.B. Roxlo, B.R. Weinberger, J. Orenstein: Band edge and neutral soliton absorption in polyacetylene: the role of Coulomb correlation. Mol. Cryst. Liq. Cryst. **117**, 275-282 (1985)

2.349 K.S. Wong, W. Hayes, T. Hattori, R.A. Taylor, J.F. Ryan, K. Kaneto, Y. Yoshino, D. Bloor: Picosecond studies of luminescence in polythiophene and polydiacetylene. J. Phys. C **18**, L843-L847 (1985)

2.350 W.J. Feast, I.S. Millichamp, R.H. Friend, M.E. Horton, D. Phillips, S.D.D.V. Rughooputh, G. Rumbles: Optical absorption and luminescence in poly(4,4'-diphenylenediphenylenevinylene). Synth. Met. **10**, 181-191 (1985)

2.351 U. Sum., K. Fesser, H. Büttner: A model with broken charge conjugation symmetry for conducting polymers. J. Phys. C **20**, L71-L75 (1987)

2.352 M. Kakano: Effects of density coupling on solitons in half-filled electron-lattice system. J. Phys. Soc. Jpn. 56, 2826-2834 (1987)

2.353 P.L. Danielsen, R.C. Ball: A theoretical study of photoluminescence quenching in cis-polyacetylene. J. de Phys. 46, 1611-1622 (1985)

2.354 U. Sum, K. Fesser, H. Büttner: Coulomb interaction and optical spectra in conjugated polymers. Solid State Commun. 61, 607-610 (1987)

2.355 A.R. Bishop, D.K. Campbell, P.S. Lomdahl, B. Horovitz, S.R. Phillpot: Breathers and photoinduced absorption in polyacetylene. Phys. Rev. Lett. 52, 671-674 (1984)

2.356 A.R. Bishop, D.K. Campbell, P.S. Lomdahl, B. Horovitz, S.R. Phillpot: Nonlinear dynamics, breathers and photoinduced absorption in polyacetylene. Synth. Met. 9, 223-239 (1984)

2.357 B. Horovitz, A.R. Bishop, S.R. Phillpot: Semiclassical formalism of optical absorption and breathers in polyacetylene. Phys. Rev. Lett. 60, 2210-2213 (1988)

2.358 M. Grabowski, D. Hone, J.R. Schrieffer: Photogenerated solitonic states in trans-polyacetylene. Phys. Rev. B 31, 7850-7854 (1985)

2.359 B.E. Kohler: The polyene 2^1A_g state in polyacetylene photoinduced photoabsorption and thermal isomerization. J. Chem. Phys. 88, 2788-2792 (1988)

2.360 E. Mulazzi, G.P. Brivio, E. Faulques, S. Lefrant: Experimental and theoretical Raman studies in trans-polyacetylene. Solid State Commun. 46, 851-855 (1983)

2.361 E. Mulazzi: Polarized resonant Raman scattering spectra from stretched trans-polyacetylene. Solid State Commun. 55, 807 (1985)

2.362 Z. Vardeny, E. Ehrenfreund, O. Brafman, B. Horovitz: Resonant Raman scattering from amplitude modes in $trans-(CH)_x$ and $-(CD)_x$. Phys. Rev. Lett. 51, 2326-2329 (1983)

2.363 Z. Vardeny, E. Ehrenfreund, O. Brafman, B. Horovitz: Classification of disorder and extrinsic order in polymers by resonant Raman scattering. Phys. Rev. Lett. 54, 75-78 (1985)

2.364 Z. Vardeny, E. Ehrenfreund, O. Brafman, B. Horovitz, H. Fujimoto, J. Tanaka, M. Tanaka: Detection of soliton shape modes in polyacetylene. Phys. Rev. Lett. 57, 2995-2998 (1986)

2.365 C.-Q. Wu, X. Sun: Effects of the $e - e$ interaction on the localized modes of solitons in polyacetylene. Phys. Rev. B 33, 8772-8775 (1986)

2.366 K. Yonemitsu, Y. Ono, Y. Wada: Correlation effects on the phonons localized around a soliton or a polaron in polyacetylene. J. Phys. Soc. Jpn. 56, 4400-4407 (1987)

2.367 J.P. Pouget: "Structural Features of Pure and Doped Polyacetylene: $(CH)_x$" in Electronic Properties of Polymers and Related Compounds ed. by H. Kuzmany, M. Mehring, S. Roth, Springer Ser. Solid-State Sci., Vol.63 (Springer, Berlin, Heidelberg 1985) pp.26-34

2.368 H. Shirakawa, T. Ito, S. Ikeda: Electrical properties of polyacetylene with various cis-trans compositions. Makro. Chem. 179, 1565-1573 (1978)

2.369 M. Ozaki, D. Peebles, B.R. Weinberger, A.J. Heeger, A.G. MacDiarmid: Semiconductor properties of polyacetylene $p-(CH)_x:n$-CdS heterojunctions. J. Appl. Phys. 51, 4252-4256 (1980)

2.370 B. Ankele, G. Leising, H. Kahlert: Optical properties of conjugated segments embedded in polyvinylidene chloride. Solid State Commun. 62, 245-248 (1987)

2.371 G.W. Bryant, A.J. Glick: Impurity states in doped polyacetylene. Phys. Rev. B 26, 5855-5866 (1982)

2.372 B.R. Bulka: Stability of the Peierls state due to impurity. Phys. Status Solidi (b) 107, 359 (1981)

2.373 D. Baeriswyl: Impurity-induced defect states in polyacetylene. J. Physique 44, C3, 381-385 (1983)

2.374 C.T. White, M.L. Elert, J.W. Mintmire: Effects of off-diagonal disorder on soliton- and polaron-like states in *trans*-polyacetylene. J. Physique **44**, C3, 481-484 (1983)

2.375 S.R. Phillpot, D. Baeriswyl, A.R. Bishop, P.S. Lomdahl: Interplay of disorder and electron-phonon coupling in models of polyacetylene. Phys. Rev. B **35**, 7533-7550 (1987)

2.376 D. Boyanovsky: Random disorder in one-dimensional electron-phonon systems. Phys. Rev. B **27**, 6763-6769 (1983)

2.377 B.-C. Xu, S.E. Trullinger: Supersymmetric treatment of random disorder in the continuum model of polyacetylene. Phys. Rev. Lett. **57**, 3113-3116 (1986)

2.378 K. Fesser: The influence of disorder on the electronic structure of conjugated polymers. J. Phys. C **21**, 5361-5368 (1988)

2.379 B.R. Bulka, B. Kramer: On the stability of the polaron in one-dimensional disorderd system. Z. Phys. B **63**, 139-147 (1986)

2.380 T. Bohr, S.A. Brazovskii: Soliton statistics for a system of weakly bound chains; mapping to the Ising model. J. Phys. C **16**, 1189 (1983)

2.381 R.H. Baughman, G.J. Moss: Interchain contribution to soliton properties in polyacetylene. J. Chem. Phys. **77**, 6321 (1982)

2.382 S. Jeyadev: Interchain Coulomb interaction in polyacetylene. Phys. Rev. B **28**, 3447-3456 (1983)

2.383 D. Baeriswyl, K. Maki: Soliton confinement in polyacetylene due to interchain coupling. Phys. Rev. B **28**, 2068-2073 (1983)

2.384 D. Baeriswyl, K. Maki: Interchain order, soliton confinement and electron-hole photogeneration in *trans*-polyacetylene. Phys. Rev. B **38**, 8135-8141 (1988)

2.385 G. Leising, O. Leitner, H. Kahlert: Structure of fully oriented crystalline *trans*-$(CH)_x$. Mol. Cryst. Liq. Cryst. **117**, 67 (1985)

2.386 S. Kivelson: Electron hopping conduction in the soliton model of polyacetylene. Phys. Rev. Lett. **46**, 1344-1348 (1981)

2.387 S. Kivelson: Electron hopping in a soliton band: conduction in lightly doped polyacetylene. Phys. Rev. B **25**, 3789-3821 (1982)

2.388 A.J. Epstein, H. Rommelmann, M. Abkowitz, N.W. Gibson: Anomalous frequency-dependent conductivity of polyacetylene. Phys. Rev. Lett. **47**, 1549-1553 (1981)

2.389 S. Jeyadev, J.R. Schrieffer: Interchain polaron tunneling in *trans*-polyacetylene. Phys. Rev. B **30**, 3620-3624 (1984)

2.390 D. Emin: Self-trapping in quasi-1-D solids. Phys. Rev. B **33**, 3973-3975 (1986)

2.391 Yu.N. Gartstein, A.A. Zakhidov: Instability of polarons and bipolarons in conducting polymers at various 3d ordering types. Solid State Commun. **62**, 213-220 (1987); Erratum, ibid. **65**, No.4, ii (1988)

2.392 D. Baeriswyl, K. Maki: Fate of solitons, polarons and bipolarons in conjugated polymers: the role of interchain coupling. Synth. Met. **28**, D507-D512 (1989)

2.393 D. Baeriswyl, G. Harbeke, H. Kiess, W. Meyer: "Conducting Polymers: Polyacetylene" in *Electronic Properties of Polymers*, ed. by J. Mort, G. Pfister (Wiley, New York 1982) pp.267-326

2.394 Z.-B. Su, L. Yu: Theory of soliton generation and lattice relaxation in polyacetylene: (I) General formalism. Comm. Theor. Phys. (Beijing) **2**, 1203-1218 (1983); (II) Non-radiative transitions. ibid. **2**, 1323-1339 (1983); (III) Radiative transitions. ibid. **2**, 1341-1356 (1983)

2.395 Z.-B. Su, L. Yu: Soliton pair generation in polyacetylene: a lattice relaxation approach..Phys. Rev. B **27**, 5199-5202 (1983); Erratum: ibid. **29**, 2309 (1984)

2.396 Z.-B. Su, Y.-X. Wang, L. Yu: Quantum fluctuation of the order parameter in polyacetylene. Mol. Cryst. Liq. Cryst. **118**, 75-79 (1985)

2.397 D. Schmeltzer, R. Zeyher, W. Hanke: Effects of quantum fluctuations on one-dimensional electron-phonon systems: the Su-Schrieffer-Heeger model. Phys. Rev. B **33**, 5141-5144 (1986)

2.398 J.P. Sethna, S. Kivelson: Photoinduced soliton pair production in polyacetylene: an instanton approach. Phys. Rev. B **26**, 3513-3516 (1982); Erratum ibid. **27**, 7798 (1983)

2.399 A. Auerbach, S. Kivelson: Large amplitude quantum fluctuations and sub-gap optical absorption in *trans*-polyacetylene. Phys. Rev. B **33**, 8171-8179 (1986)

2.400 M. Sinclair, D. Moses, D. McBranch, A.J. Heeger, J. Yu, W.-P. Su: "Instantons" as the origin of the nonlinear optical properties of polyacetylene. Phys. Scr. **T27**, 144-147 (1989)

2.401 J. Chen, A.J. Heeger: In situ electron spin resonance experiments on polyacetylene during electrochemical doping. Synth. Met. **24**, 311-327 (1988)

2.402 A. Kotani: Continuity of self-consistent solutions between commensurate and incommensurate phases of the Peierls instability. II. Numerical calculations at zero temperature. J. Phys. Soc. Jpn. **42**, 416-423 (1977)

2.403 H.-Y. Choi, E.J. Mele: Dynamical conductivity of soliton lattice and polaron lattice in the continuum model of polyacetylene. Phys. Rev. B **34**, 8750-8757 (1986)

2.404 A. Takahashi: Exact periodic solutions in the continuum models of polyacetylene. Prog. Theor. Phys. **81**, 610-632 (1989)

2.405 B.R. Bulka: Ground state of doped polyacetylene. Synth. Met. **24**, 41 (1988)

2.406 S. Kivelson, A.J. Heeger: First-order transition to a metallic state in polyacetylene: a strong coupling polaronic metal. Phys. Rev. Lett. **55**, 308-311 (1985)

2.407 S. Kivelson, A.J. Heeger: Theory of the soliton-lattice to polaron-lattice transition in conducting polymers. Synth. Met. **17**, 183-188 (1987)

2.408 D.B. Tanner, G.L. Doll, A.M. Rao, P.C. Eklund, G.A. Arbuckle, A.G. MacDiarmid: Infrared absorption in K-doped $(CH)_x$. Synth. Met. **28**, D141-D146 (1989)

2.409 E.J. Mele, M.J. Rice: Semiconductor-metal transition in doped polyacetylene. Phys. Rev. B **23**, 5397-5412 (1981)

2.410 R.H. Baughman, N.S. Murthy, G.G. Miller, L.W. Shacklette, R.M. Metzger: Structure and properties of conducting polyacetylene complexes. J. Physique **44**, C3, 53-59 (1983)

2.411 E.M. Conwell, H.A. Mizes, S. Jeyadev: Metal-insulator transition in *trans*-polyacetylene. Phys. Rev. B **40**, 1630 (1989)

2.412 S. Roth, H. Bleier: Solitons in polyacetylene. Adv. Phys. **36**, 385-462 (1987)

2.413 K.L. Ngai, R.W. Rendell: "Dielectric and Conductivity Relaxations in Conducting Polymers" in *Handbook of Conducting Polymers*, ed. by T.A. Skotheim (Dekker, New York 1986) pp.967-1039

2.414 A.B. Kaiser: "Electronic Transport in Low-Conductivity Metals and Comparison with Highly Conducting Polymers" in *Electronic Properties of Conjugated Polymers*, ed. by H. Kuzmany, M. Mehring, S. Roth, Springer Ser. Solid-State Sci., Vol. 76 (Springer, Berlin, Heidelberg 1987) pp.2-11

2.415 F.L. Carter (ed.): *Molecular Electronics* (Dekker, New York 1982)

2.416 S. Roth, G. Mahler, Y. Shen, F. Coter: Molecular electronics of conducting polymers. Synth. Met. **28**, C815-C822 (1989)

2.417 T.D. Holstein, L.A. Turkevich: Field theory for the 1-D optical polaron. I Incorporation of the Goldstone mode and interaction with internal phonons. Phys. Rev. B **38**, 1901-1937 (1988) and references therein

2.418 C. Kunz: Soliton diffusion in polyacetylene: memory-function formalism. Phys. Rev. B **34**, 3288-3296 (1986)

2.419 N. Basescu, Z.-X. Liu, D. Moses, A.J. Heeger, H. Naarmann, N. Theophilou: Long mean free path coherent transport in doped polyacetylene, in *Electronic Properties of Conjugated Polymers*, ed. by H. Kuzmany, M. Mehring, S. Roth, Springer Ser. Solid-State Sci., Vol. 76 (Springer, Berlin, Heidelberg 1987) pp.18-22

2.420 Th. Schimmel, W. Riess, J. Gmeiner, G. Denninger, M. Schwoerer, H. Naarmann, N. Theophilou: DC-conductivity of a new type of highly conducting polyacetylene, N - $(CH)_x$. Solid State Commun. **65**, 1311-1315 (1988)

2.421 P. Sheng: Fluctuation-induced tunneling conduction in disordered materials. Phys. Rev. B **21**, 2180-2195 (1980)

2.422 E.M. Conwell, H.A. Mizes: Conduction in metallic *trans*-polyacetylene. Synth. Met. **38**, 319-329 (1990)

2.423 L. Pietronero: Ideal conductivity of carbon π polymers and intercalation compounds. Synth. Met. **8**, 225-231 (1983)

2.424 S. Kivelson, A.J. Heeger: Intrinsic conductivity of conducting polymers. Synth. Met. **22**, 371-384 (1988)

2.425 D.S. Boudreaux, R.R. Chance, J.F. Solf, L.W. Shacklette, J.-L. Brédas, B. Thémans, J.M. André, R. Silbey: Theoretical studies on polyaniline. J. Chem. Phys. **85**, 4584-4590 (1986)

2.426 J.E. Hirsch: "Monte Carlo Simulation of Models for Low-Dimensional Conductors" in *Low-Dimensional Conductors and Superconductors*, ed. by D. Jérome, L.G. Caron, NATO ASI Series B, Vol. 155 (Plenum, New York 1987) pp.71-86

3. Charge Transport in Polymers

WALTHER REHWALD and HELMUT G. KIESS

On a microscopic scale polymers are characterized by a strongly anisotropic electronic structure. This is a consequence of the chain-like form of the molecules, having a strong covalent bonding along the chain and comparatively weak bonding of van der Waals type between the chains. Band-structure calculations, based on an ideal and regular crystalline polymer structure, can demonstrate this property. As an example, the bandwidth of electronic states in polyacetylene is of the order of 10 eV along the chain, but only some 0.1 eV perpendicular. However, the structure of most polymers ranges between crystalline and completely amorphous; this may give rise to considerable variations of the interchain transfer integral and thus also of the bandwidth perpendicular to the chain direction. The transfer integrals and the bandwidth parallel to the chain axis are usually less affected.

The electrical properties of a large variety of polymers have been studied. It has been found that polymers with a conjugate bonding system, i.e. an uninterrupted sequence of single and double bonds running through the whole molecule, are the most successful candidates for conducting polymers. Therefore conjugate polymers will be the subject of this article. Such polymers are known to exhibit a nearly metallic conductivity after doping with proper acceptor or donor molecules. (We will use the term "dopant molecule" for both single atoms, such as Na^+, and small groups, such as I_3^- and AsF_6^-.) The next section will deal with models for the insulating and semiconducting states of conjugated polymers, i.e. low and intermediate doping. Elementary excitations that are peculiar to these materials, such as solitons and polarons, receive special attention. In Sect. 3.2, models for charge transport are discussed. Experiments performed in this doping range are used in Sect. 3.3 to either prove or disprove models previously discussed. At high doping levels a transition takes place to a quasi-metallic state. The phenomena observed and models to explain them are the subject of Sect. 3.4.

Another point is worth mentioning at the beginning. Physicists and chemists have moved into this field from different sides. As a consequence, two names often exist for the same notion. Although this chapter on charge transport originates from a physicist's view, the corresponding chemical terms are also used whenever they are better suited to describe the phenomena.

3.1 Models for the Insulating and Semiconducting States

Conjugated polymers are often considered to originate from a one-dimensional system with one electron per carbon atom. It can be shown that such a system cannot exist as a one-dimensional metal with a half-filled band, but rather as an insulator with a gap forming at the Fermi level. The reason for this is either the Peierls instability, or electron correlation, or a combination of both. Exhibiting an energy gap of 1–3 eV, polymers seem to resemble inorganic crystalline semiconductors. The quasi-one-dimensional structure of these materials gives rise to a strong electron–phonon interaction, leading to new quasi-particles upon doping: these are solitons, polarons and bipolarons. Similar arguments hold if electron correlation effects prevail. Disorder within the chain and its environment plays an important role.

3.1.1 The Electronic Ground State

It is well known that the accessible energy levels of an electron in a crystal are grouped into bands, which may be visualized as originating from the electronic levels of the atom. The bands form by a splitting of the atomic levels when the atoms approach one another and obtain their equilibrium positions in the crystal. The bands are separated by forbidden energy ranges called gaps. The gap relevant for conductivity processes in a semiconductor separates the band highest in energy which is completely filled (valence band) from the lowest lying band which is completely empty at absolute zero temperature (conduction band).

In a metal the conduction band is partially filled, implying that a finite density of states, $N(E_F) > 0$, exists at the Fermi level. Organic polymers differ from crystalline semiconductors and metals in several respects. The anisotropy in their structure is reflected in large differences in the way the energies vary with quasi-momentum along the chain and perpendicular to it. This anisotropy has further consequences when the charge state is changed by doping, as will be discussed later.

Organic conjugated polymers are often successfully treated theoretically as one-dimensional systems. The one dimensionality may not be taken too literally in the sense of a strict geometrical linear arrangement of the atoms forming the polymer. Rather, one dimensionality has to be understood as the property that each lattice point is coupled to two neighbouring points only. Two theoretical approaches are now used to explain the formation of a gap. They take either electron–phonon interaction or electron correlation as the dominant mechanism.

If the gap is the result of the electron–phonon interaction, Peierls' argument [3.1] can be used to give a physical reason for why a chain of unsaturated carbon atoms with one conduction electron (π electron) per atom does not exhibit metallic properties: If all the atoms are spaced with equal distance a, the basic cell in reciprocal space (Brillouin zone) is the interval $-\pi/a < k < \pi/a$. With one electron per atom the band would be half filled and, hence, the chain would exhibit metallic behaviour. A periodic distortion of the chain, commensurate with the original structure, generates an n-fold super-structure and reduces the Brillouin zone to $-\pi/na < k < \pi/na$, with n being the number of atoms in the new unit cell (Fig. 3.1). The effect of the distortion is to open up a gap at the boundaries $k = \pm \pi/(na)$ of the new Brillouin zone. Therefore, if only states below the new gap are occupied with electrons, a reduction of the energy will occur and the distorted state will be more favourable, implying that the semiconducting phase is more stable than the metallic phase. It seems that Peierls' argument indicates quite fundamentally that a one-dimensional chain of unsaturated carbon atoms leads necessarily to a semiconducting state. However, as has been shown by *Froehlich* [3.2], the Peierls state is only semiconducting if the periodic lattice distortion is commensurate with the lattice. If it is incommensurate, the phase of the periodic lattice distortion can move through the lattice carrying a charge density wave. In this case the Peierls state is conducting.

A more detailed investigation of the question of lattice distortion and of bond alternation in polymeric chains has been made within the framework of the Hückel theory (for details see Chap. 2). The effective single particle Hamiltonian is represented by a sum over electron sites n and spin states s, with c_{ns}^+ and c_{ns} denoting electronic creation and annihilation operators

$$H_{\text{eff}} = \sum_n f_n - \sum_n t_{n,n+1}(c_{ns}^+ c_{n+1,s} + c_{n+1,s}^+ c_{ns}). \tag{3.1}$$

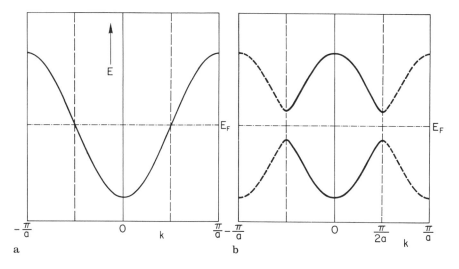

Fig. 3.1 a, b. One-dimensional electronic system with a half-filled band; band-structure before (**a**) and after the Peierls distortion (**b**)

Here f_n describes the elastic energy associated with deformations of the σ-bonds alone, and $t_{n,n+1}$ the transfer or resonance integral for π-electrons. Both parameters depend on the interatomic distances and are expanded into a series in atomic displacement coordinates u_n

$$f_n = (1/2)K_0(u_n - u_{n+1})^2 \tag{3.2}$$

$$t_{n,n+1} = t_0 - \alpha(u_{n+1} - u_n). \tag{3.3}$$

In these equations K_0 represents the elastic force constant and α the electron–phonon coupling constant. Introducing bond alternation into the Hamiltonian

$$u_n = (-1)^n u, \tag{3.4}$$

a gap equal to

$$E_g = 2\Delta_0 = 8\alpha u \tag{3.5}$$

is found for the Peierls-distorted state. In the weak coupling limit the energy gap is related to the total π-electron bandwidth $W = 4t_0$ and the parameter

$$\lambda = 2\alpha^2/(\pi t_0 K_0) \tag{3.6}$$

by the equation

$$2\Delta_0 = 4W/(\exp(1 + 1/(2\lambda)). \tag{3.7}$$

The dimerized state was shown to represent a minimum in energy and to exhibit an alternating bond-order component. The bond order is a measure for the amount of single-, double- or triple-bond characters and is closely related to the bond length [3.3].

The Peierls model neglects completely the Coulomb repulsion for an electron that is transferred to a state already occupied. In the simple Hubbard model [3.4] electron correlation is taken into account, but electron–phonon interaction is assumed to be negligible. Following this approach, the Hamiltonian in (3.1) is replaced by

$$H_{eff} = -t\sum_n (c_{ns}^+ c_{n+1,s} + c_{n+1,s}^+ c_{ns}) + U\sum_n n_{ns}n_{ns} \tag{3.8}$$

with $n_{ns} = c_{ns}^+ c_{ns}$ and U the on-site Coulomb interaction. The charge excitation spectrum which is derived from this Hamiltonian for N electrons on $N_a = N$ sites is given by

$$E_g \sim (8/\pi)(Ut)^{1/2}\exp(-2\pi t/U) \text{ in the limit } U \to 0 \text{ and}$$

$$\sim U - 4t + (8t^2/U)\ln 2 + \cdots. \quad \text{for } U \to \infty \tag{3.9}$$

This model yields a gap also in the absence of a Peierls distortion. The question arises therefore, whether the on-site Coulomb interaction or the electron–phonon interaction is the dominant mechanism. It has been shown that both interactions are of similar order of magnitude and that indeed Coulomb interaction enhances bond alternation. An interpretation of the results requires

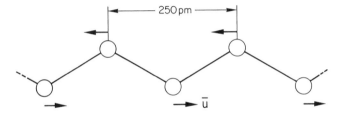

Fig. 3.2. Displacement of carbon atoms in trans-polyactylene

some care and should be based on models that take both interactions into account. The simple Hubbard model gives, for the case of an exactly half-filled band, an insulating behaviour but for the case $N \neq N_a$ however, conducting behaviour. Interestingly enough, there is no gap for the excited conducting state, and Pauli magnetism is to appear in the weakly doped material. More details on correlation effects are found in Chap. 2.

Both models can be directly applied to the simplest and most extensively studied conducting polymer, trans-polyacetylene, $(CH)_n$. It consists of a carbon backbone, each carbon atom carrying one hydrogen. The bonding is of sp^2 type, resulting in a planar molecule with bond angles of about 120°. Therefore, one electron of π character remains per carbon atom. Experiments indicate that this material is indeed semiconducting; the optical gap has been determined to be 1.4–1.8 eV, depending on how the experiments are analyzed [2.5].

The bond lengths were measured by X-ray [2.6] and NMR techniques [3.7] and found to be 144 and 135 pm, respectively, giving a bond alternation parameter of about $u = 2.6$ pm. Figure 3.2 shows the corresponding displacement vectors. Using these values together with (3.5) and (3.7) and using a bandwidth of $W = 10$ eV (from band-structure calculations) the parameters α and λ have been determined to be $\alpha = 40$–70 meV/pm and $\lambda \approx 0.2$. It is clear that the Hückel theory neglects correlation effects, or at most takes them into account in an averaged way by adjusted parameters α and λ. The observation of bond alternation by itself is no indication that correlation is negligible. Hence the parameters derived solely on the basis of electron–phonon interaction have to be taken with care and require corrections due to correlation effects.

Polyyne, a one-dimensional allotrope of carbon (carbyne) with alternating single and triple bonds, has also been treated within the one-electron tight binding description [3.8]. This polymer is expected to have the ground state of a Peierls insulator. With the parameters t_0, K_0 and λ employed for polyacetylene, a bandgap of 5 eV is calculated. However, this polymer has not been intensively studied experimentally, and the preliminary findings disagree with the theoretical predictions [3.9].

In the afore-mentioned polymers the transfer integrals between a carbon atom and its two nearest neighbours are equal, due to the symmetry of the atomic structure. In cis-$(CH)_n$, however, the hydrogen atoms take different positions at two adjacent carbon atoms; as a consequence the transfer integrals

Table 3.1 Quasi-particles, their symbolic representation, and selected representative materials

are different. This can be accounted for by writing

$$t_{n,n+1} = t_0 - (-1)^n t_1 + \alpha(u_{n+1} - u_n) \tag{3.10}$$

with t_1 representing the contribution of the hydrogen atoms. In the continuum limit and for weak coupling, the gap parameter

$$2\Delta_0 = 4W \exp((1 - \gamma)/(2\lambda) + 1) \tag{3.11}$$

depends, in addition to (2.7), on the confinement parameter $\gamma = t_1/(2\lambda t_0)$. For $\gamma = 0$ the result for *trans*-$(CH)_n$ is reproduced. In *cis*-$(CH)_n$ $\gamma \neq 0$ and, hence, the band gap is expected to be slightly larger than for *trans*-$(CH)_n$, which is in agreement with the experimental findings. Another consequence is that in *cis*-$(CH)_n$ the two possible arrangements of single and double bonds, known as *cis-transoid* and *trans-cisoid* (Table 3.1) are different in energy. There exists a single ground state in contrast to *trans*-$(CH)_n$, where the ground state is two-fold degenerate.

The Hückel type calculation and other semi-empirical methods have been extended to polydiacetylene [3.10] and to polymers with aromatic rings as building units [3.11]. With the appropriate choice of parameters the ground state configuration can be calculated astonishingly well and trends for the magnitude of the gaps can be predicted.

In the case of aromatic polymers, computations have been used to show that the bandgap values and the bandwidth of the highest occupied band decreases with increasing torsion angle of the monomeric unit from the equilibrium position, the change being small for low angles. Nevertheless, a coplanar conformation can turn out to be of importance since out-of-plane distortions can give rise to fluctuations of the band parameters and hence negatively influence the conduction process. Other effects of such distortions on the carrier transport will be discussed at a later stage.

3.1.2 The Nature of the Charge Carriers

Having discussed the ground state, conjugated polymers seem to resemble at first sight the well-known inorganic crystalline semiconductors. However, conducting polymers do by no means behave like conventional semiconductors, the important difference being that adding or removing an electron from the polymer will not generate free electrons or holes at the conduction or valence band edges. The physical reason for this is that the electronic degrees of freedom couple rather strongly to the structural deformations as a consequence of the one-dimensional coordination, so that a transfer of charge leads to a molecular distortion. For small molecules it is well known that the atomic equilibrium positions are different in the neutral and ionic state. Since in the case of long polymer chains the distortion will be limited to relatively small segments, it is legitimate to call these distorted segments defects.

Although these defects have also been treated within the framework of the Hückel theory, the continuum model developed by *Brazovskii* and *Kirova* [3.12] and by *Fesser* et al. [3.13] probably gives a more coherent description of these defects. Since the extension of the defects is usually several monomer units, the discreteness of the lattice is of minor importance and the continuum model adequate. In this limit many results can be obtained analytically, for which numerical calculations are required in the discrete model [3.14–16]. The simplest of the defects is created by adding an electron or hole to the ground state and by allowing the lattice to relax. In fact, this gedankenexperiment simulates the doping process of a polymer chain, assuming that no mixing of the wavefunctions between dopant molecule and polymer occurs. The defects generated in this way are called radical anion or radical cation in chemistry, in physics positively or negatively charged polaron. It is useful to realize that a polaron represents a combination of two defects, namely of a charged and a neutral defect or, in chemical terms, of an ion and a radical. Within the continuum theory the defect on the chain can be represented by a spatially varying gap parameter

$$\Delta(x) = \Delta_0 + 2t_0 a\kappa_0 \left[\tanh \kappa_0(x - x_0) - \tanh \kappa_0(x + x_0)\right]. \tag{3.12}$$

Here a represents the repetition length. It is easy to see from this formula that with an appropriate choice of the parameters two solutions are possible, namely either the well-separated kink and anti-kink, or a small indentation in the gap-parameter function (Fig. 3.3). Small values of x_0 characterize the latter, large values the former solution. The choice of the parameters κ_0 and x_0 is not arbitrary, but depends on the value of the confinement parameter γ or, in physical terms, on the question of whether the polymer has a degenerate ($\gamma = 0$) or non-degenerate ($\gamma = 0$) ground state.

For the purpose of discussing carrier transport it is important to know the energy of formation of the various quasi-particles; this decides which species is created. Polymers with a degenerate ground state are trans-polyacetylene and polyyne. In these materials theory predicts that polarons and solitons are possible quasi-particles. In acceptor-doped trans-$(CH)_n$, a hole in the valence band needs 0.7 eV (half the bandgap energy) for its formation [3.15, 16], a polaron only 0.65 eV [3.14]. As already mentioned, the polaron can be envisaged

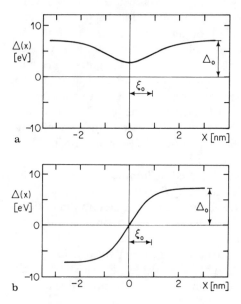

Fig. 3.3 a, b. Gap parameter for a polaron (**a**) and soliton (**b**), calculated in a continuum model. The coherence length (half-width of the excitation) is denoted by ξ_0 (from [3.13])

as a combination of a charged and a neutral defect in close proximity. When these defects are separated, the energy of formation increases to about 0.8 eV [3.15, 16] and the limiting case of $\kappa_0 x_0 \gg 1$ in (3.12) is approached. So the first charge added to a chain will lead to the formation of a polaron. Solitons can be created only in pairs for topological reasons. When a second charge is added, the two polarons disintegrate into two charged solitons (kink and antikink). It is worth noting that the existence of widely separated kinks is connected to the fact that trans-$(CH)_n$ has a degenerate ground state.

In polymers with a non-degenerate ground state, a widely separated kink–antikink pair is energetically unfavourable, because the isomer between the kink and antikink usually has a higher energy than the other. Thus, at low doping concentration, when on average only one charge per chain is found, polarons are expected; if more than one charge is transferred to the chain, correlated pairs of charged defects—bipolarons—are usually favoured due to their lower energy of formation. Bipolarons are called in chemical parlance carbodication or carbodianion. Table 3.1 gives a compilation of the excitations discussed here in terms of chemical formulae. It has to be borne in mind that these quasi-particles are extended over several repeat distances.

The difference between polymers with and polymers without degenerate ground state can be easily visualized, if the energy of the two possible structures is plotted (Fig. 3.4) as a function of the distortion parameter for a spatially

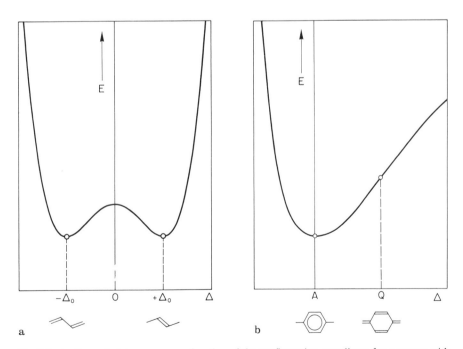

Fig. 3.4 a, b. Ground state energy as a function of the configuration coordinate for a system with a degenerate ground state, such as trans-$(CH)_n$ (**a**), and a single ground state (**b**)

uniform bond arrangement. In the case of *trans*-(CH)$_n$ the energy of the two structures is the same, whereas for e.g. poly(p-phenylene), it is not. In the latter material the quinoid state has a higher energy than the ground state with the more benzenoid (aromatic) bond sequence. Table 3.1 demonstrates that a charge residing on the chain is always accompanied by a short segment of predominantly quinoid structure and a radical defect at its opposite side. At higher doping levels, when bipolarons are formed, a second charge replaces the radical position. Obviously, if the structure between the two defects needs a higher energy, the tendency will be to minimize this polymer segment to a length which is given by the balance with the Coulombic repulsion of the charges. Thus bound pairs of defects are formed which can be considered to be single enitities, called bipolarons. The same holds for *cis*-polyacetylene and the poly-heteroaromates. As can be visualized in Fig. 3.4, an attractive potential between two defects does not exist in *trans*-(CH)$_n$, and therefore the defects, i.e. the solitons, are widely separated due to the Coulombic repulsion of their charge, unless other mechanisms come into play favouring confinement.

In the bandstructure picture the polaron forms two states in the energy gap in positions symmetrical to the centre of the gap. The lowest level of a hole polaron is singly occupied, the higher level empty. In the electron polaron the lower level is doubly occupied by electrons with antiparallel spin, whereas the upper level is singly occupied. Both states carry an unpaired spin and give rise to paramagnetism. The length of the deformed chain section is of the order of several repeat distances, e.g. about four rings in polythiophene. The soliton, on the other hand, forms an energy level at the centre of the gap (as long as electron–hole symmetry is fulfilled). This level is singly occupied if the soliton is neutral, resulting in an unpaired spin and paramagnetism. By doping with donors or acceptors the soliton can become negatively or positively charged. Both types of charged solitons are characterized by zero spin. Figure 3.5 gives a schematic view of the energetic positions of these excitations. Taking electron–electron interaction into account, the soliton level is no longer located at the centre of the gap, and differences occur whether its level is occupied or not.

The effective mass of a carrier is of particular interest to the charge transport along a polymer chain. For electrons or holes it is determined by the curvature

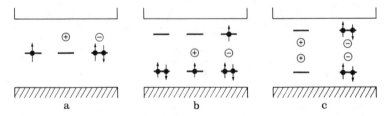

Fig. 3.5 a–c. Positions of the energy levels of (**a**) a soliton in the neutral (S^0), positively (S$^+$) and negatively charged state (S$^-$); (**b**) a polaron in the neutral (P^0), positively (P$^+$) and negatively charged state (P$^-$); (**c**) a bipolaron charged positively (B^{++}) and negatively (B^{--}). The neutral polaron does not exist.

of the respective band at the band edge. For polarons, solitons and bipolarons it is calculated from the kinetic energy [3.15, 16]. The effective masses for these quasi-particles, estimated on this basis, are of the order of 1–6 electron masses. Since these defects are also large and not localized within a unit cell, their motion will be coherent along a chain and their mobility is expected to be high. Experimental evidence for these defects comes from optical measurements (Chap. 4) and the expected change of the magnetic properties with doping (Chap. 5).

Polymers with a degenerate ground state, i.e. *trans*-polyacetylene and polyyne, are susceptible to a defect that is not present in other polymers. If the chain has an odd number of electrons, then the ground state already contains a kink with a non-bonding electron. This defect has been called *neutral soliton*, and it has evoked much interest due to its mobility and the fact that it carries spin $1/2$ in *trans*-$(CH)_n$. In polyyne the properties of solitons are expected to be even more exotic [3.8]. Since the non-bonding electron is reactive, the soliton is likely to lose its electron to the doping molecule, having generally a high electron affinity, and to be converted into a positively charged soliton during doping. This has indeed been observed [3.17]. The mobility of the neutral soliton has been experimentally established by observing motional narrowing of the ESR line width and by observation of the Overhauser effect in dynamic nuclear polarization (for details see Chap. 5). The diffusion coefficient of the solitons has been derived to be between 10^{-6} and $10^{-8}\,m^2/s$ [3.18, 19].

Of particular interest is the nature of the photo-excited state, the question for our purpose being whether excitons, electrons and holes in the bands, polarons, or solitons are generated. For *trans*-$(CH)_n$ this question has been addressed by *Su* and *Schrieffer* [3.20], who found in a computer simulation that the system relaxes in a time of the order of $10^{-13}\,s$ after photo-excitation, and the photo-excited carriers evolve into a positively and negatively charged soliton. It is surprising that the recombination of the soliton and antisoliton is suppressed despite the fact that both are located on the same chain.

If an electron or hole is injected, e.g. by a tunnelling process, theory predicts that a polaron is created. However, up to now no evidence for polarons in *trans*-$(CH)_n$ could be given. Theoretically the question of photo-excited carriers in polymers having a non-degenerate ground state has been discussed to a lesser degree. It is in general expected that, depending on the energy of the photons, excitons and polarons are the excited entities.

Very early structural and optical investigations gave evidence that these polymers, in particular polyacetylene, are in part crystalline and that inter-chain coupling cannot be neglected. It is, therefore, legitimate to ask whether the excitations discussed above in the strictly one-dimensional model without inter-chain coupling survive or whether they become destabilized in a three-dimensional system. This question cannot be answered in general, since the answer will depend sensitively on the strength of the interaction and, therefore, on details of the structure of the crystalline regions of the polymer. It has been shown in various calculations [3.21, 22] that in a strictly three-dimensional model of *trans*-(CH)$_n$, polarons cannot form. Similarly, bipolarons with a separation distance of the two charges (kink–antikink) smaller than about 6–7 lattice constants are unstable. If separated by a larger distance, bipolarons can exist, but with a small binding energy of the order kT. Since the inter-chain energy increases with separation of the charges in the bipolaron, the maximum distance between the kink and antikink should be limited to values below 50 lattice constants. These three-dimensional interactions cause a charged and a neutral soliton on adjacent chains to be strongly bound; the same holds for a pair of neutral solitons, forming a soliton molecule. As a consequence neither polarons nor bipolarons could be formed in perfectly crystalline *trans*-(CH)$_n$ upon doping, and it would be expected that injected electrons should remain free carriers in the conduction band. However, disorder is able to reduce the inter-chain coupling, so that quasi-particles might form in real materials. These findings will also have consequences when quasi-particle motion and transfer between chains is considered.

An experimental verification of these three-dimensional effects is, however, difficult since even in the material with the highest degree of crystallinity (stretched Durham-type *trans*-(CH)$_n$, see Sect. 3.4.2) the length of undisturbed chains is only about 40 CH units. In the Shirakawa material, on the other hand, many chains are in a disordered environment.

3.1.3 Disorder Along the Chains

So far the discussion of the one-dimensional model has been based on the assumption of an ideal, infinitely long chain in a homogeneous environment. In fact, the chains have a finite length, and their lengths very randomly. In poly-acetylene the structure may change between *trans* and *cis*, resulting in bends, folds, kinks, jogs and similar geometric defects. In poly(p-phenylene) torsions of monomer units out of their ideal planar position represent another type of geometric disorder. Apart from that, chemical defects are possible, such as cross-links between chains, sp^3 defects in polyacetylene and β linkage of heterocycles in a chain of α-linked monomer units. Chemical modifications at discrete points of the chain, such as oxidation, also belong to this category.

A third type of disorder results from changes in the environment of the chain, the so-called secondary structure, being described by the limiting cases of crystalline and amorphous regions. Disorder may also come from doping. At least at low dopant concentrations, the dopant molecules occupy random positions between the chains. They affect the electronic properties by their Coulomb potential and by hybridization with the polymer π orbitals. Disorder due to doping and other impurities has been studied theoretically by *Bryant* and *Glick* [3.23]. Further effects of doping will be discussed in the next section.

The main question in the context of electrical properties is how disorder influences charge carrier transport. An element of disorder may either block quasi-particle movement completely by acting as a high barrier, it may attract the quasi-particle and trap it, or its influence may only be weak. At present no definite answer is known. It is evident, however, that transport of solitons, polarons or bipolarons within unperturbed segments of the chain cannot be investigated by conductivity measurements at dc or low frequencies. Frequencies in the upper GHz range are necessary, otherwise the transport is dominated by barriers, traps and, as will be pointed out later, also by inter-chain contacts.

Various attempts were made to experimentally determine the length of unperturbed, fully conjugated chain segments. From the spin density, measured by ESR, the fraction of neutral solitons is deduced, being one soliton among 3000 carbon atoms [3.24]. The length of fully conjugated chain segments is, however, considerably smaller. In trans-polyacetylene, grown by the Shirakawa method, lengths of about 50–100 carbon atoms were observed [3.25, 26]. This value has to be compared with the size of a soliton, which is 14 carbon atoms (1.75 nm) in trans-polyacetylene, or of a polaron or bipolaron in poly(p-phenylene) or polypyrrole, which is estimated to be about four rings wide (1.5–1.7 nm). A further characteristic length is the average distance of charged quasi-particles, which is a function of the dopant concentration. At a doping level of $y = 0.05$ this distance is 20 carbon atoms along the chain.

Further elements of disorder may be introduced by doping and will be discussed in the next section.

3.1.4 Low and Intermediate Doping

Doping with acceptor or donor molecules causes a partial oxidation or reduction of the polymer molecules. Positively or negatively charged quasi-particles are created, presumably polarons in the first steps of doping. When doping proceeds, reactions among polarons take place, leading to energetically more favourable quasi-particles, i.e. a pair of charged solitons in materials with a degenerate ground state, and a bipolaron in polymers with a single ground state. Applying statistical mechanics allows calculation of the density of polarons, bipolarons and solitons, and the density of electrons and holes in band states at any temperature. The basis for this calculation is the neutrality condition and the Fermi distribution function. Furthermore it has to be assumed that the lifetime of the electrons (holes) and of the polarons is sufficiently long to treat them as quasi-particles. On this basis *Conwell* [3.27] has given an estimate of the density of free electrons at a doping level of about 5% and found a value for the free charge carrier density of about $3.10^{24}\, m^{-3}$ or less at 300 K. This is about a fraction of 1/1000 of the total dopant concentration. If the charge carriers in band states have a mobility comparable to that in inorganic semiconductors, their contribution alone would already explain the measured conductivity values. The other quasi-particles (polarons, solitons, bipolarons) exist in a much higher concentration, but have presumably a significantly lower mobility. In view of this, the contributions of the various particles to the conductivity are unclear.

Recent theoretical treatments of three-dimensional interactions in *trans*-$(CH)_n$ have suggested soliton ordering on adjacent chains, soliton confinement and the absence of polarons and bipolarons. These findings shed additional light upon this question and might explain the importance of free carriers in the conduction band. Although only outlined for $(CH)_n$, these considerations are also relevant for other conducting polymers with only polarons, electrons and holes in the bands as carriers for the electrical transport.

In the previous section we pointed out that disorder on a molecular or atomic scale gives rise to significant deviations from the ideal transport properties of charge carriers along an intact chain. However, grain boundaries and inter-fibrillar contacts are also important in this respect. In addition to this, disorder is also generated in samples doped with ions which might tend to localize the charge carriers. The electrostatic interaction with the oppositely charged dopant molecule pins the charged soliton or polaron with a binding energy, for which a value of about 0.3 eV for polyacetylene is derived from experiments, whereas calculation gives values between 0.7 and 1.3 eV [3.28]. At low and intermediate doping levels the positions of the solitons or polarons are determined by the randomly distributed dopant molecules, rendering the sample a so-called *soliton or polaron glass*. At higher doping levels the pinning potential is gradually decreased by shielding. For solitons at still higher doping levels their mutual repulsion should then become relevant, leading to a quasi-ordered *soliton lattice*. Instead of a decreasing pinning energy, an ordered array of dopant molecules can similarly lead to a soliton lattice. Such ordering is observed in the case of alkali-metal doping and is described as a phenomenon similar to stage formation in intercalation compounds. With potassium doping, for instance, the lowest observed ordered phase is a donor arrangement in linear channels, each being surrounded by four polyacetylene chains, and with a spacing equal to four carbon units [3.29]. As seen from the corresponding formula $(C_4H_4)_4K$, this leads to a relative dopant concentration of $y = 0.0625$. Such dopant ordering depends strongly on the size of the dopant. Lithium cations are smaller than the voids within the molecules and tend to be incorporated randomly, whereas in the case of larger alkali ions the resulting strain favours ordering. Following these ideas, a two-phase model was proposed to describe the moderately doped state. Above a certain threshold two phases form, one with high concentration and an ordered arrangement of dopants, the other very weakly doped. With increasing doping the ordered phase grows at the expense of the other, up to e.g. $y = 0.0625$ in this example. For anion doping a different view is taken: Due to their generally larger size anionic dopants first tend to form clusters between the planes of $(CH)_n$; at higher doping levels full sheets of dopant molecules intercalate between layers of polymer chains [3.30, 31]. Models based on inter-calation phenomena were used to interpret the rapid increase of the dc conduc-tivity with doping as a type of percolation phenomenon, but are in coflict with the vanishing Pauli susceptibility in the intermediate doping range in *trans*-polyacetylene.

3.2 Models for Transport Processes

The large number of models developed for discussing conduction in amorphous semiconductors and impurity-band conduction in crystalline semiconductors offer themselves for a first attempt to interpret electrical transport in polymers. The common feature in all these systems is disorder. This view, however, has to be modified for conjugated polymers, since quasi-particles (charged solitons, polarons or bipolarons) exist as charge carriers.

3.2.1 Conduction in Extended States

Mott and *Davis* [3.32] developed a model for amorphous semiconductors, based on the assumption that spatial fluctuations in potential, due to configurational disorder, lead to localized states whose density of states extends into the forbidden gap and forms so-called tails. A schematic density-of-states diagram is shown in Fig. 3.6. A sharp boundary E_c, called the mobility edge, separates extended states with mobile charge carriers from localized states, whose carriers are not mobile at absolute zero temperature. The position of the Fermi level is essentially determined by the charge distribution on the gap states. If the Fermi level is shifted beyond the mobility edge by doping, the material becomes metallic and its conductivity does not vanish at absolute zero. In addition, the conductivity in this case was asserted to be no smaller than the so-called minimum metallic conductivity, a value of the order of $10\,000\,\text{S/m}$.

When the Fermi level drops below the mobility edge E_c, the carrier mobility decreases by several orders of magnitude [3.33]. Only those carriers being excited into states above the mobility edge can contribute significantly to conductivity. If the Fermi level is positioned sufficiently far away from E_c, so that Boltzmann statistics can be applied, the conductivity is given by

$$\sigma = eN(E_c)kT\mu\exp(-(E_c - E_F)/kT). \tag{3.13}$$

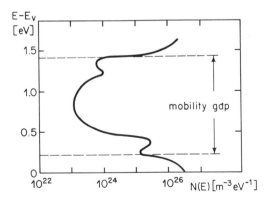

Fig. 3.6 Electronic density of states for amorphous silicon (from [3.109])

Mott estimated the mobility μ just at E_c to be of the order of 10^{-3}–$10^{-4}\,\mathrm{m^2/Vs}$. In order to separate the mobility from the charge carrier density the measurement of a second quantity, such as the Hall coefficient or the thermoelectric coefficient, is required. The latter is in general a measure for the activation energy E_c–E_F. If the observed conductivity has the same activation energy, conduction in extended states can be assumed.

3.2.2 Conduction in Localized States

Carriers in localized states can move into adjacent empty states by thermally assisted tunnelling, called hopping. The hopping probability depends on the mutual overlap of the wavefunctions, determined by the ratio of the intersite distance R_{ij} to the localization length r_0 (decay length of the wavefunction). It depends in addition on the occupation of the initial and final state and, if the jump takes place to a site j with higher energy, also on the energy difference $E_j - E_i$. An approximation proposed by *Ambegaokar* et al. [3.34] is often used for the hopping rate

$$\Gamma_{ij} = \gamma_0 \exp\left[-\frac{2R_{ij}}{r_0} - \frac{|E_i - E_F| + |E_j - E_F| + |E_i - E_j|}{2kT} \right]. \tag{3.14}$$

The prefactor γ_0 denotes a kind of attempt frequency of the order of a molecular vibration frequency, i.e. 10^{12}–$10^{13}\,\mathrm{s^{-1}}$. In equilibrium the hopping rates Γ_{ij} and Γ_{ji} in both directions are equal. An applied electric field changes the energies E_i and E_j and a net current is the result.

At sufficiently high temperatures and in diluted systems the hopping rate is dominated by the overlap. Therefore only hops between nearest neighbours contribute to the conductivity. The mobility is determined by an activated process.

If the temperature is low and if hopping takes place near the Fermi level, *Mott* [3.35] has shown that the hopping probability reaches a maximum at an optimum combination of hopping distance R_{ij} and energy difference $E_i - E_j$. If the hop goes over a large distance, more sites are available to which the carrier can hop, and as a consequence the carrier has a higher probability to find a site with lower energy than in a short hop; therefore such a process is preferred at low temperatures. For a constant density of states at the Fermi level Mott's famous variable-range hopping law results:

$$\sigma = \sigma_0 \exp(-T_0/T)^{1/4} \tag{3.15}$$

with

$$T_0 = 24/(\pi r_0^3 k N(E_F))$$

and

$$\sigma_0 = \frac{9}{4}\sqrt{\frac{3}{2\pi}}\, e^2 \gamma_0 \sqrt{\frac{r_0 N(E_F)}{kT}}.$$

A similar formula can be derived from a percolation treatment of the hopping conductivity [3.36], but only asymptotically in the low-temperature limit, as long as $T_0/T > 10\,000$. In amorphous inorganic materials, typical values of T_0, as derived from experiments, are of the order of 10^8 K. When the density of states near the Fermi level is not constant, but varies quadratically, the temperature exponent changes from 1/4 to 1/2 [3.37]. On the other hand, as *Overhof* and *Thomas* [3.38] have pointed out, other density-of-states functions also lead to an exponent 1/4 and the same variable-range hopping law. A $T^{-1/4}$ law is by no means a unique characteristic of variable-range hopping. Taking into account multi-phonon processes (instead of one-phonon processes only, as in most treatments) hopping leads to a similar $T^{-1/4}$ law, as *Emin* [3.39] has calculated. In addition, if Coulomb correlations are taken into account, exponents varying between 1/4 and 1 have been obtained.

Although many experimental data on the conductivity of disordered systems (polymers, amorphous inorganic semiconductors, impurity-band conduction) were interpreted in terms of variable-range hopping, there was always the problem of obtaining a reasonable numerical value for the prefactor. According to (3.15), T_0 and the prefactor have the variables $N(E_F)$ and r_0 in common. With r_0 and $N(E_F)$ fixed, unrealistic values result for γ_0. *Würtz* and *Thomas* [3.40] could give a remedy for curing this discrepancy by including hopping between π-like electronic states. Averaging over all directions introduces a factor $(T_0/T)^{m/4}$ with m between 2 and 10 into the prefactor that can easily account for the values of σ_0 determined from the experiment.

The thermoelectric coefficient S from carriers, hopping between localized states at the Fermi level, is low since no kinetic energy is gained in a hopping process. For variable-range hopping it is given by [2.32]

$$S = (1/2)(k^2/e)(T_0 T)^{1/2}(d \ln N/dE)_{E_F}. \tag{3.16}$$

The thermoelectric coefficient vanishes if the density of states is constant at the Fermi level. However, other expressions for S have also been given, and it was shown that a temperature-independent contribution $(k/e)\ln 2$ exists.

Calculating the ac conductivity by hopping transport, the pair approximation is often used [3.41,42]. It relies on the fact that at sufficiently high frequencies only transitions between independent pairs of sites contribute to charge transport. Each pair is represented by a relaxing electric dipole. Starting with one occupied site, the contributions of transitions to neighbouring empty sites depend on an integral of the form

$$4\pi N(E) \int \frac{r^4 \omega^2 \tau}{1 + \omega^2 \tau^2} \, dr. \tag{3.17}$$

When an exponential dependence of the reciprocal transition rate on distance is assumed

$$\tau(r) = \tau_0 \exp(2r/r_0) \tag{3.18}$$

the integrand has a sharp maximum at a distance

$$r_m = (r_0/2)\ln(1/\omega\tau_0). \tag{3.19}$$

This allows an approximate calculation of the integral, leading to

$$\sigma(\omega) = \frac{\pi}{96} e^2 kT [N(E_F)]^2 r_0^5 \omega [\ln(1/(\omega\tau_0))]^4. \tag{3.20}$$

The transport is assumed to take place here by carriers with an energy close to the Fermi energy. Therefore the temperature dependence is weak and only proportional to T. If hopping occurs in localized tail states below the mobility edges, an exponential dependence on the temperature is expected. Sufficiently below $\omega = 1/\tau_0$ (3.20) gives a frequency dependence σ proportional to ω^s with an exponent of the order of 0.8.

The above-mentioned formula considers only electron hops between pairs of sites and is therefore unable to describe correctly the ac conductivity at low frequencies and its merger into the value for the dc conductivity. A more general treatment of hopping transport by *Movaghar* et al. [3.43, 44], based on a Green's-function formalism and solved in the effective medium approximation, is able to give a unified description of ac and dc conduction that is in agreement with experimental findings. A general formula, combining dc and low- and high-frequency ac behaviour was first mentioned by *Schirmacher* [3.45]. Similar results were obtained by *Summerfield* and *Butcher* [3.46] within their extended pair approximation, the dc limit of which yields Mott's variable range hopping law, provided that $T_0/T > 10\,000$. The extension of the pair approximation considers the environment of a single pair and takes it into account in a mean-field description. Recently a scale-invariant "universal" law for $\sigma(\omega)$ was derived, with the localization radius r_0 and a numerical model-dependent factor A as parameters [2.47]:

$$\sigma(\omega)/\sigma(0) = f\{A\omega e^2/[kTr_0\sigma(0)]\} \tag{3.21}$$

and $f(x)$ being approximated by $f(x) = 1 + x^{0.725}$.

This formula describes the frequency dependence in a series of different materials rather well. Characteristic of the last three approaches is a constant conductivity at low frequencies and a rise with a power slightly less than 1 at higher frequencies. The crossover frequency shifts with increasing dc conductivity $\sigma(0)$.

3.2.3 Transport in One Dimension

The conductivity in one-dimensional systems was treated theoretically by *Alexander* et al. [3.48]. It is extremely sensitive to any perturbation. Impurities or other defects present barriers to the motion of carriers, which in three dimensions are easily circumvented. The interrupted-strand model with completely blocking barriers predicts a zero dc conductivity. Models, taking a distribution of barrier heights (or equivalently trap depth) into account, yield a non-vanishing dc conductivity and a frequency dependent ac conductivity due to a sufficient spread in hopping or release times.

An exponential barrier height (Δ) distribution is often assumed

$$g(\Delta) = g_0 \exp(-\Delta/kT_m) \tag{3.22}$$

for Δ above a threshold value Δ_0, resulting in a complex conductivity

$$\sigma(\omega) = c(T)(-i\omega)^\nu \tag{3.23}$$

with a temperature-dependent frequency exponent

$$\nu(T) = (T_m - T)/(T_m + T). \tag{3.24}$$

To our knowledge these equations have never been applicable to polymers. Measurements on hollandite, a one-dimensional fast ionic conductor, gave frequency exponents of 0.2 at room temperature, rising to 0.8 at 90 K in agreement with theoretical predictions. Such a strong temperature dependence of the frequency exponent has not been observed in polymers.

Since the optical absorption along the chain direction is much higher than that perpendicular to it, optically excited carriers give, in principle, information on one-dimensional transport. Calculations [3.20] have shown that in *trans*-(CH)$_n$ photo-excited carriers should lead to negatively and positively charged solitons within about 10^{-13} s after excitation. Recombination in (CH)$_n$ is expected to be fast due to the finite chain length and the Coulomb attraction of the carriers, i.e. geminate recombination is expected to be dominant and, hence, the quantum efficiency for free-carrier or soliton generation to be small. Geminate recombination implies that the quantum efficiency depends on both the electric field and the wavelength of the exciting light [3.49]. The electric field reduces the Coulomb capture radius and increases the escape probability from recombination. Typically, the electric fields at which this occurs are above 10^7 V/m. Similarly, for carriers highly excited up into the bands, the probability for the carriers to diffuse beyond the Coulomb capture radius during relaxation and to escape geminate recombination becomes greater than for excitations just across the bandgap. Therefore, the quantum efficiency should rise with increasing electric field and with increasing photon energy. Measurements of the dc photo-currents in Schottky depletion layers allow one in principle to obtain the dependence of the quantum efficiency on the electric field and on the wavelength of the exciting light. The experimental results reported so far are, however, conflicting [3.50]. This might be due to different electric fields in the Schottky depletion layer and possibly also due to different degrees of crystallinity in the material used for these experiments.

The decay of the photocurrents after excitation with short light pulses has been studied in the time domain between 10^{-12} and 10^{-1} s [3.51, 52]. These experiments were performed with ohmic and not with Schottky contacts and the electric fields were usually below 5×10^6 V/m. Excitation with ps light pulses gives a photoconductive signal, which decays initially as rapidly as it is formed, i.e. with a time constant of the order of 1 ps. After this initial rapid decay the photocurrent decreases sluggishly and is still observable after milliseconds. Surprisingly, in highly oriented samples the photocurrent is larger if the exciting light is polarized perpendicularly to the chain direction. This is an indication that primarily carriers, being excited on different chains, give rise to photo-conductivity. If this conclusion is correct, the charge carriers responsible for photoconduction must be free electrons, holes or polarons, which might in part be converted into charged solitons, if they encounter a neutral soliton during their lifetime.

In nondegenerate polymers photoconductive phenomena should be absent [3.53]. Photoexcitation leads to an exciton-like state, due not only to Coulomb interaction but also to a lattice configuration such as that within a polaron. Indeed, experiments in cis-$(CH)_n$ demonstrate that photoconductivity is not observable, but luminescence is.

3.2.4 Transport by Quasi-Particles

The models developed for transport in localized or extended states of amorphous semiconductors have been widely applied to the conductive properties of polymers. In a conjugated polymer, however, and as long as a one-dimensional model is valid, the addition of charge leads to the formation of polarons, bipolarons or charged solitons. These quasi-particles are expected to be quite mobile along intact segments of the polymer chain. For electrical transport through real macroscopic samples, however, these quasi-particles have to move finally from one chain to a neighbouring one. This section will deal with the question, of the extent to which dc and ac conductivity are affected by such inter-chain transfer processes.

The inter-chain transitions of polarons, bipolarons or solitons are difficult to assess, because not only a charge but also the molecular distortion has to be transferred. Electrons can tunnel, but how is the deformation transmitted to the neighbouring chain? In the case of transfer of a soliton in the middle of a chain, half of the chain would have to change its structure in both the initial and final molecule. Two models have been proposed and will be discussed subsequently: one for bipolaron transitions, which is *mutatis mutandis* also applicable to a pair of confined charged solitons, and a model considering electron hopping between two solitons.

Chance et al. [3.54] calculated the transfer rate for a bipolaron. It is determined by the combined probability for finding a bipolaron on one chain and a sufficiently long segment free of any excitation on the neighbouring chain, and by the appropriate Franck–Condon factor f. Denoting the length of a bipolaron by l (expressed in monomer units) and the molar fraction of dopant molecules by y, the probability for an inter-chain transfer is given by

$$P_b = (y/2)(1 - ly/2)^{l+2}, \tag{3.25}$$

as long as interactions between bipolarons can be neglected. The transition rate, and hence the conductivity, peak at $y = 0.05$–0.07, assuming a bipolaron length $l = 4$–5. More sophisticated treatments are required if Coulomb interaction between bipolarons plays an important role at higher dopant concentrations.

The same authors also considered inter-chain transfer of solitons by a similar mechanism. Two charged solitons on a chain approach each other, being separated by a distance l, and perform a transition to an adjacent chain in the same way as described for a bipolaron. Taking $\exp(-\beta l)$ for the Franck–Condon factor, the total transfer rate is obtained

$$P_s = y \sum P(l) = \text{const} \{ y^2(1 - y) \} / \{ 1 - (1 - y)^2 \exp(-\beta l) \}. \tag{3.26}$$

The transition probability and hence the conductivity as a function of y are s-shaped, levelling off at higher doping concentrations.

The above calculation implies that a hop of two solitons to an adjacent chain is only highly probable if the region of "wrong" bond arrangement between them is small, and only a relatively small molecular distortion has to be transferred together with the charge carrier. This mechanism for a transition is no longer possible in doping ranges, at which doping proceeds via conversion of neutral solitons into charged ones, and at which the soliton gas is so diluted that only single, charged solitons may be found on the chains. Single solitons are unlikely to hop due to the high energy required for structural reorganization. *Kivelson* [3.55] envisaged for this doping range a mechanism for the transition, involving the charged soliton generally pinned to a doping molecule, and a neutral soliton on a neighbouring chain. The unpaired electron of the neutral soliton can tunnel to the positively charged soliton, if the electronic levels of both excitations nearly coincide. This requires that the neutral soliton is also close to an ionized impurity, otherwise an energy equal to the binding energy of the charged soliton would have to be overcome. Under such conditions the charge can tunnel without the need to transfer a large amount of molecular deformation. The resulting formula

$$\sigma(T) = 0.45 \frac{e^2}{kT} x_0 (1 - x_0) \frac{\gamma(T)}{N R_0} \frac{r_0}{R_0} \exp(-2.78 R_0 / r_0) \tag{3.27}$$

depends on the fraction of neutral (x_0) and charged solitons ($1 - x_0$), the average chain length N, the ratio of the average localization length r_0 to the average acceptor distance R_0, and on a frequency factor $\gamma(T)$, which is assumed to vary with temperature like T^n, with $n = 10$–12, a rather unphysical dependence. This formula predicts, firstly, a strong exponential increase of the conductivity with doping, provided that the fraction of neutral solitons remains constant and, secondly, a power dependence of σ on temperature. Considering three-dimensional effects [3.21, 22], this process becomes unlikely, since solitons on adjacent chains are bound to one another forming a soliton molecule.

The calculation of the ac conductivity in Kivelson's model follows closely that of the pair approximation and is described by a formula similar to (3.20).

The thermoelectric coefficients S is a measure for the entropy transported per charge carrier. Consequently in diluted systems S is high, whereas in dense and ordered systems, such as metals at low temperatures, S vanishes in general. In the case of a system of highly diluted charge carriers, i.e. in the low-doping regime, S is generally interpreted by *Heikes'* formula [2.56]

$$S = \pm (k/e)\ln[(1-x)/x] \tag{3.28}$$

with x denoting the fraction of sites occupied by particles or quasi-particles. In general the transport entropy and hence S contain in addition a kinetic energy term, depending on the transport and scattering mechanism. In Kivelson's theory for intersoliton hopping [3.55] the temperature exponent $n = 10$–12 in the hopping rate gives such an additive contribution of the form $(k/e)(n+2)/2$. In both cases a thermoelectric coefficient more or less independent of temperature is to be expected. *Emin* and *Ngai* [3.57], however, strongly critisize Kivelson's calculation of S.

Interaction of the quasi-particles with the oppositely charged dopant molecule has not been taken into account so far. This interaction leads to the so-called pinning, resulting in bound particles. It should give rise to an activation energy in the conductivity, even for transport along the chain, equal to the binding energy E_b. The same quantity is expected to determine the thermoelectric coefficient

$$S = (k/e)(E_b/kT). \tag{3.29}$$

The binding energy is defined as the difference in energy between a quasi-particle on a chain closest to a dopant molecule and in its most widely separated position. It is mainly due to Coulomb interaction. The binding energy is expected to decrease with doping, since the Coulomb forces are shielded by more charge carriers, and a single quasi-particle comes under the influence of more than one dopant molecule.

Mele and *Rice* [3.58] have investigated this in a numerical study. Due to the disorder introduced by impurities, there are states at the Fermi energy which are expected to be localized. Assuming that transport at the Fermi level is by variable range hopping, they find that the conductivity increases sharply with concentration in the range below 2%; above this impurity concentration the conductivity rises less steeply. From a computer simulation they conclude that the sharp increase of σ is caused by hopping between quasi-metallic regions due to statistical fluctuations of the impurity density. These fluctuations are purely statistical and must not be confused with metallic islands which might originate from inhomogeneous aggregations of dopant molecules. Conductivity induced by metallic islands embedded in an insulating matrix has been investigated theoretically by *Sheng* [3.59]. His formula interpolates between quantum mechanical tunnelling (temperature independent) and thermally activated transport, and has often been used to interpret results on the temperature dependence of the conductivity. The problem of inhomogeneous doping continuously turns up in the literature and will be discussed in its various contexts.

The question of the conductivity increase with doping and of the binding energy of quasi-particles has also been discussed by *Fukutome* and *Takahashi*

[3.60] within their theory of polyelectrolytes. They consider a model in which, at low doping levels, the dopants are adsorbed at the surface of the fibres, thus giving rise to an accumulation of quasi-particles there. With increasing doping level the electrostatic forces between dopant and oppositely charged quasi-particles start to overcome the van der Waals forces between the chains, so that intercalation and doping of the bulk material becomes effective. A strong increase of the conductivity with dopant concentration is predicted due to a decreasing pinning energy.

3.3 Experiments in the Insulating and Semiconducting State

In this section the electrical transport properties of materials in the low and intermediate doping range will be discussed, i.e. from unintentionally doped samples up to doping levels of the order of 5%. Experiments comprise measurements of the dc and ac conductivity as a function of temperature, doping level and dopant material, and measurements of the thermoelectric coefficient. When necessary, other properties are also considered, such as the magnetic susceptibility.

A vast amount of data has been published on both isomers of polyacetylene, but less on other conjugated polymers. Therefore a whole section will be devoted to $(CH)_n$. The main goal is to specify the transport mechanisms and the relevant charge carriers.

3.3.1 Polyacetylene

Probably the most prominent feature to be observed is the rapid increase of conductivity with doping. In Fig. 3.7, the dc conductivity at room temperature is plotted as a function of the relative dopant concentration y. The conductivity increases by many orders of magnitude in the doping range between $3 \cdot 10^{-4}$ and 0.003. Below $y \approx 3 \cdot 10^{-4}$ the data for trans-polyacetylene depend less on y and level off at a value of $3 \cdot 10^{-4}$ S/m. In cis-polyacetylene the conductivity versus doping curve looks similar, but extends to lower conductivity values of the order of $5 \cdot 10^{-8}$ S/m. The thermoelectric coefficient (measured at a fixed temperature) decreases in the range where the conductivity rises with doping (Fig. 3.8). The drop, also occurring around $y = 0.001$, is very pronounced: S drops from values of the order of 1400 μV/K to below 100 μV/K.

Some authors interpret the temperature dependence of their conductivity data in terms of an activated process [3.24, 61], others in terms of variable-range hopping [3.62, 63]. Characteristic in all cases is the fact that the temperature dependence of σ becomes weaker with increasing doping level, which results either in a decreasing activation energy, or in a decreasing T_0-value (3.15). In

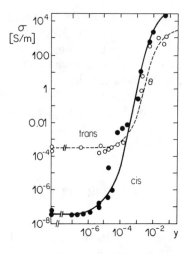

Fig. 3.7 Dc conductivity of iodine (I_3^-)-doped *trans-* and *cis*-polyacetylene as a function of relative dopant concentration y (from [3.71])

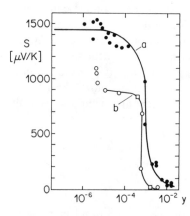

Fig. 3.8. Thermoelectric coefficient S of trans-polyacetylene as a function of doping with I_3^- (a) and AsF_5 (b) (from [3.71])

the case of variable-range hopping, an exponent of $1/2$ was found for doping levels below 0.05 [3.62], which can be taken as evidence for hopping in one dimension at a constant density of states at the Fermi energy, or for an $N(E)$ varying quadratically around E_F [3.37] in a three-dimensional model. An exponent of $1/4$ was found in a fit to low-temperature data of iodine-doped polyacetylene with y ranging between 0.025 and 0.045 [2.64]. The density of states obtained at the Fermi level was of the order of several $10^{26}\,\text{m}^{-3}\,\text{eV}^{-1}$. Deviations at higher temperatures were explained by a density of states that varies with energy at or close to the Fermi level. An interpretation in terms of hopping at the Fermi level, however, suffers from all the objections discussed before, namely that it does not uniquely pinpoint the assumed mechanism.

Moses et al. [3.65] interpret their data on pressure dependence of the conductivity of as-grown samples in terms of intersoliton hopping between chains (3.27). The conductivity ratio $\sigma(p)/\sigma(p=0)$ was found to be independent of temperature. This is an argument in favour of Kivelson's theory and against variable-range hopping. The samples were estimated to have a doping level of $y = 3 \cdot 10^{-5}$. In such not intentionally doped samples the density of neutral solitons is relatively high ($1.5 \cdot 10^{25}$ m^{-3}), and intersoliton hopping as a transport mechanism may happen. On the other hand, *Scott* and *Clarke* [3.66] concluded from studies of the nuclear spin-lattice relaxation rate that in polyacetylene only a small fraction of the chains contains solitons; this would make intersoliton hopping rather unlikely. In addition, in crystalline parts of $(CH)_n$, binding of charged and neutral solitons by three-dimensional interactions into a soliton molecule would make soliton hopping inoperative.

The thermoelectric coefficient S was found to be high and temperature independent for weak or no doping (Fig. 3.8). Its decrease at higher doping levels runs parallel with a more and more pronounced temperature dependence. Above $y \approx 0.01$ the thermoelectric coefficient S is smaller than $50\,\mu V/K$ at room temperature, and shows a strong temperature dependence [3.67]. The observed non-linear variation (Fig. 3.9) is described by a superposition of a linear law, characteristic for metals (3.4.2), and a square-root dependence, as derived for variable-range hopping (3.16). At doping levels above 0.05 the thermoelectric coefficient has decreased further and its temperature dependence becomes linear (Fig. 3.15). This behaviour is characteristic for metals and will be discussed in the next section.

The high and constant value for S in the low-doping regime is generally interpreted by (3.28), where x is taken as the ratio of the numbers of charged to neutral solitons [3.65]. An evaluation of the observed data gives an x of the order of $1 \cdot 10^{-7}$ in Heikes' formula, and $1 \cdot 10^{-4}$ if Kivelson's temperature exponent is taken into account. From such a general treatment, however, no conclusions can be drawn with respect to the conduction mechanism and the nature of the charge carriers involved.

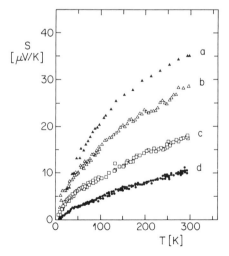

Fig. 3.9. Theremoelectric coefficient vs. temperature for FeCl$_3$-doped *cis*-polyacetylene; doping levels are (a) $y = 0.0106$; (b) $y = 0.0179$; (c) $y = 0.0378$; (d) $y = 0.0593$ and 0.0612 (from [3.67])

Frequency-dependent conductivity measurements were performed and show in a restricted frequency range the characteristic features represented by the extended pair-approximation (3.21): a frequency-independent conductivity at low frequencies, equal to the dc value, is followed by a monotonous increase with a frequency exponent $s < 1$ at higher frequencies. The low-frequency branch varies with temperature in the same way that dc conductivity does, whereas the high-frequency branch varies in general considerably less. This causes the cross-over frequency to vary with temperature and to shift to lower values at lower temperatures. *Epstein* et al. [3.68] found that this effect is more pronounced in *cis*-$(CH)_n$ than in the trans-isomer. Fig. 3.10 shows the behaviour of trans-polyacetylene. The authors describe their results as intersoliton hopping and use Kivelson's high-frequency formula, which is based on the pair approximation and can describe only the high-frequency branch. *Ehinger* et al. [3.64] successfully interpret these data on the basis of the extended pair approximation. An observed Drude-like behaviour at much higher frequencies is connected with quasi-metallic behaviour and will be mentioned later.

Before discussing the main question, namely the nature of the charge carriers and the physical processes responsible for the drastic changes in σ and S at $y \approx 0.001$, some properties of the dopant molecule and its incorporation into the polymer have to be summarized. Currently used acceptors are iodine, incorporated as triiodide anion I_3^-, or higher polyiodides such as I_5^-, and arsenic pentafluoride AsF_5 built in as AsF_6^-. $SbCl_5$, SbF_5, ClO_4^- and $FeCl_3$ are also used as acceptors. Donor doping is accomplished by introducing alkali metals from alkali naphtalide. The size and shape of the dopant molecule seem to have an effect on the process of doping and, as a result, also on the final arrangement of the dopant molecules between the polymer chains. Noticeable differences in

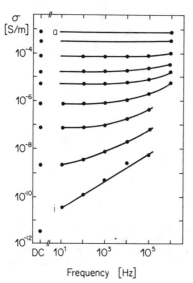

Fig. 3.10. Frequency and temperature dependence of the conductivity in *trans*-polyacetylene. The temperatures vary from 294 K (a) to 84 K (i) (from [3.68])

homogeneity are reported, depending on the speed of doping [3.70–72]. Alkali metals as dopants are relatively small and can be considered as spheres with ionic radii ranging from 95 pm for Na to 169 pm for Cs. Triiodide forms a linear molecule, while the other acceptors have the shape of a regular tetrahedron or octahedron and are considerably larger in overall size. From an evaluation of X-ray diffractograms, an ordered arrangement of dopant molecules has been postulated, forming columns in the case of alkali metals [3.73], or layers for iodine doping [3.74]. In both cases a generalized stage formation should occur.

Another unsolved question is the role which interfibrillar contacts play. This is especially important for polyacetylene, grown by the Shirakawa method from gaseous acetylene in a catalyst solution. This material is known to consist of a random fleece of fibrils with an average diameter of 50 nm or less [3.75]. Interfibrillar contacts can become blocking or ohmic upon doping and, therefore, the measured dc conductivity does not necessarily describe the conductivity change in the bulk of the fibrils. In this context measurements of the thermoelectric coefficient are expected to be less influenced by these phenomena and should yield results more characteristic for the bulk of the material.

The prime phenomenon in the doping range discussed here is the sharp increase in conductivity and an accompanying drop in the thermoelectric coefficient at doping levels around $y = 0.001$. This effect has often been called a semiconductor-metal transition, but we will reserve this name for a second transition, occurring at $y = 0.06$, where a steeply rising Pauli susceptibility clearly indicates the appearance of a finite density of states at the Fermi level. Both transitions have to be clearly separated; the latter will be discussed in the subsequent section. The conductivity rise can neither be explained by a mere increase in the number of charge carriers nor by a decreasing average distance R_{ij} between hopping sites in (3.14). Some kind of transition must take place here, involving a new mechanism that increases the effective mobility drastically.

Two different views of this phenomenon have developed: the melting of the soliton glass and the two-phase model, based on an ordered array of dopant molecules in one of the phases. As described in Sect. 3.1.4, the charge carriers, predominantly charged solitons in the case of *trans*-$(CH)_n$, are electrostatically bound to their dopant molecules. Increasing the dopant concentration lowers the binding energy of pinned solitons and leads to gradual depinning. This model of "a melting of a soliton glass" has been introduced by *Su* et al. [3.76] and is supported by the decrease of the apparent activation energies for the conductivity in the doping range around 0.001 [3.61]. Such a decrease in activation energy can also explain the variation of the thermoelectric coefficient in a simple way (3.29). The model proposed by *Fukutome* and *Takahashi* [3.60] uses a similar picture, although based on a different mechanism.

The alternative two-phase model relies on the existence of an ordered, stoichiometric "compound" of polymer chains and dopant molecules. In polyacetylene, doped with sodium, the lowest composition corresponds to $y = 0.0625 = 1/16$. At lower dopant concentrations the dopant molecules would, for energetic reasons, form ordered and thus highly conducting regions. In this model the observed steep conductivity increase would be the result of a percolation.

At the moment it is not possible to give a clear answer to the interpretation of this phenomenon from electric transport properties alone. Important in this context is that *trans*- and *cis*-polyacetylene exhibit the transition at the same dopant concentration, and that the functional dependence of σ and S on doping is nearly indistinguishable for both materials. Taking into account that solitons can exist only in *trans*-$(CH)_n$, difficulties arise in interpreting the observed transitions in terms of soliton effects alone. This fact is, however, to some extent obscured by the observation that doping *cis*-polyacetylene leads to gradual isomerization. Since the trans-isomer regions are doped more easily than the remaining *cis* regions, inhomogeneous doping is the consequence. The question of up to which doping level *cis*-polyacetylene retains its configuration is still controversial. If inhomogeneous doping is assumed, the interpretation of the rise in conductivity depends strongly on details of the dopant molecule and the way it is incorporated into the polymer.

3.3.2 Other Polymers

The electrical transport properties of aromatic conducting polymers such as poly(p-phenylene) or poly(phenylenesulfide) and heterocyclic polymers such as polypyrrole, polythiophene etc., show a common behaviour, which is in many respects similar to polyacetylene. Although verified on one material only, conductivity as a function of relative dopant concentration exhibits a similar

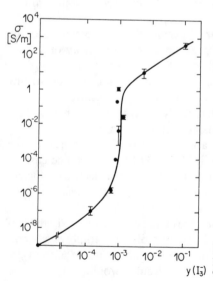

Fig. 3.11. DC conductivity of iodine (I_3^-)-doped polythiophene as a function of relative dopant concentration y (from [3.77])

steep and sharp increase as polyacetylene. Figure 3.11 shows the results on iodine-doped polythiophene, where the doping was done carefully and the dopant concentration determined accurately [3.77]. The rise occurs at $y = 0.001$ and is even steeper than in polyacetylene. A similar drop in the apparent activation energy is also found. The thermoelectric coefficient changes from high, temperature-independent values around $1700\,\mu V/K$ below $y = 0.001$ to values below $100\,\mu V/K$, measured at room temperature (Fig. 3.12), with S now depending linearly on temperature [3.78].

The temperature dependence of the conductivity also closely resembles that of polyacetylene. Various authors have tried to fit their data to (3.15) for variable-range hopping with an exponent of $1/4$ [3.79], assuming that hopping is the relevant process for carrier transport. In practically all cases encountered, the data cannot be interpreted by a simple variable-range hopping mechanism below a room-temperature conductivity of $10\,S/m$. The general behaviour of the frequency dependence of the conductivity is the same as in polyacetylene, as can be seen from Fig. 3.13, where the results on poly(3-methylthiophene) are plotted. These data can be described by the extended pair-approximation and can be scaled into a common universal curve corresponding to (3.21) [3.80].

Since the polymers considered here have no degenerate ground state, the quasi-particles formed at low concentrations are expected to be polarons, whereas at higher doping bipolarons are often formed. Experimental evidence comes from the optical absorption and from ESR experiments and is discussed in the chapters on magnetic and optical properties. Comparing the transport properties with that of polyacetylene demonstrates the great similarity between soliton and polaron transport. One important difference, however, exists in the magnetic properties. A small number of structural defects gives rise to a small Curie-type susceptibility in undoped material. Upon doping a strong increase in the susceptibility (generally determined by ESR) indicates the formation of a spin-carrying species, identified as polarons [3.81]. At higher doping levels,

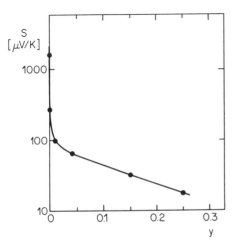

Fig. 3.12. Thermoelectric coefficient of BF_4^--doped polythiophene as a function of doping (from [3.78])

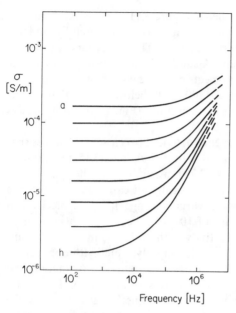

Fig. 3.13. Frequency and temperature dependence of the conductivity in poly(3-methylthiophene), synthesized electrochemically and being nearly completely reduced [3.80]. The temperature varies from 290 K (a) to 200 K (h) in steps of 10 K

where spinless bipolarons should form, the magnetic behaviour is not unique: either a decrease [3.82], or a levelling-off [3.83], or a more complicated behaviour [3.77, 78] has been reported.

Recently, polyaniline has gained great interest as a polymeric conductor. In this novel material the electrical properties can be changed not only by adding or subtracting electrons (reduction, oxidation) but also by adding or subtracting protons (protonation, deprotonation). This additional degree of freedom allows doping of the material by changing the number of protons without any change in the number of electrons. Generally a 1:1 mixture of the fully oxidized (poly(phenylene-imine)) and fully reduced species (poly(phenylene-amine)), the so-called emeraldine base, is taken as the starting material. By exposing it to a protonic acid, such as hydrochloric acid, the polymer chain is protonated at the imine positions, and at the same time an equal amount of chloride anions is incorporated. During protonation the conductivity increases from about 10^{-8} S/m to 500 S/m [3.84], where a state is reached with all the imine positions being protonated ($y = 0.5$).

3.4 The Semiconductor-Metal Transition and the Metallic State

In the last section the steep conductivity increase, occurring around a doping level of 0.001 in all the materials studied, was discussed. Although the conductivities obtained are quite high, there is no safe indication of a metallic behaviour,

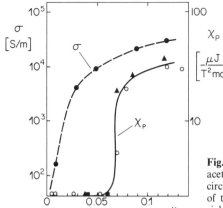

Fig. 3.14. Onset of the Pauli susceptibility in poly-acetylene, doped with AsF_6^- (triangles and open circles) and I_3^- (squares), compared with the variation of the conductivity with doping (broken lines and right-hand scale) (from [3.110])

unless the doping level increases beyond a certain threshold concentration. This concentration is $y \approx 0.06$ in trans-polyacetylene (Fig. 3.14). At this doping level the magnetic susceptibility rises rather sharply from very small, practically un-measurable values to values of the order of some $10\,\mu J/T^2\,mol$ within a narrow y range [3.69]. This dramatic rise in the magnetic susceptibility is in no way reflected in the conductivity. Being temperature independent, the susceptibility must be due to Pauli paramagnetism. The thermoelectric coefficient, which has dropped to small values already at $y \approx 0.001$, decreases further with doping and finally becomes linearly dependent on temperature [3.67]. The dc conductivity does not vanish in the limit of low temperatures, but retains finite and constant values that persist down to several mK [3.24, 85]. Whereas the Pauli para-magnetism develops rather abruptly in polyacetylene, in others, e.g. polythiophene, a gradual increase with doping is observed. In all cases a finite density of states at the Fermi level develops, leading to a behaviour that is in many respects characteristic for a metal. It is, therefore, legitimate to call this transition a semiconductor-metal transition.

3.4.1 Models for the Highly Doped State

Common to all the observations discussed above is that they require a finite electronic density of states at the Fermi level. For a highly degenerate electron gas, the thermoelectric coefficient is given by

$$S = \pi^2 k^2 T/(3e)\left(\frac{\partial \ln \sigma(E)}{\partial E}\right)_{E_F} \tag{3.30}$$

which becomes

$$S = \pi^2 k^2 T N(E_F)/(3en) \tag{3.31}$$

in the case of an energy-independent mean free path and parabolic bands. Here $N(E_F)$ denotes the density of states at the Fermi level for both spin orientations and n the total density of charge carriers.

Other quantities depending on $N(E_F)$ are the Pauli susceptibility

$$\chi_P = \mu_B^2 N(E_F) \tag{3.32}$$

$(\mu_B = 9.274 \cdot 10^{-24} \text{ J/T} = \text{Bohr magneton})$

and the electronic contribution to the molar heat capacity

$$\Delta C = \pi^2 k^2 T N(E_F)/3. \tag{3.33}$$

Generally $N(E_F)$ is referred to the density of carbon atoms in a conjugated sequence, which is $n_c = 5.2 \cdot 10^{28} \text{ m}^{-3}$ for an ideally compact sample of polyacetylene. Values of the order of $N(E_F)/n_c = 0.1 \text{ eV}^{-1}$ are typical.

A model capable of interpreting the semiconductor-metal transition has to explain the sudden appearance of a finite density of states at the Fermi level and the fact that it develops in a doping range, where the dc conductivity has already attained rather high values, e.g. for polyacetylene values above 10^4 S/m (Fig. 3.14).

Increasing disorder leads to the formation of band tails and a gradual filling of the gap with density of states, but it cannot explain the sudden appearance and the high value of $N(E_F)$ in a simple way. A second model, also related to doping, is based on a commensurate-incommensurate transition. The Peierls-insulator state, originating from a one-dimensional exactly half-filled band, is equivalent to a commensurate ordered state. Changing the number of π electrons per carbon atom by heavy doping tends to make the ordering incommensurate. As in other incommensurate structures, the incommensurability is taken up first by discommensurations or phase solitons [3.86]. With an increasing number of discommensurations the "plane-wave region" is finally attained with a sinusoidal modulation of the order parameter. *Mele* and *Rice* [3.58] investigated this model and found a gradual widening of the soliton band, but no closing of the gap. The addition of disorder and three-dimensional interactions between the chains causes the gap between valence and soliton band to close and leads to a density of states $N(E_F)/n_c = 0.11 \text{ eV}^{-1}$.

A third model, valid for *trans*-polyacetylene only, interprets the semiconductor-metal transition in terms of a transition from a soliton lattice to a polaron lattice and the formation of a polaron metal. Due to the degenerate ground state, solitons have a smaller energy of formation than polarons. *Kivelson* and *Heeger* [3.87] claimed that this difference can be decreased to zero and become even negative, if the polaron bands widen sufficiently by adding dopant molecules. Then polarons would become more stable, and the half-filled polaron band would show metallic properties.

Recent calculations by *Hicks* et al. [3.88] shed some doubt upon this interpretation. According to their results, a polaron lattice in the limit of high doping should have a vanishingly small oscillator strength in the doping-induced infrared absorption line at 1390 cm^{-1} (41.7 THz) as compared to a soliton lattice. The spectra taken together, however, do not show any significant jump in the oscillator strength [3.89].

It is worth noting that the insulator-metal transition takes place at a doping level, at which the average distance between charge carriers approaches the mean length of the corresponding quasi-particles. When doping is increased beyond this point, the lattice can no longer relax in the same manner as at low dopant concentrations, and new effects are expected.

3.4.2 Experiments in the Highly Doped State

The most convincing evidence for the semiconductor-metal transition comes from the abrupt increase of the Pauli susceptibility in polyacetylene when the doping level increases above $y = 0.06$ (Fig. 3.14). Values of the order of 20–30 μJ/T^{2}mol were finally reached, corresponding to a density of states $N(E_F)/n_c = 0.06$–0.09 eV^{-1} [3.69, 90]. The magnetic susceptibility of other polymers behaves differently. A high value for the temperature-independent Pauli susceptibility was reported for polythiophene, doped with 24% AsF$_5$: $\chi_p = 300$ μJ/T^{2}mol [3.91, 92], whereas for lower doping, $y < 0.14$, only a Curie component was measured. From the measured χ_p value, a density of states of $N(E_F)/n_c = 0.23$ eV^{-1} is calculated, which is considerably higher than that been found in other polymers. In poly(3-methyl-thiophene), doped with BF$_4^-$, *Schärli* et al. [3.81] observed in ESR measurements a density of mobile spins that increases linearly with doping up to a concentration of about $y = 0.12$, where it saturates. A Pauli susceptibility, rising approximately linearly with doping, is also observed in polyaniline [3.93]. In poly(p-phenylene) a gradual increase of the Pauli susceptibility with doping was observed by *Kume* et al. [3.83], reaching $\chi_p = 15$ μJ/T^{2}mol at $y = 0.2$ and leading to a density of states $N(E_F)/n_c = 0.012$ eV^{-1} for the highest doping. For the same material *Peo* et al. [3.94] reported no measurable Pauli susceptibility after heavy doping. In polypyrrole also no Pauli susceptibility could be detected [3.82, 95]; it was interpreted as formation of spinless bipolarons.

The thermoelectric coefficient is low in this doping range and strongly temperature dependent. Only in polyacetylene, doped with 14.8% AsF$_5$, is a strictly linear variation with temperature observed; if other dopants are used, a nonlinear contribution is seen (Fig. 3.15) that makes an evaluation in terms of $N(E_F)$ difficult [3.67]. For the AsF$_5$-doped sample, a value $N(E_F)/n_c = 0.18$ eV^{-1} is calculated, based on the assumption of a complete charge transfer, i.e. $n = yn_c$. A linear temperature dependence is also reported for polythiophene doped with 25% BF$_4^-$ [3.78]; evaluating these data under the same assumption yields $N(E_F)/n_c = 0.19$ eV^{-1}.

An important fact is that the electrical conductivity in polyacetylene has already attained high values long before the semiconductor-metal transition takes place; at the semiconductor-metal transition itself, no break or jump

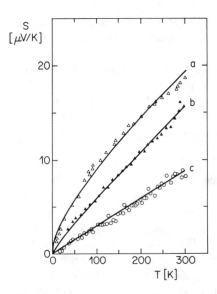

Fig. 3.15. Thermoelectric coefficient vs. temperature for *trans*-polyacetylene, doped with (a) I_3^-, $y = 0.073$; (b) $FeCl_4^-$, $y = 0.061$ and (c) AsF_6^-, $y = 0.148$ (from [3.67])

occurs. At low temperatures signs of a metallic behaviour can be seen. In earlier experiments [3.24] a nonvanishing conductivity was observed as soon as the doping level increased above 0.03. This y value is smaller by a factor of two compared to the doping level, where the Pauli susceptibility becomes measurable. Later investigations [3.85], performed on stretched samples of polyacetylene doped with 9.7% AsF_5, found a levelling-off to a constant conductivity value between 1 and 0.01 K, but a further decrease in the temperature range below this. These findings were explained as an effect of interfibrillar contacts. In the meantime, a new method of synthesis has been developed by a group at Durham University [3.96]. It starts from a precursor polymer from which, by an elimination reaction, cis-polyacetylene is obtained. Samples prepared in this way are reported to show no fibrillar structure, but to be compact. Initial investigations on AsF_5-doped samples gave room-temperature conductivities of the order of 50 000–80 000 S/m, whereas in iodine-doped samples only 1000–10 000 S/m could be achieved [3.97]. By stretching the precursor polymer the chain molecules can be oriented to a high degree. If such a non-fibrillar structure is verified, future conductivity measurements at very low temperatures will be able to contribute to the understanding of metallic conductivity.

An important point is the maximum conductivity obtained by heavy doping. Generally the conductivity increases only weakly with dopant concentration, as compared to the range around $y = 0.001$; it sometimes even saturates. Synthesis conditions also play an important role here, since they dominate the ultimate morphology of the sample [3.98]. In general values between 10^4 and 10^5 S/m can be obtained in Shirakawa-type polyacetylene and polypyrrole [3.82], poly(p-phenylene) [3.83] and polythiophene [3.99]. Stretching the sample results in a further increase in σ_{max} and also in the anisotropy ratio $\sigma_\parallel / \sigma_\perp$ [3.100].

In 1987 *Naarmann* and *Theophilou* reported a novel method of synthesizing polyacetylene in a catalyst containing a reducing agent [3.101]. After heavy doping with iodine a conductivity as high as $8 \cdot 10^6$ S/m was measured. This new material shows the highest conductivity ever measured in a polymer. *Schimmel* et al. [3.102] measured the temperature dependence of the conductivity in such a material and found a steep decrease at temperatures below 0.5 K, in agreement with the findings of [3.85], suggesting also that here interfibrillar contacts determine the low-temperature behaviour. Later a different synthesis, using a solvent-free catalyst, was described by *Akagi* et al. [3.103], resulting in a dense material and conductivities in the same range.

Protonation of polyaniline also gives a highly conducting polymer with conductivities around 500 S/m. This material exhibits a metal-like thermoelectric coefficient and a Pauli susceptibility that increases approximately in proportion to the amount of doping (protonation in this case) [3.93]. From this it is concluded that heavily doped polyaniline is a two-phase material, consisting of metallic, fully protonated regions and insulating more or less undoped regions. The highest magnetic susceptibility reached is $1100 \, \mu J/T^2$ mol, leading to a density of states $N(E_F)/n_c = 0.34 \, eV^{-1}$; here the nitrogen atoms also have to be included in the sequence of conjugation.

Frequency-dependent conductivity does not contribute much to the understanding of the semiconductor-metal transition and the general behaviour in the highly doped regime. The conductivity stays high at a constant value for all frequencies extending into the GHz range. At higher frequencies, a decrease in conductivity is reported for polyacetylene and interpreted as Drude-like free-carrier absorption [3.104]. In the infrared *Kuivalainen* et al. [3.105] observed a plasma edge in strongly $FeCl_3$-doped poly(p-phenylene) at a frequency of about 270 THz. From this an "optical" conductivity of 10^5 S/m was calculated. Similar values were reported for polyacetylene [3.5].

Several attempts were made to determine the mobility of the charge carriers. Hall-effect measurements did not succeed, but magneto-resistance measurements look more promising, except that they need a specific model for their interpretation. From the magneto-resistance measured on a Corbino disc, values for the hole mobility of the order of $0.01-0.02 \, m^2/Vs$ were obtained for heavily doped polyacetylene ($y > 0.07$) [3.106]. Under the assumption that each dopant molecule generates one charge carrier, mobilities of such an order of magnitude result in conductivities of several 10^6 S/m, values that had been reached recently [3.101-103]. The cited values are considered to be those of the bulk material not influenced by interfibrillar contacts.

An important question remains to be answered, namely whether the semiconductor-metal transition is limited to trans-polyacetylene or whether it is a property common to all conjugated polymers. The safest indicator for the appearance of a metallic state would be a non-vanishing conductivity at low temperatures. The experiments, however, are incomplete and the conclusions drawn on a weak basis. Therefore in general the Pauli paramagnetism is taken as indicator for the appearance of a metallic (or quasi-metallic) state. An abrupt increase in χ_P is found only in poly-acetylene; on the other hand, polythiophene, poly(p-phenylene) and polyaniline show a more gradual increase with doping, whereas in polypyrrole no Pauli paramagnetism was found. The question is still open.

3.5 Summary

Conjugated polymers in their undoped state are insulating, like an anisotropic semiconductor. Doping with acceptors or donors, which corresponds chemically to partial oxidation or reduction, leads to conductive materials and a series of interesting new phenomena. Contrary to inorganic crystalline semiconductors, where charge is transported in general by electrons in the conduction band and holes in the valence band, in doped conjugated polymers charged solitons, polarons, and bipolarons can act as charge carriers. These quasi-particles arise from a strong interaction between a charge put on the chain (electron, hole) and the molecular structure. Such behaviour is typical for conjugated polymers. Solitons depend on a degenerate ground state and can develop only in trans-polyacetylene (and polyyne). Polarons and bipolarons correspond to a pair of confined solitons (charged–neutral and charged–charged, respectively) and are the general type of charge carriers. If three-dimensional interactions are taken into account, quasi-particles may no longer be stable.

In the discussion of electric transport properties, reference is made to the electric conductivity and its dependence on temperature, frequency, dopant concentration and material, and to the thermoelectric coefficient. In many cases the rate-limiting step is the charge transfer between chains and, if the material has a fibrillar structure, also between fibrils.

Two important phenomena occur when the doping concentration is varied: Firstly, at a relative concentration of the order $y = 0.001$ the dc conductivity increases by many orders of magnitude within a decade of y. This behaviour is common to all materials studied, to both isomers of polyacetylene and to poly-aromates and polyheterocycles. In the same range the thermoelectric coefficient drops from values above $1000\,\mu V/K$ to below $100\,\mu V/K$. These effects depend on the fact that at low dopant concentrations the charge carriers (solitons, polarons) are bound to dopant molecules. Around the critical concentration the pinning forces are reduced, and a pronounced increase in conductivity occurs.

Secondly, at $y \approx 0.06$ a semiconductor-metal transition is observed in poly-acetylene. This transition is not yet fully understood. Characteristic features are the sudden appearance of Pauli paramagnetism, a metallic behaviour of the thermoelectric coefficient, and a nonvanishing dc conductivity at low temperatures.

Another important point in the discussion was the ordering of dopant molecules. Evidence for this has come from diffraction studies, but implications on the interpretation of electric transport properties are to be expected.

Despite of the success achieved in the field of conducting polymers within the last decade, a great amount of work still lies ahead. Material preparation and characterization need considerable improvement, the understanding of several phenomena is still unsatisfactory, new and interesting materials are synthesized and need evaluation, and some of the theoretical concepts wait for their experimental verification. Within the last years the highest room-temperature conductivity obtained by doping has been pushed up and is now only one order of magnitude below that of copper. Where the ultimate value for a polymer material is to be expected, is still a matter of debate, but the guesses range from twice [3.107] to 25 times [3.108] that of metallic copper.

Acknowledgement

We gratefully acknowledge the support and valuable comments we received from Dr. S. Roth, MPI, Stuttgart.

References

3.1.   R.E. Peierls: *Quantum Theory of Solids* (Clarendon, Oxford 1955) p. 108
3.2.   H. Froehlich: Proc. Roy. Soc. (London) Ser. **223**, 296 (1954)
3.3.   L. Pauling: *The Nature of the Chemical Bond* (Cornell Univ. Press, Ithaca 1948) p. 171–175
3.4.   J. Hubbard: Proc. Roy. Soc. **A276**, 238 (1963)
3.5.   C.R. Fincher, M. Ozaki, M. Tanaka, D. Peebles, L. Lauchlan, A.J. Heeger, A.G. MacDiarmid: Phys. Rev. **B20**, 1589 (1979)
3.6.   C.R. Fincher, C.-E. Chien, A.J. Heeger, A.G. MacDiarmid, J.B. Hastings: Phys. Rev. Lett. **48**, 100 (1982)
3.7.   T.C. Clarke, R.D. Kendrick,, C.S. Yannoni: J. de Phys. Coll. **44**, C3-369 (1983)
3.8.   M.J. Rice, A.R. Bishop, D.K. Campbell: Phys. Rev. Lett. **51**, 2136 (1983)
3.9.   K. Akagi, M. Nishiguchi, H. Shirakawa, Y. Furukawa, I. Harada: Synth. Metals **17**, 557 (1987)
3.10. N.A. Cade, B. Movaghar: J. Phys. **C16**, 539 (1983)
3.11. J.L. Brédas, R. Elsenbaumer, R.R. Chance, R. Silbey: J. Chem Phys. **79**, 5656 (1983)
3.12. S.A. Brazowski, N.N. Kirova: Pis'ma Zh. Eksp. Teor. Fiz. **33**, 6 (1981) [Engl. transl. JETP Lett. **33**, 4 (1981)]
3.13. K. Fesser, A.R. Bishop, D.K. Campbell: Phys. Rev. **B27**, 4804 (1983)
3.14. J.L. Brédas, R.R. Chance, R. Silbey: Phys. Rev. **B26**, 5843 (1982)
3.15. W.P. Su, J.R. Schrieffer, A.J. Heeger: Phys. Rev. **B22**, 2099 (1980)
3.16. W.P. Su, J.R. Schrieffer, A.J. Heeger: Phys. Rev. Lett. **42**, 1698 (1979)
3.17. A.J. Heeger, A.G. MacDiarmid: In *Physics in One Dimension*, ed. by J. Bernasconi, T. Schneider, Springer Ser. Solid-State Sci. Vol. **23** (Springer, Berlin, Heidelberg 1981) p. 179
3.18. F. Devreux, K. Holczer, M. Nechtschein, T.C. Clarke, R.L. Greene: In *Physics in One Dimension*, ed. by J. Bernasconi, T. Schneider, Springer Ser. Solid-State Sci. Vol. **23** (Springer, Berlin, Heidelberg 1981) p. 194
3.19. N.S. Shiren, Y. Tomkiewicz, H. Thomann, L. Dalton, T.C. Clarke: J.de Phys. Coll. **44**, C3-223, (1983)
3.20. W.P. Su, J.R. Schrieffer: Proc. Natl. Acad. Sci. USA **77**, 5626 (1980)
3.21. D. Baeriswyl, K. Maki: Synth. Metals **28**, D507 (1989)
3.22. P. Vogl, D.K. Campbell, O.F. Sankey: Synth. Metals **28**, D513 (1989)
3.23. G.W. Bryant, A.J. Glick: Phys. Rev. **B26**, 5855 (1982)
3.24. Y.W. Park, A.J. Heeger, M.A. Druy, A.G. MacDiarmid: J. Chem. Phys. **73**, 946 (1980)
3.25. H. Kuzmany, P. Knoll: Mol. Cryst. Liq. Cryst. **117**, 385 (1985)

3.26. E. Mulazzi, G.R. Brivio, R. Tiziani: Mol. Cryst. Liq. Cryst. **117**, 343 (1985)
3.27. E.M. Conwell: Phys. Rev. **B33**, 2465 (1986)
3.28. I.A. Howard, E.M. Conwell: Synth. Metals **10**, 297 (1985)
3.29. L.W. Shacklette, N.S. Murthy, R.H. Baughman: Mol. Cryst. Liq. Cryst. **121**, 201 (1985)
3.30. J.P. Pouget, J.C. Pouxviel, P. Robin, R. Comes, D. Begin, D. Billaud, A. Feldblum, H.W.
       Gibson, A.J. Epstein: Mol. Cryst. Liq. Cryst. **117**, 75, (1985)
3.31. R.H. Baughman, N.S. Murthy, G.G. Miller, L.W. Shacklette, R.M. Metzger: J. de Phys. Coll.
       44, C3-53 (1983)
3.32. N.F. Mott, E.A. Davis: *Electronic Processes in Non-Crystalline Materials* (Clarendon, Oxford
       1979), p. 7-64
3.33. E.A. Davis: "Electronic Conduction in Non-Crystalline Systems" in: *Conduction in Low-
       Mobility Materials*, Proc. 2nd Int. Conf., Eilat, 1971, ed. by N. Klein, D.S. Tannhauser
       (Taylor and Francis, London 1971) p. 175
3.34. V. Ambegaokar, B.I. Halperin, J.S. Langer: Phys. Rev. **B4**, 2612 (1971)
3.35. N.F. Mott: Phil. Mag. **19**, 835 (1969)
3.36. P.N. Butcher: Phil. Mag. **B42**, 799 (1980)
3.37. A.J. Epstein, R.W. Bigelow, H. Rommelmann, H.W. Gibson, R.J. Weagley, A. Feldblum,
       D.B. Tanner, J.P. Pouget, J.C. Pouxviel, R. Comes, P. Robin, S. Kivelson: Mol. Cryst. Liq.
       Cryst. **117**, 147 (1985)
3.38. H. Overhof, P. Thomas: Proc. 6th Internat. Conf. on Amorphous and Liquid Semiconductors,
       Leningrad, 1975, ed. by B.T. Kolomiets (Izdatel'stvo Nauka, Leningrad 1976) p. 107
3.39. D. Emin: Phys. Rev. Lett. **32**, 303 (1974)
3.40. D. Würtz, P. Thomas: Phys. Status Solidi **B88**, K73 (1978)
3.41. M. Pollak, T.H. Geballe: Phys. Rev. **122**, 1742 (1961)
3.42. P.N. Butcher: *Linear and Nonlinear Electron Transport in Solids*, in: NATO Adv. Study
       Institute, Antwerpen, 1975, ed. by J.T. Devreese, V.E. van Doren (Plenum, New York 1976)
       p. 341-381
3.43. B. Movaghar, B. Pohlmann, W. Schirmacher: Phil. Mag. **B41**, 49 (1980)
3.44. B. Movaghar, B. Pohlmann, G.W. Sauer: Phil. stat. sol **B97**, 533 (1980)
3.45. W. Schirmacher: Solid State Commun. **39**, 893 (1981)
3.46. S. Summerfield, P.N. Butcher: J. Phys. **C15**, 7003 (1982) and: J. Phys. **C16**, 295 (1983)
3.47. S. Summerfield: Phil. Mag. **B52**, 9 (1985)
3.48. S. Alexander, J. Bernasconi, W.R. Schneider, R. Orbach: Rev. Mod. Phys. **53**, 175 (1981)
3.49. F. Galluzzi, M. Schwarz: Chem. Phys. Lett. **105**, 95 (1984)
3.50. H. Kiess, R. Keller, G. Wieners: Mol. Cryst. Liq. Cryst. **117**, 235 (1985)
3.51. H. Bleier, S. Roth, H. Lobentanzer, G. Leising: Europhys. Lett. **4**, 1397 (1987)
3.52. H. Bleier, K. Donovan, R.H. Friend, S. Roth, L. Rothberg, R. Tubino, Z. Vardeny, G. Wilson:
       Synth. Metals **28**, D189 (1988)
3.53. S. Etemad, A.J. Heeger, L. Lauchlan, T.-C. Chung, A.G. MacDiarmid: Mol. Cryst. Liq.
       Cryst. **77**, 43 (1981)
3.54. R.R. Chance, J.L. Brédas, R. Silbey: Phys. Rev. **B29**, 4491 (1984)
3.55. S. Kivelson: Phys. Rev. **B25**, 3798 (1982) and: Phys. Rev. Lett. **46**, 1344(1981)
3.56. R.R. Heikes: In *Thermoelectricity*, ed. by R.R. Heikes, W.U. Ure (Interscience, New York
       1961) Chap. 4
3.57. D. Emin, K.L. Ngai: J. de Phys. Coll. **44**, C3-471 (1983)
3.58. E.J. Mele, M.J. Rice: Phys. Rev. **B23**, 5397 (1981)
3.59. P. Sheng: Phys. Rev. **B21**, 2180 (1980)
3.60. H. Fukutome, A. Takahashi: Synth. Metals **13**, 135 (1986)
3.61. H. Kiess, W. Meyer, D. Baeriswyl, G. Harbeke: J. Electron. Mater. **9**, 763 (1980)
3.62. K. Ehinger, W. Bauhofer, K. Menke, S. Roth: J. de Phys. Coll. **44**, C3-115 (1983)
3.63. A.J. Epstein, H. Rommelmann, R. Bigelow, H.W. Gibson, D.M. Hoffmann, D.B. Tanner:
       Phys. Rev. Lett. **50**, 1866 (1983)
3.64. K. Ehinger, S. Summerfield, W. Bauhofer, S. Roth: J. Phys. **C17**, 3753 (1984)
3.65. D. Moses, J. Chen, A. Denenstein, M. Kaveh, T.-C. Chung, A.J. Heeger, A.G. MacDiarmid,
       Y.W. Park: Solid State Commun. **40**, 1007 (1981)
3.66. J.C. Scott, T.C. Clarke: J. de Phys. Coll. **44**, C3-365 (1983)
3.67. Y.W. Park, W.K. Han, C.H. Choi, H. Shirakawa: Phys. Rev. **B30**, 5847 (1984)
3.68. A.J. Epstein, H. Rommelmann, M. Abkowitz, H.W. Gibson: Phys. Rev. Lett. **47**, 1549 (1981);
       and: Mol. Cryst. Liq. Cryst. **77**, 81 (1981)
3.69. S. Ikehata, J. Kaufer, T. Woerner, A. Pron, M.A. Druy, A. Sivak, A.J. Heeger, A.G.
       MacDiarmid: Phys. Rev. Lett. **45**, 1123 (1980)

3.70. M. Audenaert: Phys. Rev. **B30**, 4609 (1984)
3.71. J.C.W. Chien, J.M. Warakomski, F.E. Karasz, W.L. Chia, C.P. Lillya: Phys. Rev. **B28**, 6937 (1983)
3.72. D. Moses, A. Denenstein, J. Chen, A.J. Heeger, P. McAndrew, T. Woerner, A.G. MacDiarmid, Y.W. Park: Phys. Rev. **B25**, 7652 (1982)
3.73. R.H. Baughman, L.W. Shacklette, N.S. Murthy, G.G. Miller, R.L. Elsenbaumer: Mol. Cryst. Liq. Cryst. **118**, 253 (1985)
3.74. R.H. Baughman, N.S. Murthy, G.G. Miller, L.W. Shacklette: J. Chem. Phys. **79**, 1065 (1983)
3.75. D.C. Bott, C.K. Chai, D.L. Gerrard, R.H. Weatherhaed, D. White, K.P.J. Williams: J. de Phys. Coll. **44**, C3-45 (1983)
3.76. W.P. Su, S. Kivelson, J.R. Schrieffer: In *Physics in One Dimension*, ed. by J. Bernasconi, T. Schneider Springer Ser. Solid-State Sci. Vol. **23** (Springer, Berlin, Heidelberg 1981), p. 201
3.77. Sh. Hayashi, K. Kaneto, K. Yoshino, R. Matsushita, T. Matsuyama: J. Phys. Soc. Jpn. **55**, 1971 (1986)
3.78. K. Kaneto, S. Hayashi, S. Ura, K. Yoshino: J. Phys. Soc. Jpn. **54**, 1146 (1985)
3.79. F. Devreux, F. Genoud, M. Nechtschein, J.P. Travers, G. Bidan: J. de Phys. Coll. **44**, C3-621 (1983)
3.80. W. Rehwald, H. Kiess, B. Binggeli: Z. Phys. **B68**, 143 (1987)
3.81. M. Schärli, H. Kiess, H. Harbeke, W. Berlinger, K.W. Blazey, K.A. Müller: Synth. Metals **22**, 317 (1988)
3.82. J.C. Scott, P. Pfluger, M.T. Krounbi, G.B. Street: Phys. Rev. **B28**, 2140 (1983)
3.83. K. Kume, K. Mizuno, K. Mizoguchi, K. Nomura, Y. Maniva, J. Tanaka, M. Tanaka, A. Watanabe: Mol. Cryst. Liq. Cryst. **83**, 285 (1982)
3.84. J.-Ch. Chiang, A.G. MacDiarmid: Synth. Metals **13**, 193 (1986)
3.85. C.M. Gould, D.M. Bates, H.M. Bozler, A.J. Heeger, M.A. Druy, A.G. MacDiarmid: Phys. Rev. **B23**, 6820 (1981)
3.86. W.L. McMillan: Phys. Rev. **B14**, 1496 (1981)
3.87. S. Kivelson, A.J. Heeger: Phys. Rev. Lett. **55**, 308 (1985)
3.88. J.C. Hicks, J.T. Gammel, H.-Y. Choi, E.J. Mele: Synth. Metals **17**, 57 (1987)
3.89. D.B. Tanner, G.L. Doll, A.M. Rao, P.C. Eklund, G.A. Arbuckle, A.G. MacDiarmid: Synth. Metals **28**, D141 (1989)
3.90. F. Moraes, J. Chen, T.-C. Chung, A.J. Heeger: Synth. Metals **11**, 271 (1985)
3.91. F. Moraes, D. Davidov, M. Kobayashi, T.-C. Chung, J. Chen, A.J. Heeger, F. Wudl: Synth. Metals **10**, 169 (1985)
3.92. J. Chen, A.J. Heeger, F. Wudl: Solid State Commun. **58**, 251 (1986)
3.93. J.M. Ginder, A.F. Richter, A.G. MacDiarmid, A.J. Epstein: Solid State Commun. **63**, 97 (1987)
3.94. M. Peo, S. Roth, K. Dransfeld, B. Tieke: Solid State Commun. **35**, 119 (1980)
3.95. F. Genoud, M. Guglielmi, M. Nechtschein, E. Geniers, M. Salmon: Phys. Rev. Lett. **55**, 118 (1985)
3.96. J.H. Edwards, W.J. Feast: Polymer **21**, 595 (1980)
3.97. D.C. Bott, C.S. Brown, C.K. Chai, N.S. Walker, W.J. Feast, P.J.S. Foot, P.D. Calvert, N.C. Billingham, R.H. Friend: Synth. Metals **14**, 245 (1986)
3.98. M. Satoh, K. Kaneto, K. Yoshino: Synth. Metals **14**, 289 (1986)
3.99. W. Kobel, W. Rehwald, H. Kiess: Unpublished results
3.100. Ying-Chung Chen, K. Akagi, H. Shirakawa: Synth. Metals **14**, 173 (1986)
3.101. H. Naarmann, N. Theophilou: Synth. Metals **22**, 1 (1987)
3.102. Th. Schimmel. G. Denninger, W. Riess, J. Viot, M. Schwoerer, W. Schoepe, H. Naarmann: Synth. Metals **28**, D11 (1989)
3.103. K. Akagi, M. Suezaki, H. Shirakawa, H. Kyotani, M. Shinomura, Y. Tanabe: Synth. Metals **28**, D1 (1989)
3.104. L. Genzel, F. Kremer, A. Poglitsch, G. Bechthold, K. Menke, S. Roth: Phys. Rev. **B29**, 4595 (1984)
3.105. P. Kuivalainen, H. Stubb, H. Isotalo, P. Yli-Lahti, C. Holmström: Phys. Rev. **B31**, 7900 (1985)
3.106. W. Röss, A. Philipp, K. Seeger: J. de Phys. Coll. **44**, C3-127 (1983)
3.107. J. Voit: Synth. Metals **28**, D495 (1989)
3.108. L. Pietronero: Synth. Metals **8**, 225 (1983)
3.109. P.G. LeComber, W.E. Spear: In *Amorphous Semiconductors*, ed. by M.H. Brodsky, Topics Appl. Phys. **36** (Springer, Berlin, Heidelberg 1979) p. 251
3.110. M. Peo, H. Förster, K. Menke, J. Hocker, J.A. Gardner, S. Roth, K. Dransfeld: Mol. Cryst. Liq. Cryst. **77**, 103 (1981)

4. Optical Properties of Conducting Polymers

Helmut G. Kiess and Günther Harbeke

4.1 Elementary Considerations

The optical constants of solids reflect their electronic and vibronic structure since the electromagnetic field of the light wave interacts with all fixed and mobile charges, i.e. with free electrons of the solid, with polarons, bound electrons, ions, polar phonons, plasmons etc. In general the response to light is described by the complex dielectric function $\varepsilon = \varepsilon_1(\omega) + i\varepsilon_2(\omega)$. For a degenerate semiconductor with Drude behaviour at small frequencies, a vibrational mode in the IR and a bandgap in the visible, the dielectric function has a frequency dependence of the form depicted in Fig. 4.1. Conversely, if the dielectric function is known, one can obtain information on the nature of the interactions or quasi-particles themselves and their optical transitions.

The dielectric function is obtainable from optical transmission and reflection measurements, the results of which are often expressed in terms of the absorption coefficient K and the reflectivity R. The propagation of electromagnetic waves in a medium is governed by Maxwell's equation and, as already stated, by the physical properties of the medium which are described phenomenologically by the magnetic permeability μ, the dielectric tensor ε_1, and the conductivity tensor σ. Maxwell's equations relate the electric and magnetic vector of the light wave by

$$\text{curl } E = -\mu\mu_0 \partial H/\partial t \tag{4.1}$$

$$\text{curl } H = \varepsilon_1\varepsilon_0 \partial E/\partial t + \sigma E. \tag{4.2}$$

Maxwell's equation can be transformed into two equivalent wave equations

$$c^{-2}\partial^2 E/\partial t^2 = \Delta E \tag{4.3}$$

$$c^{-2}\partial^2 H/\partial t^2 = \Delta H. \tag{4.4}$$

Here is $c = c_0(\varepsilon\mu)^{1/2}$, c_0 being the speed of light. The imaginary part of the dielectric tensor is given by $\varepsilon_2 = \sigma/(\omega\varepsilon_0)$.

In an isotropic medium a solution to these equations is the plane wave with

$$E = E_0 \exp[i(\omega t - k \cdot r)]. \tag{4.5}$$

The complex wave vector k is given by

$$k^2 = \frac{\omega^2}{c^2}\varepsilon\mu. \tag{4.6}$$

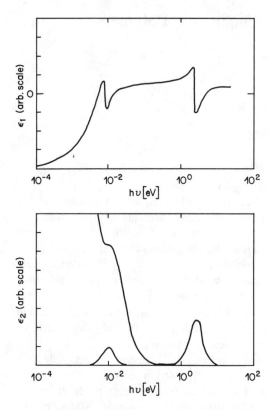

Fig. 4.1. Schematic diagrams of the contribution of the valence electrons (band gap at 2 eV), of phonons (at 10^{-2} eV) and of free carriers to the dielectric function $\varepsilon = f(\omega)$. The small peak at 10^{-2} eV shows separately the magnitude of the phonon contribution to ε_2, giving rise to the shoulder in the curve

Introducing a complex refractive index $N(=(\varepsilon)^{1/2}) = n + i\kappa$, relations between the complex dielectric constant and the refractive index can be established:

$$\varepsilon_1 = n^2 - \kappa^2 = \mathrm{Re}\,\{\varepsilon\} \qquad (4.7)$$

$$\varepsilon_2 = 2n\kappa = \frac{\sigma}{\omega\varepsilon_0} = \mathrm{Im}\,\{\varepsilon\}. \qquad (4.8)$$

The experimentally determined absorption coefficient K and reflectivity R are connected with the absorption index κ and the refractive index n through the relations [4.1]

$$K = \frac{2\omega\kappa}{c} = \frac{4\pi\kappa}{\lambda} \qquad (4.9)$$

and

$$R = \frac{(n-1)^2 + \kappa^2}{(n+1)^2 + \kappa^2}. \qquad (4.10)$$

Having determined both R and K, (4.9, 10) serve to obtain the refractive and the absorption indices n and κ and, hence, also the complex dielectric function.

If R can be measured over a sufficiently large spectral range, a Kramers–Kronig analysis also allows computation of K [4.1]. In these circumstances reflectivity measurements alone suffice in obtaining the dielectric function.

Electron energy loss spectroscopy (EELS) is a further experimental tool to acquire information on the electronic and vibronic properties of solids. In contrast to optics, EELS [4.2] measures the dielectric loss function which is defined as the imaginary part of $-1/\varepsilon$. Via momentum-dependent measurements, EELS also yields information on the dispersion of the electronic bands of a solid which is not accessible using optical methods.

4.2 Dielectric Response Function and Band Structure

We have claimed that the dielectric function $\varepsilon(\omega)$ is related to the electronic, ionic or dipolar properties of the solid. In this section we will limit the discussion to a frequency range, in which $\varepsilon(\omega)$ is determined by the transitions of electrons from valence bands into conduction bands and by free carriers in the bands. In time-dependent perturbation theory the probability per unit time and unit volume for a direct transition of an electron from the valence band into the conduction band can be calculated to be [4.1]

$$P_{vc} = 4\pi\hbar(e/m)^2 A^2 \int d^3q(1/4\pi^3)|M_{vc}|^2 \delta(E_c(k+q) - E_v(q) - \hbar\omega). \tag{4.11}$$

A is the vector potential, M_{vc} is the matrix element for the transition and q is the wave vector of the electron. The δ function represents the first selection rule based on energy conservation. The transition probability is only different from zero if the energy difference between empty state $E_c(k+q)$ and occupied state $E_v(q)$ is equal to the energy of the photon $\hbar\omega$. The second selection rule, based on conservation of momentum, follows from a discussion of the matrix element M_{vc} which is only different from zero when the wave vectors of the initial and final states differ by the wave vector of the radiation, k. Further specific selection rules are established through group-theoretical considerations depending on the symmetry of the initial and final states.

The energy loss per unit volume and time due to absorption at the energy $\hbar\omega$ is given by $\hbar\omega P_{vc}$. This loss can also be described phenomenologically by $(1/2)\sigma|E|^2$. Using (4.8), the desired relationship between the imaginary part of ε and the expression (4.11) can be established by equating the losses

$$\varepsilon_2 = \frac{8\pi\hbar^2 e^2}{\varepsilon_0 m^2 \omega^2}\int d^3q(4\pi^3)^{-1}|M_{vc}|^2 \delta(E_c(k+q) - E_v(q) - \hbar\omega). \tag{4.12}$$

Since M_{vc} is in general a slowly varying function, this equation can be approximated by

$$\varepsilon_2 = \text{const } J_{vc}|M_{vc}|^2 \tag{4.13}$$

with $J_{vc} = \int (4\pi^3)^{-1} (\nabla_k (E_c(\boldsymbol{q} + \boldsymbol{k}) - E_v(\boldsymbol{q})))^{-1} dS$. The integral is performed over a constant energy surface. The dielectric function is thus essentially determined by J_{vc}, the joint density of states. J_{vc} varies rapidly whenever, in the expression $\text{grad}_q (E_c(\boldsymbol{q} + \boldsymbol{k}) - E_v(\boldsymbol{q}))$, each term separately or the difference $\text{grad}_q E_c - \text{grad}_q E_v$ equals or is approximately zero. Such points are called critical points. In the neighbourhood of such points M_{vc}/ω^2 can be regarded as constant and hence ε_2 has the same dependence on the frequency as J_{vc}. It is often sufficient to consider the absorption coefficient K to be representative of J_{vc} assuming a constant refractive index, but one has to keep in mind that it is ε_2 and not K that is the quantity which is really related to J_{vc}.

Up to this point we have assumed a filled valence band and an empty conduction band. This is no longer true at high doping concentrations, higher temperatures or under intense illumination, where free holes and electrons have to be taken into account. The presence of free carriers has two effects. On the one hand, the intensity of inter-band transitions is decreased due to the Pauli exclusion principle, and on the other the free carriers give rise to intra-band transitions which may appear as a Drude-like contribution at low frequencies. At high photo-excitation levels the free carrier density may be sufficiently changed to shift the plasma frequency and the occupation of the states in the valence band and conduction band significantly. In these conditions the refractive and absorption indices become a function of the excitation intensity, i.e. the optical constants become nonlinear. If electron–hole interaction is taken into account, photo-excitation can give rise to bound electron–hole pairs and nonlinear optical effects are then usually enhanced above those due only to the increase of the free carrier concentrations [4.3, 4]. The influence of structural relaxations on the magnitude of nonlinear optical effects is not yet clear (see also Sect. 4.5.2c).

4.3 Band Gap and Band Structures of Undoped Conjugated Polymers

4.3.1 Results of Band Structure Calculations

The theory of the optical band gap is still dominated by the discussion of whether the gap is due to electron–phonon or electron–electron interaction or, if both are responsible, about the relative strengths of these interactions. The calculations refer usually to idealized one-dimensional models. If only electron–phonon interaction is taken into account, then relatively simple independent electron theories result. This is not true if models incorporate the electron–electron interaction; accurate results are then much harder to obtain. The difficulty of deciding on the relative importance of the interactions stems from this theoretical problem and also from the difficulty of obtaining reliable molecular parameters. Since these questions are discussed in detail in Chap. 2, we will give here only a brief overview of the essential results. In practice, the structures of conducting polymers are far from the idealized models. Changes of the band gap may be due to varying bond lengths, non-planarity of the molecules and different conformations; results hereof are also presented.

Probably the most complete calculations of the band gap and band structure have been performed for polyacetylene. The band gap E_g and the structure of the π bands within the framework of the Su-Schrieffer-Heeger (SSH) Hamiltonian [4.5] (see also Chap. 2) are determined by the electron–phonon coupling constant α, the transfer integral t_0 and the elastic constant K_0

$$E_g = 2\Delta = 16t_0/\exp(1 + 1/2\lambda) \tag{4.14}$$

with $\lambda = 2\alpha^2/\pi t_0 K_0$. The conductivity tensor based on the SSH Hamiltonian gives expressions for the imaginary part of the dielectric function [4.6]. Results of these calculations are shown in Fig. 4.2, where ε_{yy} is the dielectric function

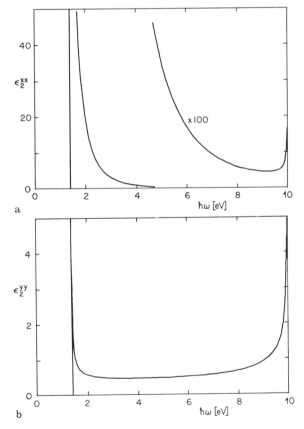

Fig. 4.2 a, b. Imaginary part of the dielectric function of $(CH)_n$ parallel and perpendicular to the chain direction (after [4.6])

perpendicular and ε_{xx} parallel to the chain axis, both being in the plane of the macromolecule. ε_{xx} and ε_{yy} both show a square root singularity at E_g and $4t_0$.

Within the framework of the SSH Hamiltonian, electron–electron interaction is implicitly taken into account by adjusted parameters α, K_0 and t_0. If unbiased parameters derived from small molecules had been taken [4.7], and assuming that only electron–phonon interaction prevails, then the optical gap of trans-(CH)$_n$ is calculated to be only 0.1 eV in contrast to the measured gap of 1–2 eV. Realizing this, Ovchinnikov et al. [4.8] had argued that correlation effects alone would give rise to the optical gap in (CH)$_n$ and that electron–phonon interaction would be insignificant. Based on this assumption and neglecting bond alternation, they calculated an optical gap of about 2.2 eV for a chain of infinite length. However, since the experimentally observed bond alternation requires electron–phonon coupling, the electron–electron interaction alone cannot be responsible for the optical gap in (CH)$_n$. Conversely, the existence of bond alternation does not imply that electron–electron interaction is negligible. Indeed, it has been shown by various authors [4.9–11] that electron–electron interaction enhances bond alternation in (CH)$_n$ if it is smaller than or comparable to the electron–phonon interaction. In contrast to these investigations, it has been argued [4.12] that bond alternation is reduced if off-diagonal terms in the electron–electron interaction are taken into account. These terms describe physically the interaction of electrons on the bonds. Despite these differences in theoretical prediction, it is generally agreed that both electron–phonon and electron–electron interactions must be taken into account to describe the properties of conjugate polymers correctly. Nevertheless, discussion of their relative importance continues. Equation (4.14) for the gap is then no longer valid as the determination of the coupling parameters becomes involved and somehow uncertain.

Grant and Batra [4.13, 14] and Ashkenazi et al. [4.15] have used density functional theory to compute the band structure and density of states of trans- and cis-polyacetylene assuming a $P2_1/n$ symmetry. Results for trans-(CH)$_n$ are depicted in Fig. 4.3 and may be compared with the 1D results in Fig. 4.2. As can be seen, inter-chain interaction is weak, so that one-dimensional models are appropriate to describe many features of (CH)$_n$. The electronic structure of (CH)$_n$ depends critically on the bond alternation. Differences in band structure also arise if the dimerization of adjacent chains is assumed to be in phase or in antiphase ($P2_1/a$ or $P2_1/n$ symmetry). The question of crystal structure has not been solved satisfactorily by experiments; however, calculations by Vogl et al. [4.16] indicate that the $P2_1/a$ structure is slightly more stable than $P2_1/n$. Disturbances of the crystalline order will sensitively affect the electronic structure of (CH)$_n$, and unless true single crystals are available, all experimental materials will be plagued by these and other disturbances; thus comparisons of calculated band structures with optical results are difficult.

Utilizing the valence effective Hamiltonian method, Brédas et al. [4.17] have obtained essentially the same band structure and density of states for the occupied bands as did Grant and Batra. The density of states can be probed

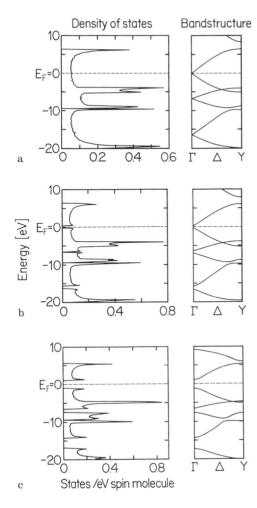

Fig. 4.3 a–c. Density of states and band structure of $(CH)_n$. (a) no, (b) weak, and (c) strong bond alternation (after [4.14])

with photo-electron spectroscopy and there seems to be a satisfactory agreement between experiments and these theoretical models [4.18].

The band gaps of other polymers also depend critically on the bond lengths assumed for the computations. Of particular interest is the influence of the bond that links the monomer units. In the case of polypyrrole, the optical gap is theoretically expected to decrease from 3.2 eV to 2.6 eV if this bond length is reduced from 1.49 Å to 1.4 Å [4.19]. The threshold for absorption is, in addition, a strong function of the number of monomer units used for the calculation.

Calculations usually assume the polymers to be strictly planar. It has been found that twisting the monomer units of a pyrrole dimer around the linking single bond does not change the density of states significantly as long as the twisting angle is below 10°. However, the positions of the computed optical transitions change significantly [4.19]. The optical transition is expected to shift by as much as 0.1 eV to lower energies for twisting angles of 10°. With a twist angle of 90°, conjugation is destroyed and concomitantly, the density of states significantly altered. Therefore, an assumed helicity of the pyrrole polymer might be expected to reduce the band gap and rotational disorder to give rise to a broadening of the optical absorption feature at the band gap. In the case of cis-polyacetylene, helix formation would give rise to a dramatic increase of the band gap, from 1.5 eV in the planar conformation to 4.3 eV in the helical isomer [4.20]. Helical disorder would hence smear out the square root singularity of the absorption and broaden the absorption peak at E_g.

Ab initio self-consistent-field restricted Hartree–Fock calculations performed on polyacetylene [4.21, 22] and on poly(diacetylene) overestimate the band gaps of these polymers considerably. The effects of correlation and electron–hole interaction have also been taken into account. Correlation effects were shown to reduce the Hartree–Fock energy band gap of polyacetylene from 5 eV to about 2.5 eV [4.23].

4.3.2 Experimental Results

a) Polyacetylene

The absorption coefficient of trans-polyacetylene uncorrected for reflection is plotted in Fig. 4.4. It comprises results obtained on unoriented Shirakawa-type material and on highly oriented material prepared from a precursor polymer [4.24, 25]. The band gaps obtainable by extrapolation are independent of the preparation method of the material. The width of the first absorption band deviates strongly from that of the ideal infinitely long chain; in contrast to an expected singularity at E_g, a pronounced tail is observed below E_g. Impurities, morphological disorder and varying chain lengths are possible reasons for the broadening of the band. Quantum fluctuations could also cause the tailing; the

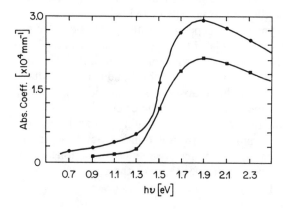

Fig. 4.4. Optical absorption of oriented (●) and unoriented (■) $(CH)_n$ ((●) after [4.25])

absorption below E_g would then be intrinsic to the material and not a result of defects and disorder.

The anisotropy of the absorption parallel and perpendicular to the fibre orientation of oriented $(CH)_n$ is quite pronounced [4.26]. Spectra obtained by EELS on the same material show that the anisotropy persists to high energies, though the anisotropy ratio decreases from 170 at 2 eV to about 2 at 10 eV [4.27]. The real and imaginary parts of the dielectric function ε_1 and ε_2 obtained by EELS are depicted in Fig. 4.5. Results of ε_2 obtained with unoriented samples [4.28] are shown in the same figure. There are essentially two electronic transitions which cause the peaks in ε_2 at 2.0 eV and at about 12 and 17 eV. By inspection of the band structure of $(CH)_n$, one is able to identify two separate bands to which these transitions can be assigned. These are inter-band transitions between $\pi - \pi^*$ states and $\sigma - \sigma^*$ states. Though the energy loss of the

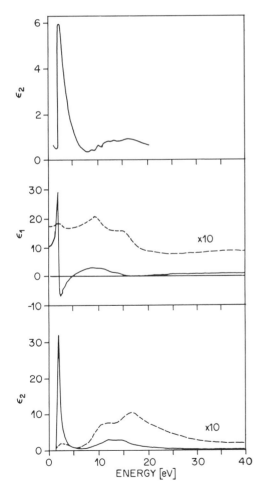

Fig. 4.5. Dielectric function ε_1 and ε_2 of oriented $(CH)_n$ obtained from EELS data between 0 and 40 eV (after [4.27]). The dashed lines are representative results for perpendicular, the full lines for parallel orientation. ε_2 of unoriented $(CH)_n$ obtained by optical measurements between 0 and 20 eV are depicted in the upper graph (after [4.28])

first absorption peak is momentum dependent, the absorption threshold itself has been found to be independent of momentum transfer. This has been attributed to excitonic effects [4.29, 30], although this is not the only possible explanation.

Comparing the real part of the dielectric function obtained by optical measurements [4.26, 31, 32] with EELS measurements [4.30] shows a remarkably similar complex dielectric function.

b) Aromatic and Heterocyclic Conjugated Polymers

The absorption spectra of the different undoped materials indicate a significant variance of the onset of intrinsic absorption, ranging from about 3 eV for poly(paraphenylene) down to 1.0 eV for poly(isothianaphthene). Thus, depending on the position of the band gap, these materials can be transparent, coloured or opaque in the reduced state. Results of absorption measurements are summarized in Fig. 4.6. Analyses of data in terms of band structure and dielectric function have been made up to now only with polypyrrole [4.33], polythiophene [4.34] and poly(p-phenylene) [4.35] using EELS as the experimental tool. The momentum dependence of the dielectric loss function of polypyrrole is shown in Fig. 4.7. The peak at 3.2 eV is assignable to a $\pi - \pi^*$ transition. It shows a strong dispersion, indicative of a large width of the π band. The shoulders at about 4.5–5.5 eV and the transition at 8 eV are caused by excitations from a second occupied π band to the lowest unoccupied π band, and from the highest occupied σ band to the lowest unoccupied π-band respectively. Since no dispersion is observed, these bands are practically flat. A transition at 14 eV (not shown in Fig. 4.7) is caused by $\sigma-\sigma^*$ transitions. From a comparison of the data with calculations on oligomers it had been concluded that undisturbed segments of at least 5 pyrrole rings must exist in these samples.

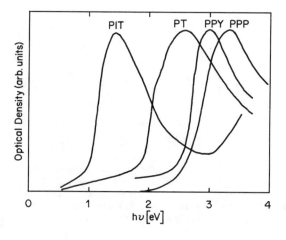

Fig. 4.6. Absorption spectra of poly(isothianaphthene) (PIT), polythiophene (PT), polypyrrole (PPY) and of poly(paraphenylene) (PPP)

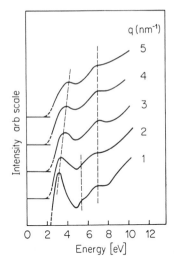

Fig. 4.7. Momentum dependence of the dielectric loss function of polypyrrole between 2 and 10 eV. The dispersion of the peak at ≈ 3.2 eV is indicative for a large width of the π bands (after [4.33])

4.4 Photon–Phonon Interaction

4.4.1 General Remarks

The interaction of electromagnetic radiation with lattice vibrations in solids results in two main effects: absorption or emission of the photon due to the creation or annihilation of phonons, respectively, and scattering of the electromagnetic wave by phonons.

The basic selection rules for optical absorption due to lattice vibrations are, as in the case of photon–electron interaction, based on the conservation of energy and momentum. Since the energies of phonons are generally below 0.5 eV, the spectral region for the measurements of these excitations is in the infrared. The wave vector of radiation, k, is small compared to the Brillouin zone dimensions, hence only phonons with wave vector $q_{ph} = 0$ can participate in one-phonon processes. Moreover, only lattice vibrations having certain types of symmetry give rise to infrared absorption. Such vibrations are said to be infrared active. For two- or multiphonon processes the sum of the momenta of the involved phonons must vanish. This leads again to critical points in the combined density of states and to corresponding structures in the spectra. In terms of absorption mechanisms, the first-order dipole moment in heteropolar crystals gives rise to very strong one-phonon absorption lines. For homopolar crystals, where the first-order dipole moment vanishes, two- or multiphonon processes can occur due to anharmonic terms in the potential or due to higher order dipole moments.

Inelastic light scattering is due to changes in the electronic polarizability induced by the lattice vibrational modes. Traditionally, the distinction is made between Raman scattering from optical phonons and Brillouin scattering from acoustic phonons. The part of the scattered light with lower frequency than the incident light is called the Stokes component, while that with higher frequency is called the anti-Stokes component. Only lattice vibrations having certain types of symmetry are Raman active. In centrosymmetric crystal structures, Raman activity and infrared activity are mutually exclusive for a given lattice vibration. In other crystal structures a phonon can be simultaneously Raman and infrared active. As in infrared absorption, we will find one-, two- and multiphonon scattering with similar features determined again by energy and momentum conservation.

In a one-phonon Raman scattering event, the incoming photon of frequency ω_i is annihilated, a scattered photon of frequency ω_s is created and in the case of a Stokes component, a phonon of frequency ω is created such that $\omega_i = \omega_s + \omega$. The scattering mechanism involves therefore virtual electronic transitions due to the absorption of the incident photon ω_i, the creation of the phonon ω and the emission of the photon ω_s. Since energies E_i of the virtual intermediate states are involved, the Raman matrix elements consist of terms which contain frequency-dependent denominators $(\hbar\omega_i - E_i)$ and $(\hbar\omega_s - E_i)$. These terms can thus diverge as the frequency $\hbar\omega_i$ or $\hbar\omega_s$ approaches an energy eigenvalue E_i. The resonant denominators are the origin of the resonance Raman scattering. While the ordinary Raman effect already gives information on the vibrational excitations of the system, resonance Raman scattering is mainly used to study electronic transitions and the electron–phonon interaction as it expresses itself in the scattering mechanism.

4.4.2 Calculations of Vibrational Spectra of Polymers

Calculations of the infrared and Raman-active vibrations of polymers are usually based on structural models taking the bond lengths and angles between the constituents from the structural data. Usually these models have to be three dimensional, even if one-dimensional models are appropriate in describing with sufficient accuracy the electronic transitions and the related band structure. For polymers the force fields between neighbouring atoms are often derived from appropriate small organic complexes. In this sense the vibrations of polymers are described by structural units which are linked together without essential changes. In the case of conjugated polymers, however, additional interaction with the vibrations comes from the delocalized π electrons. A question then arises concerning the reliability of the force fields derived from small units.

Trans-polyacetylene is the simplest of the conjugated polymers for which, therefore, the most extensive computations of the vibrational spectra have been performed. The C_{2h} symmetry of *trans*-$(CH)_n$ allows for four Raman- and two infrared-active modes. *Mele* and *Rice* [4.36] first analyzed the problem by constructing a force field consisting of two parts. The nearest-neighbour bond-stretching and bond-bending forces were taken from double and single C–C bonds of small organic molecules and extrapolated to the corresponding bond lengths of the infinitely long polyene chain. From these fields the so-called bare phonons of the chain can be calculated. Interaction of the atomic displacements with the extended π electrons was taken into account utilizing the electron–phonon coupling constant α. The calculation is thus a dynamical extension of the SSH model. The electron–phonon coupling constant was adjusted to bring the C–C and C=C stretching modes into agreement with the principal Raman lines at 1470 and 1080 cm^{-1}. The coupling constant was found to be 8 eV/Å, implying a Peierls gap of 1.3 eV. The measured Raman lines were utilized here to determine the unknown parameters in the SSH theory, rather than to calculate strictly the active modes. If the gap of polyacetylene is mostly due to electron–electron interaction, then the parameters tacitly imply this interaction also and, therefore, do not reflect their bare values.

A different approach was taken by various other authors [4.37–39]. *Zannoni* and *Zerbi* used a valence force field, whose parameters are derived from the vibrational spectra. For conjugated systems, delocalized force constants have to be included in contrast to the case of saturated molecules. The principal Raman line at 1470 cm^{-1} was found to be essentially a C=C stretch vibration whereas the lines at 1290 cm^{-1} and 1080 cm^{-1} are strongly coupled to C–H bending vibrations. The Raman line at 3060 cm^{-1} is a pure out-of-phase C–H stretching vibration.

Rakovic et al. [4.40, 41] have computed spectra and intensities of infrared absorption lines of *cis*- and *trans*-polyacetylene using a single periodic chain model within a harmonic potential approximation. The force constants and dipole moments including their derivatives were obtained from *cis*- and *trans*-hexatriene, respectively. A similar calculation of the infrared and Raman active lines was also performed for poly(p-phenylene) [4.42]. Here, the parameters for the force field and the dipole moments were transferred from the toluene molecule.

A microscopic approach similar to that of Mele and Rice (i.e. taking the π electrons explicitly into account) was used by *Horovitz* [4.43]. He considered the continuum version of the SSH Hamiltonian, where the electronic gap is directly coupled to the lattice dimerization. Thus oscillations of the dimerization amplitude give rise to oscillations of Δ. Since in this model not all the molecular degrees of freedom are represented, *Horovitz* et al. [4.44, 45] extended the continuum model and introduced a multi-component order parameter $\Delta_n(x)$, with $\Delta(x) = \Sigma\Delta_n(x)$. In this sum three of the normal modes coupled to π electrons are included. This theory is usually referred to as amplitude mode formalism. The Raman-active modes of the defect-free chain can be obtained as solutions of the equation

$$D(\omega) = \sum(\lambda_n/\lambda)\omega_{n,0}^2(\omega^2 - \omega_{n,0}^2)^{-1} = 1/(2\lambda - 1) \tag{4.15}$$

where $\omega_{n,0}$ are the bare frequencies, i.e. the normal mode frequencies without

π electrons, λ_n is a dimensionless coupling parameter belonging to mode n, and $\lambda = \Sigma \lambda_n$. Equation (4.15) has been solved graphically. For each bare mode $\omega_{n,0}$ coupled to the π electrons one root is obtained, and thus one Raman-active mode ω_n^R with $\omega_n^R < \omega_{n,0}$.

The models so far described do not reflect real polymers which are far from the theoretical model of a perfect infinitely long chain. Real polymers have various types of defects such as chain breaks, cross links, sp^3 defects, impurities etc. The imperfections have been globally modelled either in terms of a distribution of conjugation lengths [4.49] or of the coupling parameters [4.50, 51].

If disorder shifts the coupling parameter λ_n, the Raman lines will shift correspondingly. Disorder can be introduced into this model by a distribution function $P(\lambda)$ which transforms into a distribution of the gap energies through a relationship $\lambda = f(\Delta)$; compare e.g. with (4.14).

The influence of electron–electron interaction has also been considered and incorporated into this general formalism. A similar analysis can be carried out for a defected chain, the defect being described as a local perturbation of the phonon propagator $D(\omega)$ (see Sect. 4.5.3). Thus Raman data and defect-induced local vibrations can be traced back to the same set of parameters.

The relation between the amplitude mode theory and the phenomenological valence force method has been discussed by *Zerbi* and coworkers [4.46–48]. As already pointed out by *Mele* and *Rice* [4.36], delocalized force constants originate from interactions of atomic displacements with the extended π-electronic states. Since the delocalized force constants change with increasing conjugation length, Zerbi et al., made a correction to the local force constants by an additive term ΔK. ΔK was shown to be related to λ and to play the same role in the valence force model as λ in the amplitude mode theory. Therefore the two models for treating imperfections are not mutually exclusive but they both reflect the delocalized π-electron system.

4.4.3 Experimental Results

The IR spectra of not intentionally doped *trans*-(CH)$_n$—according to the symmetry, a total of 3 modes—can be attributed to CH stretching, wagging and bending vibrations. The absorption bands of these vibrations are located at 3013, 1290 and 1015 cm^{-1}, the intensity of the 1290 band being very weak whereas the other two are strong. Since these transverse modes are also coupled to the π-electron density, the intensity of the lines is influenced by the mobile charge along the backbone. This is particularly evidenced by the in-plane CH wagging band at 1290 cm^{-1} which is very weak compared to the out-of-plane mode at 1015 cm^{-1}. The in-plane wagging is expected to induce electron motion screening the ionic charge, whereas the out-of-plane mode does not, thus explaining the difference in intensity.

A comparison of experimental IR spectra of a series of oligomers of thiophene with that of polythiophene allows one to identify the lines of the spectrum of polythiophene: The lines at 790 and 690 cm^{-1} are due to CH out-of-plane vibrations, whereas those at 1441, 1453 and 1491 cm^{-1} are due to symmetric and antisymmetric C=C stretch vibrations [4.52]. Infrared spectra of PMT are also characterized by a wealth of lines. The bands are shifted to slightly higher energies for PMT.

Raman scattering directly probes even parity vibrations of a chain. Due to the electron–phonon coupling, certain Raman lines are resonantly enhanced for photon energies above the band gap. Therefore resonant Raman scattering is also suited to investigate the π-electron system. The principal Raman lines of *trans*-(CH)$_n$ at 1500 and 1100 cm^{-1} are doubly peaked, one of the peaks being significantly shifted with exciting wavelength, as indicated in Fig. 4.8. This dispersion of the Raman bands has been the subject of considerable discussions.

Mele [4.53] suggested that exciting carriers high up into the bands will give rise to a structural relaxation of the lattice which is accompanied by the emission of a phonon. This interpretation corresponds to the emission of optical phonons by hot carriers. Radiative emission into the ground state after phonon emission would give rise to one of the double peaks. This emission of phonons would be in competition with the regular resonance Raman scattering process. Since this suggested process is intrinsic by its very nature, it is not directly related to the quality of the sample.

The changing profile has also been interpreted as being due to the inhomogeneity of the polymer [4.54–57]. It is known that both the optical excitation energy and the vibrational frequencies of the Raman active modes (e.g. the C=C

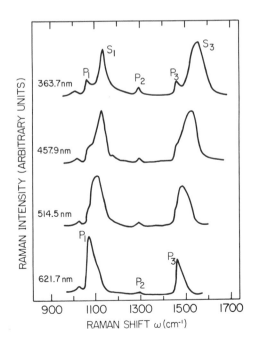

Fig. 4.8. Raman lines *trans*-(CH)$_n$ between 900 cm^{-1} and 1700 cm^{-1} at different exciting wavelengths. The lines marked with P_1, P_2, P_3 show no shift, those marked with S_1 and S_3 are shifted with exciting wavelength (after [4.45])

stretch vibration) depend on the conjugation length. It has therefore been concluded that different exciting wavelengths cause selective excitation of different
parts of the chains, which can be envisaged to differ with respect to their conjugation length. Hence, varying the optical excitation energy will give rise to shifts
of the Raman frequencies. In order to fit the experimental data, a double-peaked
distribution of chain lengths was assumed, corresponding to a distribution peaking at chain lengths of about 200 C atoms and 10–20 C atoms respectively.

In the multi-component amplitude mode theory, the Raman frequencies
depend sensitively on the electron–phonon coupling parameter λ. If a distribution $P(\lambda)$ with a single peak is introduced, then the amplitude mode theory
allows one to describe the observed doubly peaked Raman lines [4.58]. Since
the gap parameter depends exponentially on λ, $P(\lambda)$ is narrow. The primary line
of the doubly peaked structure corresponds to the peak in $P(\lambda)$, the secondary
line to the resonance condition. Thus the primary line *should not*, and the
secondary line *should* shift with increasing frequency of the exciting light in
agreement with experiment (Fig. 4.8). Further support for the amplitude mode
theory is provided by the opposite shifts of the primary and secondary line with
temperature [4.58].

Since both the Raman frequencies and Δ are function of λ, the relationship
between these quantities allows one to obtain values for the electron–phonon
and electron–electron coupling parameters [4.58]. Within the Peierls–Hubbard
model these values were estimated to be $\lambda = 0.12$ and $U = 3.7$ eV for the on-site
Coulomb interaction. The σ-bond force constant for the bare C–C stretching

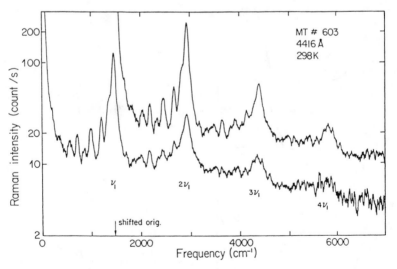

Fig. 4.9. Raman spectrum of oxidized PMT at 298 K. The upper curve represents the same spectrum
to show the exact registry of overtones. For this purpose it has been shifted laterally by 1468 cm^{-1}
and vertically offset by a factor of two. The Raman intensity is plotted on a logarithmic scale (after
[4.60])

mode, was also derived from Raman data and found to be $K = 46\,eV/\text{Å}^2$. A similar analysis has been performed for *cis*-polyacetylene.

Raman studies on polythiophene (PT) have been reported with modes at 700, 1045, 1458 and 1500 cm^{-1} [4.59]. The mode at 1458 cm^{-1} was assigned to the C=C antisymmetric stretching vibrations, the one at 1045 cm^{-1} to the CH in-plane bending and the one at 700 cm^{-1} to ring deformation vibrations. In poly(3-methylthiophene) (PMT) the C=C stretch vibration is shifted to 1470 cm^{-1}. The PMT Raman spectra (Fig. 4.9) are characterized by overtones and combination bands of the lower frequency modes with the fundamental band at 1470 cm^{-1} [4.60]. This has been taken as evidence for a high degree of ordering in this material, since disorder destroys this effect. The mode at 1470 cm^{-1} shifts linearly to higher frequencies with the frequency of the exciting laser light, in qualitative agreement with experiments on polyacetylene.

4.5 The Study of Elementary Excitations in Conjugated Polymers

4.5.1 General Considerations

The electronic excitations in quasi-one-dimensional materials are expected to be different from those in conventional semiconductors. Assuming that a linear chain model without inter-chain interactions can be applied to polymers with degenerate ground state, two types of nonlinear excitations are theoretically expected, namely the one-dimensional domain walls or solitons and the more common and ubiquitous polarons. In the case of polymers with non-degenerate ground state, polarons and bipolarons are the possible solutions of the SSH Hamiltonian. In the following, these nonlinear excitations will also be collectively called quasi-particles, if the particle aspect is emphasized, or bond-alternation defects (or simply defects) having the local change of the bond alternation in mind. If, as an alternative to the linear chain model, it is assumed that the material is crystalline and that hence inter-chain interaction has also to be taken into account, then polarons may be unstable; in this case the excited carriers are expected to be electrons or holes as in conventional semiconductors. Similarly, in crystalline (CH)$_n$ individual solitons cannot exist but are rather confined within kink–antikink pairs. The subtle energetics determining the stability of the quasi-particles is an interesting starting point for the discussion of the optical experiments; these can, in principle, identify such quasi-particles.

The energy for creation of these quasi-particles is of the order of the band gaps. Since the band gaps of these polymers are usually large (about 1–4 eV), the thermal equilibrium density of the quasi-particles turns out to be too low for experimental investigations at temperatures to which these materials can be exposed without destroying them. However, they can be generated either by doping or photo-excitation in the undoped or weakly doped material. Therefore, we discuss first the electronic and vibrational states induced by doping or optical excitation. We will then elaborate on experimental results, with emphasis on photo-induced absorption measurements which are less difficult to interpret than absorption bands caused by doping. The results of photo-induced absorption have helped to clarify various controversial points.

A further point of interest has been whether the quasi-particles contribute to or are responsible for charge transport. However, this question needs further study, even if the existence of quasi-particles in the various materials can be proven by optical means. The question of movement of quasi-particles along the chain or their transfer to an adjacent chain cannot be investigated optically unless their density is sufficiently high, allowing the observation of a possibly modified Drude behaviour (see also Chap. 3).

4.5.2 The Electronic States of the Quasi-Particles

a) Electron–Phonon Interaction: The SSH Model

Within the framework of the Hückel theory, bond-alternation defects of polyenes with degenerate ground state were discussed a long time ago. In particular their high mobility, spatial extension and electronic state at midgap was predicted by various authors [4.61]. In the case of polyenes with two-fold degenerate ground states, two stationary configurations have been identified; the domain wall which interpolates between the two alternation patterns (the soliton) and a localized deviation from homogeneous ground state configuration (the polaron). For the non-degenerate case, the solution for the polaron remains qualitatively unchanged; however, individual solitons cannot exist and are confined within pairs. Depending on the degree of confinement, these configurations either look like weakly bound soliton-antisoliton pairs or like doubly charged polarons, a so-called bipolaron.

Trans-$(CH)_n$, a polymer with a two-fold degenerate ground state, may or may not contain a soliton, depending on the number of carbon atoms. The ground state of a finite $(CH)_n$ with an even number of carbon atoms obviously does not contain a soliton, whereas in odd-numbered chains for topological and energtic reasons, a domain wall (a neutral soliton) is incorporated [4.61]. Thermal excitations will create additional solitons, but only in pairs on a chain. The density of neutral solitons in trans-$(CH)_n$ is expected to be equal to the density of chains or chain segments in the polymer, having an odd number of carbon atoms. With repect to the chain without soliton, the density of states of the conduction and valence band is equally reduced by one half-state per spin and per band. These missing states appear as localized electronic states at midgap. The neutral state of these chains at midgap level is singly occupied; thus these neutral solitons carry spin 1/2.

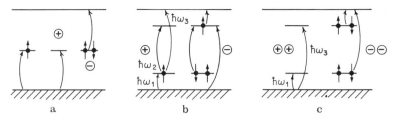

Fig. 4.10 a–c. Energy levels and possible optical transitions associated (**a**) with neutral, positively and negatively charged solitons, (**b**) with hole and electron–polaron, (**c**) with hole and electron–bipolaron

Upon doping either charged solitons or polarons are formed initially due to charge transfer from the dopant to odd chains and even numbered chains respectively. Since the formation of two charged solitons is energetically favoured over that of two charged polarons, only an odd number of charges on the polymer can give rise to polarons. Therefore, practically all quasi-particles on *trans*-(CH)$_n$ are solitons. The electronic levels associated with neutral and charged solitons are depicted in Fig. 4.10a. Only one additional optical absorption line at half the energy of the band gap is predicted.

Doping polymers with non-degenerate ground state leads to polarons or bipolarons. The polaron can be viewed as being composed of two defects, namely a charged one and a neutral one. Creating two separated defects would require more energy than the formation of one paired defect. The reason for this can be easily visualized e.g. for poly(paraphenylene) (Fig. 4.11). The quinoidal form of poly(paraphenylene) is higher in energy than the benzenoidal form, so the local separation of the radical and cation is minimized on the chain [4.62]. Thus in chemical parlance a radical cation, or in physical terms a polaron, is generated. Even if such a defect is doubly charged, it does not disintegrate (in contrast to what happens in trans-polyacetylene) but rather forms a bipolaron. The formation of a bipolaron is also energetically favoured over that of two isolated polarons. The charge of the bipolaron is either doubly positive or negative.

The density of band states for polarons and bipolarons is reduced by one state per spin and per band with respect to the chain without charge added. This is consistent with two localized states in the gap, appearing at energies $\varepsilon = \pm(\Delta^2 - h^2\kappa^2 v_F^2)^{1/2}$ symmetrically to the midgap, with Δ being the band gap parameter $(2\Delta = E_g)$, v_F the Fermi velocity and κ a parameter of the order of the inverse coherence length. The level fillings of the positively and negatively charged polaron are shown in Fig. 4.10b. There are two transitions possible from the bands to these levels and one transition between them, so that in total

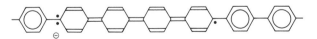

Fig. 4.11. Structure associated with electron–polaron in poly(paraphenylene)

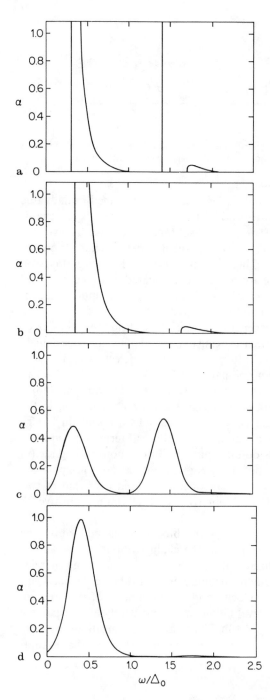

Fig. 4.12 a–d. Optical absorption intensity for a polaron (**a**) and a bipolaron (**b**). For $\omega/\Delta = 1/\sqrt{2}$ as assumed here, the high energy optical transition is significantly weaker than the low energy transition. If convoluted with a Gaussian (**c, d**), the high energy transition becomes invisible and only two and one transitions remain for the polaron and bipolaron respectively (after [4.63])

three optical absorption bands can be expected at ω_1 and ω_3 for the transitions involving band states, and at ω_2 for the transition between the two levels in the gap. Since electron–hole symmetry is implicit in the SSH Hamiltonian, the levels are, as already mentioned, located symmetrically with respect to midgap. It follows that $\hbar\omega_1 + \hbar\omega_2 = \hbar\omega_3$ and $\hbar\omega_1 + \hbar\omega_3 = E_g$. In the case of the bipolaron, only two transitions from the levels to the respective bands are possible, as depicted in the energy level diagram of Fig. 3.10c. For the bipolaron, the sum of the threshold energies of the two transitions should equal the energy of the band gap: $\hbar\omega_1 + \hbar\omega_3 = E_g$.

In addition to the number of optical transitions and their position in energy, it is important to know the structure and intensity of the optical absorption bands. The optical absorption coefficients due to a charged soliton, a polaron and a bipolaron are a function of the ratio ε/Δ [4.63]. The results obtained for the SSH Hamiltonian are summarized in Fig. 4.12. The transition matrix elements and a possible broadening of the bands are critical in determining the detectability of the absorption bands. Thus, if a Gaussian broadening is assumed, a polaron may in practice yield only two bands and a bipolaron only one band, the others being too weak to be detected (Fig. 4.12).

b) Electron–Electron Interaction and Inter-Chain Coupling

Up to this point we have considered a single polyene chain using simple Hückel theory. It has been shown by many authors that three-dimensional effects and electron–electron interaction may substantially modify Hückel results. The stability of quasi-particles is not destroyed by weak and intermediate electron–electron interaction (Chap. 2). However, inter-chain coupling can easily destabilize polarons and bipolarons [4.64, 65]. Therefore, one has to keep in mind that the physical properties of highly crystalline polymers with sufficient inter-chain interaction might differ significantly from those of randomly oriented, fibrous polymers.

Electron–electron interactions have significant input on the optical transitions involving defect levels in the gap [4.66]. With electron–electron interaction, the threshold for optical absorption by neutral solitons is shifted to higher energies by $U/3\xi$, and that for charged solitons to lower energies by the same amount (ξ is the soliton width). These perturbative results are in qualitative agreement with calculations for finite polyenes which lead to an allowed transition [4.67–69] slightly below E_g for the Pariser–Parr–Pople model, as compared to $E_g/2$ for the SSH model.

For polarons the optical transition at $\hbar\omega_3$ is predicted to split into three bands [4.70]. For bipolarons, both absorption thresholds are shifted to lower energies. Probably not only the on-site Coulomb repulsion but also nearest neighbour and possibly longer range Coulomb forces must be taken into account for a correct computation of the transition energies.

In addition to the number of states and their position, electron–electron interaction may also significantly affect the intensities of the transitions [4.71]. An assignment of optical absorption bands to specific types of quasi-particles requires, therefore, some care and additional evidence for their existence should be provided from other experiments, e.g. from electron spin resonance measurements.

c) Photo-Induced Changes and Nonlinear Optical Properties

It has been shown by *Su* and *Schrieffer* [4.72] within the framework of the SSH model that by optical excitation charged solitons are formed in $(CH)_n$ within a time of about 10^{-13} s, which is about the inverse phonon frequency. Since the formation of solitons is connected with a change of the density of states in the conduction and valence band and the appearance of a level at midgap, it is expected that the photo-excitation leads to a bleaching at and above the band gap and to a new absorption feature below the band gap. Under these conditions, the complex dielectric function will depend sensitively on the light intensity if the concentration of solitons and the transition probabilities are high. If the lifetime of the photo-excited solitons is short, the dielectric function will also be significantly affected within this time domain.

If electron–electron interactions are taken into account, covalent excitations exist corresponding to a neutral soliton–antisoliton structure [4.69]. The transition energy of this neutral excitation relative to the optical gap depends on the strength of the Coulomb interaction. In strongly correlated systems the charged excitations have higher energies than the lowest neutral excitations. Since neutral excitations in $(CH)_n$ are not accessible to one-photon processes because of selection rules, two-photon absorption experiments should yield the desired information on the relative importance of the electron–electron and electron–phonon interaction.

Phenomenologically, nonlinearities are usually expressed in terms of the induced polarization which is expanded in a power series of the electric field:

$$P = \chi_{ij}^{(1)}(E_j) + \chi_{ijk}^{(2)}(E_j E_k) + \chi_{ijkl}^{(3)}(E_j E_k E_l). \tag{4.16}$$

In systems with inversion symmetry $\chi_{ijk}^{(2)}$ vanishes. For conducting polymers we have, therefore, to focus on the third-order susceptibility $\chi_{ijkl}^{(3)}$ which is related to the nonlinear refractive index n_2, defined as

$$n = n_1 + n_2 I \tag{4.17}$$

with I being the light intensity. The relation between the two coefficients is

$$n_2 = 16\pi^2 \chi^{(3)}/cn_1^2 \tag{4.18}$$

(or $n_2 = 5.26 \cdot 10^{-6} \chi^{(3)}/n_1^2$ for χ in esu and n_2 in cm^2/W).

Nonlinear effects have been extensively studied with the Hückel model both for polydiacetylenes [4.73] and polyacetylenes [4.74, 75]. However, in poly-diacetylenes the optical absorption in a wide range (typically about 0.5 eV) above its onset is generally attributed to excitonic effects [4.76], i.e. to electron–hole pairs bound by the Coulomb attraction. *Green* et al. [4.77] have given a simple physical interpretation of the nonlinear optical response for one-dimensional systems based on the theory developed for 3D and 2D systems [4.78]. As already mentioned, strong excitonic absorption lines are observed in polydiacetylenes below the single particle gap. In essence it is argued that the optically generated exciton population reduces the oscillator strength for the exciton transition since states forming exciton wavefunctions are no longer available. In one-dimensional systems the saturation density N_S for excitons is simply determined by the inverse of the exciton dimension in the chain direction. The exciton density under resonant and stationary conditions is given by

$$N = K(1 - R)I\tau_r/(hv), \tag{4.19}$$

where K is the absorption coefficient, R is the reflectivity, τ_r is the recombination time and I is the spectral light intensity in W/cm^2 at frequency v. Using the space-filling argument for excitons [4.77], the relative change in refractive index $\delta n/n$ is related to the relative exciton density N/N_S by

$$\delta n/n = - N/(2N_s). \tag{4.20}$$

Together with (4.17), one finds

$$\delta n/I = n_2 = (1 - R)K\tau_r/(2nN_Shv). \tag{4.21}$$

On the basis of pulsed experiments, *Green* et al. have determined an exciton extent of 33 Å in good agreement with available theoretical and experimental estimates [4.76]. The extrapolation to stationary conditions on the basis of (4.21) allows one to calculate n_2 at the exciton resonance if the recombination time τ_r is known. (Using such a procedure, Green obtained for $n_2 = 3 \times 10^{-8} \, cm^2/W$ for $hv = 1.97 \, eV$ in PTS).

The derivation of these formulae and also calculations of [4.73–75] are based on rigid band models and do not take into account lattice relaxations, which are expected to occur in conjugated polymers. *Sinclair* et al. [4.79] argued that in $(CH)_n$ photo-excitation produces solitons and no excitons. Fluctuations of the order parameter give rise to breathers and thus to sub-gap levels. If pumped into these levels, the excited states are claimed to evolve into charged, separated soliton pairs. Since solitons remove band states, the optical absorption is bleached, as with excitons. This again would lead to nonlinear optical co-efficients, although the lattice relaxation leading to solitons requires a much longer time than does the formation of excitons.

4.5.3 The Vibrational State of the Quasi-Particles

A soliton represents a domain wall across which the bond alternation sequence is reversed. This makes a chain with a soliton different from the ideal chain: firstly, the soliton breaks the inversion and translational symmetries of the chain; secondly, the bond lengths within the domain wall are significantly altered from their value outside the wall and, hence, also the force constants. In the case of charged solitons, delocalization of the charge increases the polarizability within the domain wall. These considerations hold for the neutral and charged soliton also, but only charged domain walls contribute significantly to the infrared conductivity due to the strong polarizability of the charged defect. Similar arguments hold for polarons and bipolarons. The question is then whether the local structure and charge distribution give rise to characteristic infrared active lines which allow one to distinguish the various quasi-particles. The continuum theory developed by Horovitz and used to explain the Raman data, can also be applied to infrared absorption induced by bond-alternation defects [4.80]. The result corresponds to that briefly outlined in Sect. 4.4.2, the optical conductivity being given by

$$\sigma(\omega) = i\omega(\rho/M\omega_0^2)D(\omega)/(1 + (1-p)D(\omega)), \tag{4.22}$$

where $D(\omega)$ is given by (4.15), ρ is the average charge density, M is the kinetic mass per unit charge, $\omega_{n,0}$ is the nth bare frequency, $\omega_0^{-2} = \Sigma(\lambda_n/\lambda)(\omega_{n0})^{-2}$ and p is the pinning parameter due to the Coulomb potential of the dopant ion. The poles of (4.22), i.e. the relation $D(\omega) = (p-1)^{-1}$, yield the frequencies of the additional infrared active modes. These modes are associated with the centre of mass motion of the charge and are independent of the particular structure of the defect. The only information about the quasi-particle causing the IR modes is contained in M. Thus, within the continuum theory, solitons, polarons and bipolarons all give rise to the same modes and a distinction between them is only possible by comparison of the value of M obtained from theory with that derived from experiment. In discrete models additional non-universal modes occur, which may allow one to distinguish the various particles (Chap. 2).

4.5.4 Experimental Results

a) Absorption Features of Weakly Doped Trans-Polyacetylene

Since trans-polyacetylene has a two-fold degenerate ground state, it is expected that doping proceeds via soliton formation, except at low doping concentrations at which polaron formation might preferentially occur according to the considerations outlined in Sect. 4.5.2a. However, the as-prepared material is known to contain about 10^{19} spins per cm^3, generally attributed to neutral solitons. Hence it is reasonable to assume that at first doping proceeds via removing or adding an electron to the neutral solitons, thus transforming them into charged solitons.

It has been found that upon doping new strong absorption bands appear at 0.7–0.8 eV, and in the infrared vibrational excitations at 900, 1280 and 1390 cm^{-1} (100–110, 160 and 175 meV) with a concomitant decrease of the intrinsic absorption band at about 2 eV (Fig. 4.13). These bands are independent of both the chemical nature and the charge of the dopant used. *Horovitz* [4.81] was the first to assign the band at 100 meV to the translational mode and the bands at 160 and 175 meV to the internal vibrations of the charged defect. In principle the translational mode should be at $\omega = 0$, therefore, the mode is argued to be shifted by pinning to 100–110 meV. Such a large shift can easily be explained in terms of the attractive Coulomb potential of the dopant ion [4.80].

Since there is no Coulomb interaction between dopants and neutral solitons, attempts were made to measure the absorption due to the relatively abundant neutral solitons (10^{19} per cm^3) in *trans*-(CH)$_n$. The striking result was that, despite the enormous sensitivity of the measuring technique used, no absorption band near midgap could be detected [4.82, 83]. It was concluded that the absorption threshold for the neutral soliton must be at energies above 1.3 eV.

The optical transition energy of the charged soliton with about 0.7–0.8 eV has approximately the value predicted by the SSH theory, in view of the uncertainty in the value of the gap and of the width of the midgap absorption. However, as will become clear below, this apparent agreement seems to be due to the compensation of two effects, namely the electron–electron interaction, which tends to shift the absorption line to lower energies, and the Coulomb attraction between the dopant and the localized electron charge, which moves it to a higher energy.

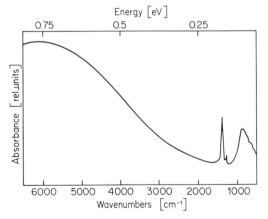

Fig. 4.13. Difference spectra between doped and undoped (CH)$_n$ sample. These spectra are independent of the dopant used

b) Photo-Induced Absorption in Polyacetylene

Computer simulations of the photo-excitation process show that electron–hole pairs decay into separated charged solitons and antisolitons within 10^{-13} s [4.72]. Since the solitonic states are supposed to appear at midgap accompanied by a decrease of the density of states in the bands, this prediction stimulated experiments to probe these states with light having energy below that of the exciting beam.

A summary of the results of photo- and doping-induced absorption is shown in Fig. 4.14. Doping gives rise to one band at 0.75 eV, assignable to an electronic transition, and to vibrational bands at 175 meV, 160 meV and about 110 meV. Upon illumination with bandgap light, two bands appear at 1.35 eV and 0.45 eV caused by electronic transitions, and vibrational bands at 175 meV, 160 meV and 60 meV [4.84–86]. None of the peaks induced by doping or photo-excitation and assignable to an electronic transition is precisely at midgap, so the identification of these absorption features was not straightforward.

The appearance of IR modes indicates that photo-generated charge carriers must be located on the chain. The lowest lying mode is significantly shifted from about 110 meV (900 cm^{-1}) to about 60 meV (500 cm^{-1}), indicating that the photo-induced carriers are less pinned than those generated by doping. The photo-excited carriers are, however, not expected to be as strongly bound as those induced by doping since the oppositely charged dopant is missing. It is surprising that the pinning frequency remains so large for photo-generated defects [4.87].

The relative change of photo-induced transmission $\Delta T/T$ and the phenomena which are related to it, depend strongly on the time domain of measurement

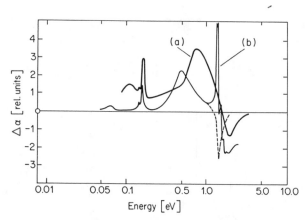

Fig. 4.14. Difference spectra induced by doping (a) and by illumination (b) with bandgap light. Spectrum (a) was taken at 300 K, spectrum (b) at 10 K. The dotted line represents the changes of the photo-induced absorption spectrum if measured at 300 K. At temperatures above about 200 K the absorption peak at ≈ 1.4 eV turns into a bleaching feature peaking at about the same energy

and temperature [4.88, 89]. We first discuss results for a delay of 10^{-2}–10^{-3} s. In this time domain the transmission change of the peak at 0.45 eV is about -10^{-5}–-10^{-4} for samples of about 0.1 μm thickness. The 0.45 eV peak and the IR modes vary with temperature and with the exciting light intensity in the same way, whereas the absorption peak at 1.35 eV behaves differently (it disappears above about 200 K). The IR modes and the absorption feature at 0.45 eV arise, therefore, from the same charged defect. The absorption band at 1.35 eV is independent of that at 0.45 eV and not coupled to the IR modes. It has therefore been assigned to neutral excitations.

If the absorption at 0.45 eV is due to solitons (or other defects), then theory predicts that it should be accompanied by bleaching in the intrinsic absorption region. Since the experimental results depend on both the time domain and the temperature of the measurements, this prediction of theory was difficult to verify. If measured after a delay of 10^{-2} s and above 200 K, the photo-induced absorption at 0.45 eV was accompanied by bleaching at 1.4 eV [4.90] (cf. Fig. 4.14). Therefore, bleaching does not occur in the intrinsic absorption region, but rather in the region characteristic for neutral solitons. It was concluded that in these experiments the charged defects are solitons obtained by a transfer of the photo-generated charges to neutral solitons [4.91].

If the 0.45 eV peak is assignable to solitons, then the position of the optical absorption band is not in agreement with SSH theory, which places the optical transition to e.g. 0.7–1.0 eV. Electron–electron interaction has been considered as a possible reason for the shift of the transition by about 0.4 eV [4.92] (see also Chap. 2). Based on this assumption, an estimate of the on-site electron–electron interaction can be given; the on-site Hubbard U was found to be of the order of 10 eV. However, since the result is based on perturbation theory requiring that the electron–electron interaction is small compared to the electron–phonon interaction, the large value of U is actually in conflict with the assumptions of the theory used. Further calculations have to be performed to allow an exact determination of the interaction energy. The difference in optical transition energies between the charged solitons induced by photo-excitation and doping is assumed to be due to the pinning of the solitons to dopant ions. Thus, the position of the absorption peak of charged solitons is determined by the electron–phonon, the repulsive electron–electron interaction and, if applicable, the attractive Coulomb attraction between dopant and charged soliton.

A different explanation of results obtainable by photo-induced absorption is given by *Flood* and *Heeger* [4.93] for their samples. Their interpretation is based on ESR experiments under photo-excitation. No decrease of the spin density was observed, which would have been expected if photo-excited carriers are trapped by neutral solitons. Trapping was, therefore, assumed not to occur and the absorption peaks at 0.4 eV and 1.4 eV were assigned to direct photo-generated solitons. According to [4.93], the destiny of the photo-excited species depends strongly on the sample quality. ESR experiments under illumination are extremely difficult and, hence, the results and their interpretation remain controversial [4.94, 95].

We turn now to experiments on photo-induced absorption in the picosecond time domain. It was found that the absorption band at 0.45 eV appears with a relative change in transmission of $\Delta T/T$ of about -6×10^{-2} [4.96]. This change is thus about two to three orders in magnitude larger than after a delay of about 10^{-2} s. In addition, the absorption at 0.45 eV is accompanied by an increase of the absorption at 1.4 eV with relative change of about 0.1 and by bleaching in the intrinsic absorption region [4.97]. Measurements of the IR modes after a delay of 10^{-12} s do not exist. Nevertheless, the peak at 0.45 eV observed in these experiments is most likely also due to the same charged defect found in the time domain of 10^{-2} s.

The interpretation of the 1.4 eV absorption peak (not bleaching) is still not clear. In view of the high change in transmission this absorption must be due to an intrinsic excitation of the chain. As discussed by *Bishop* et al. [4.98–100], the transition at 1.4 eV could be the manifestation of a breather which forms due to the energy release of the decay of an electron–hole pair into a soliton–antisoliton pair. The total energy of the moving soliton and antisoliton was found to be $1.3\,\Delta$, leaving about $0.7\,\Delta$ in vibrational energy on the chain, which gives rise to a well-localized oscillatory breather. The lattice displacement in turn causes fluctuations of the bandgap. Thus breathers could explain the 1.4 eV absorption band, though it is not clear whether the intensity of the absorption band at 0.45 eV, assigned to solitons, and that of the absorption at 1.4 eV can be explained consistently in terms of photo-generated solitons.

Models involving electron–electron interactions have also been proposed. These interactions shift the $2^1 A_g$ state, which can be viewed as a state with two alternation reversals or two neutral solitons, below the $1^1 B_u$ state in polyenes [4.101]. The $2^1 A_g$ state can be populated either by a two-photon absorption or by a dipole-allowed absorption into the $1^1 B_u$ state followed by a rapid relaxation into the $2^1 A_g$ state [4.102]. The photo-induced absorption occurs then from the $2^1 A_g$ state into some higher state to which the transition is allowed by symmetry. In octatetraene the lifetime of the $2^1 A_g$ state was found to decrease from 10^{-7} s at 10 K to 5×10^{-9} s at 300 K. If these results obtained for molecules of relatively short chain length can be transferred to *trans*-$(CH)_n$ with finite chain segments, then the photo-induced absorption peak at 1.4 eV would have to be traced back to photo-generated neutral solitons. A similar picture has been proposed by *Su* [4.103].

c) Nonlinear Optical Properties

The nonlinear optical coefficients of conducting polymers are among the highest that have been measured. A summary of the coefficents for the third harmonic generation $\chi^{(3)}(-3\omega, \omega, \omega, \omega)$ is given in Table 4.1. From the point of view of application, $\chi^{(3)}$ should have values $> 10^{-9}$ esu and, in addition, the attenuation losses of the materials should be less than 1 dB/cm [4.104]. The poly(diacetylenes)

Table 4.1

Material	$\chi^{(3)}(-3\omega, \omega, \omega, \omega)$	Wavelength
	esu	μm
Silica	2.4×10^{-14}	1.907
Ge	4.0×10^{-10}	10.6
GaAs	4.0×10^{-10}	10.6
Polythiophene	3.5×10^{-10}	1.907
trans-(CH)$_n$	4.0×10^{-10}	1.060
PDA-TCDU*	7.0×10^{-11}	1.890
PDA-PTS*	8.5×10^{-10}	1.890

* Polydiacetylenes with different sidegroups

are promising materials for optical processing, since they fulfil both requirements [4.105].

The nonlinear optical properties can also give information on the relative strength of the electron–electron and electron–phonon interaction. As has been discussed, the transition from the ground state 1^1A_g to 1^1B_u states is dipole-allowed, however, that to the 2^1A_g state is optically forbidden. Energetically the forbidden transition is below the allowed transition provided that the system is strongly correlated. The forbidden transition can be accessed by two-photon processes so that its energetic position can be determined under high excitation levels. From the relative positions of the two states, information on the strength of the electron–electron interaction can be derived. The experimental curves for the third harmonic generation $\chi^{(3)} = f(\omega)$ for (CH)$_n$ [4.106] show a peak at 0.91 eV which is not visible in $\chi^{(1)}$. In general, $\chi^{(3)}$ shows one- and three-photon resonance enhancement, but two-photon enhancement if states of the same symmetry are coupled. The 0.91 eV peak is, therefore, assigned to a two-photon resonance enhancement, the two-photon accessible 2^1A_g state being located at 1.82 eV. Since this energy is close to the single-particle gap the electron–electron interaction energy U is concluded to be smaller in (CH)$_n$ than in finite polyenes [4.75]. (For a further discussion of this subject see Chap. 2).

d) Polymers with Nondegenerate Ground State

Practically all conjugated conducting polymers except trans-(CH)$_n$ are polymers with a nondegenerate ground state. Most experimental work has been done with systems having aromatic or heterocyclic monomer units. Optical absorption spectra of these polymers have been investigated as a function of the doping level or under illumination, with the intention of exploring whether polarons or bipolarons are formed in these materials.

Optical spectra of ClO_4^--doped polypyrrole at low stages of oxidation show three absorption peaks at 0.7, 1.4 and 2.1 eV below the inter-band $\pi–\pi^*$ transition at 3.2 eV [4.107]. With increasing oxidation the peak at 1.4 eV disappears, and the other two are slightly shifted. At low doping levels the rule for transitions between polaronic states $\hbar\omega_1 + \hbar\omega_2 = \hbar\omega_3$ and $\hbar\omega_1 + \hbar\omega_3 = E_G$ is fulfilled for the three peaks, and since the ESR intensity also varies correspondingly with doping, these absorption features were assigned to polarons [4.107]. At higher doping both the 1.4 eV absorption and the ESR signal disappear, indicating the formation of bipolarons [4.108]. The calculations of *Brédas* et al. [4.62] position the electronic levels of the quasi-particles asymmetrically to midgap. There is no experimental evidence for such an asymmetry.

Polythiophene has also been investigated by various groups [4.109–111]. Two absorption bands are induced at 0.8 eV and 1.8 eV for doping levels above 2 mol% independent of the dopant used. The sum rule $\hbar\omega_1 + \hbar\omega_3 = E_G$ is fulfilled. The two transitions have been consequently assigned to bipolarons. *Chung* et al. [4.110] have deduced a confinement parameter of 0.1–0.2 which they found to be high enough to exclude polaron formation. However, at lower levels of BF_4^- and ClO_4^- doping, a third absorption feature appears at 1.2 eV which remains visible as a broad band even at the lowest dopant concentrations, at which the bands at 0.8 and 1.8 eV are no longer visible. The absorption band might be due to polarons; however, the band is too broad to draw a definite conclusion.

Poly(3-methylthiophene) also exhibits two absorption bands upon doping with BF_4^- above 2 mol% at 0.5 eV and 1.7 eV [4.111, 112]. The methyl group probably influences the π-electron system and shifts the energy levels with respect to those of poly(thiophene). Three bands become visible at lower concentrations of BF_4^- (Fig. 4.15). For the three bands and the two bands, the sum rule for polarons and bipolarons is fulfilled respectively. It has therefore been suggested that, similarly to poly(pyrrole), polarons would give rise to the bands at low

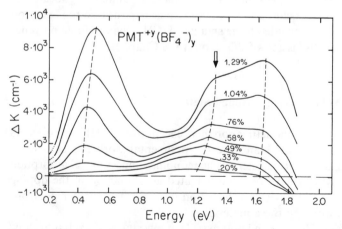

Fig. 4.15. Doping-induced difference spectra of BF_4^--doped poly(3-methylthiophene). The doping concentrations in mol% are given in the figure (after [4.111])

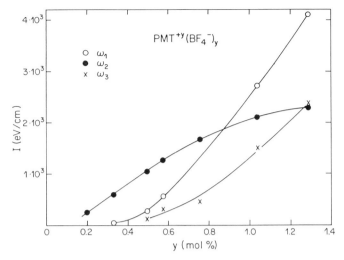

Fig. 4.16. Integrated intensity of the transition at 1.2 eV (●), 1.6 eV (x), and at ≈0.4 eV (○) as a function of doping concentration

doping levels and bipolarons to those at higher levels. However, the intensity of the bands does not increase as expected with dopant concentration: at low dopant levels the concentration of polarons (and hence also the intensity of the absorption bands) should increase linearly and that of bipolarons quadratically with dopant concentration. However, one observes that below approximately 0.8% doping (Fig. 4.16), only the intensity of the 1.2 eV band increases linearly, whereas that of the 0.5 and 1.7 eV bands increases quadratically with doping. At higher concentrations the 1.2 eV band saturates and only the two other bands continue to grow. ESR measurements indicate a linear increase of spin concentration with doping [4.113] over the whole concentration range, so that the formation of spinless bipolarons can be doubted. The dopant-induced absorption bands would then have to be due to polarons though the dependence of the intensity of the bands on the dopant concentration would then be in disagreement with such a conclusion.

The absorption bands at 0.8 and 1.8 eV induced by doping in polythiophene are accompanied by four intense vibrational peaks at 1053, 1125, 1205 and 1340 cm^{-1} (corresponding to 130.6, 139.5, 149.4 and 166.2 meV respectively) and four weak peaks at 790, 730, 680, and 640 cm^{-1} (corresponding to 98, 90.5, 84.3 and 79.4 meV respectively) [4.114–116]. The chemical structure of the quasi-particles in heterocyclic and aromatic polymers suggests a relation of these bands to the quinoidal rings (Fig. 4.11). The strong peaks between 1053 and 1340 cm^{-1} are in the range of the stretch frequencies of C–C, C=C and C–S bonds, the others may be of the ring-bending type. A broad absorption band peaking at 370 cm^{-1} (45.9 meV) has been interpreted as an impurity-pinned translational mode of the quasi-particles [4.114].

Photo-induced absorption measurements in poly(thiophene) show two absorption bands at 0.45 eV and 1.25 eV, and four strong and four weak infra-red bands [4.117, 118]. The optical transitions at 0.45 eV and 1.25 eV are significantly shifted with respect to the bands at 0.8 eV and 1.8 eV, obtained by doping. Since only two peaks have been observed, they are assigned to photo-generated bipolarons. The shift of the photo-induced bipolaron bands is explained by the missing impurity binding energy. One notices that, in contrast to the doped case, the sum rule $\hbar\omega_1 + \hbar\omega_3 = E_G$ is no longer fulfilled. This can be attributed to electron–electron interactions, as for $(CH)_n$.

4.6 Highly Conducting Conjugated Polymers

4.6.1 General Considerations

The light doping regime of the polymers has been characterized by sufficient dilution of the dopant so that interaction between quasi-particles and/or dopants can be neglected. At higher doping concentration this will be no longer true and the question arises as to whether agglomerates become observable, as in silicon [4.119]. There an increase of the dopant concentration gives rise to optical absorption lines which can be successively attributed to single impurities, dimers, trimers, etc. Finally, an impurity band will become observable, concomitant with an insulator-metal transition. The dc electrical conductivity at $T = 0$ K would then be unmeasurably small in the insulator range below, and metallic above the transition.

Above the insulator–metal transition, the classical model of Drude for free charge carriers is commonly used to describe the optical behaviour. The dynamical conductivity $\sigma(\omega)$ is then given by

$$\sigma = \sigma(0)/(1 + \omega^2\tau^2) + i\sigma(0)\omega\tau/(1 + \omega^2\tau^2) \tag{4.23}$$

where $\sigma(0)$ is the dc conductivity and τ the relaxation time. The scattering in this formula is usually assumed to be independent of frequency, though it is in reality better described by a frequency-dependent relaxation time $\tau(\omega)$ [4.120]. For comparison with optical data, it is often convenient to express $\sigma(\omega)$ in terms of reflectivity $R(\omega)$ and/or the absorption coefficient $K(\omega)$, being optical data which are readily obtained from measurements. The result of these elementary calculations is depicted in Fig. 4.17. Optical absorption bands, e.g. those caused by a resonator such as a bound electron or phonon, would have to be superimposed on the absorption and reflectivity curves of Fig. 4.17. Obviously, if the oscillator strength of the resonator is weak it would be invisible or barely visible. The electron gas in the Drude model gives rise to a characteristic, sharp rise in reflectance at the plasma frequency ω_p; below ω_p the reflectivity approaches the value of unity. The absorption coefficient K is proportional to $\sqrt{\omega}$ for small ω, goes through a plateau and drops to small values above the plasma frequency ω_p.

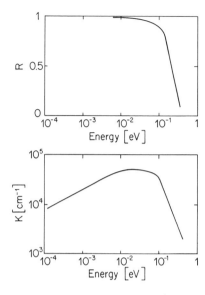

Fig. 4.17. Reflectivity and absorption coefficient as a function of energy for free electrons

4.6.2 The Highly Conducting Phase of Trans-Polyacetylene

In order to obtain a better understanding of the optical properties of $(CH)_n$ at intermediate and high doping concentrations, the results on electrical and structural properties will be briefly outlined. The electrical behaviour of trans-polyacetylene is characterized by two abrupt changes [4.121] (see also Chap. 3), the first at approximately 0.01–0.1 mol% doping level at which the conductivity rises abruptly, concomitant with a corresponding drop of the thermopower and a decrease of the Curie susceptibility to nearly unmeasurably small values. The second change occurs at 5–6% doping concentration, at which the Pauli susceptibility rises abruptly from negligibly small values to values close to those of metals. This transition occurs at about the dopant concentration at which the further transfer of charge to the chain cannot occur under complete relaxation of the lattice since the width of the soliton to be created would be larger than the space left for its accommodation on the chain. If charge transfer to the chains continues to take place, various evolutions can be envisaged: The system may undergo a continuous cross-over from a soliton lattice to a simple incommensurate Peierls distortion. It has also been proposed that a polaron lattice would be energetically favoured over a soliton lattice at higher concentrations, where a first order transition would occur [4.122]. Alternatively, the Coulomb potential of randomly distributed ions can give rise to a redistribution of the states and suppress the bond alternation, so that the Fermi energy becomes located in a continuum of states [4.123].

Above $\approx 1\%$ doping, structural investigations often indicate the appearance of a superstructure and intercalation of the dopant. An investigations of the structure of Na-doped $(CH)_n$ shows evidence for the coexistence of ordered and disordered phases in the doping range between 0.4 and 6% [4.124, 125]. The ordered phase contains 6.6%, the disordered phase 0.4% Na. Thus, the ordered phase grows at the expense of the disordered phase with increasing dopant concentration. It has been argued that a soliton lattice is formed in the ordered regions. The soliton lattice would manifest itself in a soliton band and would allow optical transitions from the soliton band to the conduction band with a first peak at $0.64 \, \Delta_0$ and a second peak at $1.58 \, \Delta_0$ [4.126]. The positions of the peaks are shown in Fig. 4.18. For a two-phase system, the optical absorption would be a superposition of the absorption due to the volume fraction of the region with the soliton lattice and the remaining volume fraction of the weakly doped regions. Thus, Maxwell Garnett's or Bruggeman's theory on optical properties of composite material would be applicable with appropriate modification. Above 6.6% doping concentration, the aforementioned restriction for the accommodation of solitons would apply and the formation of a new phase is to be expected. Since the soliton lattice gives rise to an insulating state, the question arises of how to account for the continued increase of conductivity in the doping range in which the insulating, solitonic phase grows at the expense of the other phase.

The question of the homogeneity of doping has been repeatedly discussed. It seems that the nature of the dopant and/or the different doping procedures influence the order or disorder finally obtained in the conducting polymer. Thus a phase separation, as discussed before in the case of $(CH)_n$, has not been observed for dopants other than Na and some other alkali metals. For iodine-doped $(CH)_n$, a superstructure found at higher doping might be also suggestive for phase separation. In other cases the dopant molecules are incorporated statistically and disorder is assumed to play a major role. Materials with strong disorder become gapless (at high doping) with a finite but reduced density of

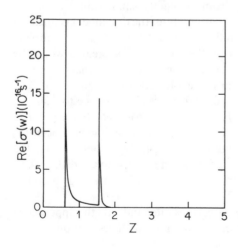

Fig. 4.18. Optical conductivity of a soliton lattice. The two singularities are due to transitions from the top and bottom of the soliton band to the bottom of the conduction band. Z is the normalized energy $Z = \hbar\omega/\Delta_0$ (after [4.126])

states at the position of the Fermi level [4.123]. There is certainly no unique and universal behaviour for all the various systems and each case should be analyzed separately.

4.6.3 Polyacetylene: Experimental Results

Results of absorbance measurements [4.127–129] in the range between 0.1–4.4 eV for polyacetylene doped with Na, I_2, SbF_5 and AsF_5 are depicted in Fig. 4.19. Na is a donor and is incorporated into polyacetylene in separate phases in the doping range shown. I_2, SbF_5 and AsF_5 form acceptors and are less likely to induce phase separation in $(CH)_n$. With Na and AsF_5 as dopants the band at 0.7 eV, associated with the formation of solitons at low concentrations, remains at its position, whereas with increasing dopant level of I_2 and of SbF_5, it shifts to a value of about 0.5 eV. The existence of a peak in absorption indicates the presence of a gap at this energy or a peak in the joint density of states. The value of the extrapolated gap is in astonishingly good agreement with calculations which put the gap between the soliton band and conduction band to 0.45 eV [4.126]. However, at these highest concentrations the material shows a Pauli susceptibility, independent of the dopant. The question remains, therefore, of whether a polaron lattice has formed, the peak in absorption being due to a transition between the two sub-bands [4.122], or whether the Fermi level is

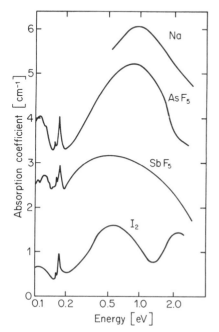

Fig. 4.19. Optical absorption coefficient of $(CH)_n$ doped to saturation for the dopants indicated in the figure. Dopant concentrations are $[I_3^-] = 9.3\%$, $[AsF_6^-] \approx 20\%$, $[Na^+] = 14.6\%$. The absorption coefficients of the broad peaks at 0.5–1.0 eV are irrespective of the dopant of the order of 1.5 to $1.8 \times 10^4\,mm^{-1}$, the curves for SbF_5, AsF_5 and Na being vertically offset in this figure

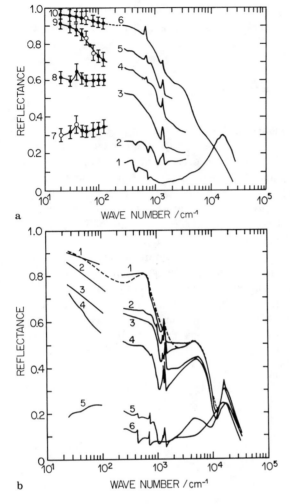

Fig. 4.20 a, b. Reflectance data of (a) AsF_5 and (b) I_2 doped $(CH)_n$. The dopant level is not specified for the AsF_5-doped sample; the conductivity of sample #6 in (a) is 10^5 S/m. The $(CHI_y)_n$ sample #1 has a concentration $y = 0.25$ ($\sigma = 7 \times 10^4$ S/cm); #2, $y = 0.214$ ($\sigma = 2 \times 10^4$ S/cm); #3, $y = 0.139$ ($\sigma = 6.6 \times 10^3$ S/cm); #4, $y = 0.10$ ($\sigma = 6 \times 10^3$ S/m); #5, $y = 0.024$ ($\sigma = 300$ S/m) (after [4.128, 131])

positioned in a continuum of states which formed between the bands due to disorder. Clearly, the onset of a Drude absorption is not visible in this range and the question arises as to whether it is visible at still smaller energies.

Reflectivity measurements performed between 20 and 3×10^4 cm^{-1} (2.4×10^{-3} eV and 3.7 eV) also show distinct differences between iodine- and AsF_5-doped $(CH)_n$: the peak at high doping levels observable for iodine-doped $(CH)_n$ at 1.6×10^4 cm^{-1} (2 eV) is absent in AsF_5-doped material (Fig. 4.20). This peak at high concentration has been argued either to be due to absorption of the incorporated I_5^- or caused by a $\pi-\pi^*$ transition [4.130]. Since the transition is barely shifted with respect to pristine $(CH)_n$, the interpretation in terms of a $\pi-\pi^*$ transition seems to be more attractive. The pronounced shoulder at 5.2×10^3 cm^{-1} (0.65 eV) of the iodine system is not visible for the AsF_5 doped $(CH)_n$ and instead a shoulder becomes observable at approximately $3.2 \times 10^3 \, cm^{-1}$ (0.4 eV). A second pronounced shoulder of the iodine-doped material at $800 \, cm^{-1}$ (0.1 eV) is shifted to $400 \, cm^{-1}$ (0.05 eV) for AsF_5 doping. Finally, a peak in reflectivity at 725–$760 \, cm^{-1}$ (0.09–0.095 eV), easily visible with AsF_5 and not discernible with iodine, is due to the dopant molecule. Surprisingly for both dopants, the vibrational structure at $1370 \, cm^{-1}$ (0.17 eV) persists up to the highest conductivity level of about 7×10^4–10^5 S/m. In a recent paper *Tanaka* and coworkers [4.131] have made similar investigations on highly oriented $(CH)_n$ and performed a Kramers–Kronig transformation of the reflection spectra. The shoulders in reflectivity appear as well-defined peaks in $\sigma(\omega)$ in addition to a clear Drude contribution below approximately $300 \, cm^{-1}$ (0.037 eV) for $(CHI_y)_n$ with $y > 0.1$. The origin of the peaks remains to be clarified. In unoriented $(CH)_n$, the reflectivity in the far infrared and at intermediate doping concentrations levels off at rather low values of about 50–60%. Only at the highest doping concentration achievable does the reflectivity increase to 95%. However, with highly oriented material the reflectivity of light polarized in the chain direction approaches 1 for $\omega \to 0$ even at lower concentrations [4.131]; at the same time it remains low for the other orientation. The result suggests that well-oriented $(CH)_n$ shows metallic behaviour in the chain direction, while it remains semiconducting perpendicular to the chains.

4.6.4 Highly Conducting Polymers with Nondegenerate Ground State

In this section we will discuss the results of optical measurements of polypyrrole/perchlorate (PPY/ClO_4^-) [4.132, 133] and poly(3-methylthiophene)/hexafluorophosphate (PMT/PF_6^-) [4.134]. These two materials seem to be prototypes of polymers with nondegenerate ground state in that PPY/ClO_4^- shows no paramagnetism in the highly conducting state [4.135], whereas highly doped PMT does show Pauli susceptibility [4.113]. Theory predicts that bipolarons form at intermediate doping levels in PPY/ClO_4^-. The bipolaron states are located asymmetrically to midgap and at the highest doping level of 0.33 mol%, bipolaron bands are formed. In the case of polythiophene, bipolaron states at intermediate doping levels are also predicted to be positioned asymmetrically in the gap and to widen to bands at the highest doping concentrations [4.17]. The theoretical results for polythiophene are not necessarily valid for PMT since the methyl group supposedly influences the electronic states of the heterocycle.

Reflectivity data of PPY/ClO_4^- and PMT/PF_6^- films having comparable electrical conductivity of the order 2×10^4–4×10^4 S/m show several remarkable features in their spectra: A high reflectance of about 90% in the far infrared in the range between 20 and $100 \, cm^{-1}$ (2.5×10^{-3} eV and 1.24×10^{-2} eV), a wealth of lines in the region of the vibrational spectra of both polymers, and finally the existance of several shoulders in the reflectance between 10^2 and $10^4 \, cm^{-1}$ (1.24×10^{-2} eV and 1.24 eV). For discussion it is advantageous to present the

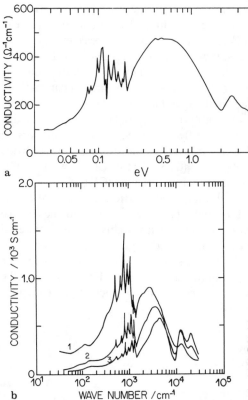

Fig. 4.21 a, b. Optical conductivity of as grown samples of poly-pyrrole perchlorate (**a**) and of poly(3-methyl-thiophene) hexafluorophosphate (PMT(PF$_6$)$_y$) (**b**). Doping levels of PMT(PF$_6$)$_y$ are $y = 0.29$ (1), $y = 0.16$ (2) and $y = 0.12$ (3) (after [4.133, 134])

data in terms of optical conductivity, which has been obtained by Kramers–Kronig analysis from the experimental data, and in the case of PPY/ClO$_4^-$ from transmittance data covering the same spectral range as that of the above-presented reflectance measurements. These results are shown in Fig. 4.21. The optical conductivity peaks at 800, 4000 and 2.17×10^4 cm^{-1} (0.1, 0.5 and 2.7 eV) in the case of PPY/ClO$_4^-$ and 725, 3225 and 1.29×10^4 cm^{-1} (0.09, 0.4 and 1.6 eV) for PMT/PF$_6^-$. Below 800 cm^{-1} (0.1 eV) the conductivity corresponds to the dc conductivity. The data have been analyzed in terms of simple harmonic oscillators. It was concluded that free carriers determine the conductivity in the far infrared. The free carriers were speculatively identified with bipolarons, the peaks at 725 cm^{-1} (0.09 eV) for PMT/PF$_6^-$ and 800 cm^{-1} (0.1 eV) for PPY/ClO$_4^-$ representing the pinned mode of the quasi-particles. The maxima at higher energies were not identified. The peaks in optical conductivity of PPY/ClO$_4^-$ at 4000 and 2.17×10^4 cm^{-1} (0.5 and 2.75 eV) could be due to transitions between the bipolaron bands, and the shoulder at 2.9×10^{-4} cm^{-1} (3.6 eV) due to an interband transition. Since ESR measurements in PMT indicate at high doping concentration the presence of paramagnetism, the optical transitions at 3225 cm^{-1} (0.4 eV) and 1.29×10^4 cm^{-1} (1.6 eV) can be assigned to polaron bands. The fact that for BF$_4^-$ doping the peak at 1.6 eV disappears and only one transition remains observable at the highest achievable dopant concentrations [4.136, 17], can be rationalized by assuming that the polaron bands merge with the π band. This would simultaneously explain the disappearance of the inter-band transition. Further investigations are, however, required to substantiate such an interpretation.

4.6.5 Concluding Remarks

Optical transmission and reflectance measurements, and electron energy loss spectroscopy have proven to be powerful tools for the investigation of the physical properties of polyconjugated materials. Results obtained in the intrinsic absorption region of the polymers gave information on the band structure and anisotropy of the polymers. Though the measurements allowed determination of the numerical values of the gaps, their origin remained unclear. Raman measurements and results of defect-induced changes of the IR spectra allowed us, however, to draw conclusions on the relative importance of electron–phonon and electron–electron interactions and thus on the origin of the gap. First estimates of the parameters determining the strength of these interactions could be derived, though the discussion is still on-going.

Recently, research has focused on the highly photo-excited state. Experiments showed that third-order nonlinear susceptibilities are observed which are comparable to or even higher than those of other organic or inorganic materials. The underlying physics of the strong nonlinear optical coefficients is not quite clear. The nonlinear optical coefficients of PDA–PTS and $(CH)_n$ are of similar magnitude, which is probably related to the fact that both materials are one dimensional and π conjugated with about the same band gap.

Extension of the reflectivity measurements to the far infra-red on highly conducting materials gives additional insight into the problem of the metallic state. Drude behaviour indicates that free charge carriers are generated in the highly doped states in contrast to quasi-particles (i.e. solitons, polarons or bipolarons) which are observed at low doping levels. The transition to the metallic state is nevertheless not really understood theoretically, presenting a challenge for future work in this field.

Overall, the field has matured and new impetus in the exploration of the properties is expected to come from improved materials and their possible applications in nonlinear optics.

Acknowledgements

It is a pleasure to thank D. Baeriswyl for his continued interest in this work and in particular for his advice and suggestions for improving the manuscript. We are also indebted to G. Zerbi for carefully reading the text.

References

4.1.   See e.g. D.L. Greenaway, G. Harbeke: *Optical Properties and Bandstructure of Semiconductors* (Pergamon, Oxford, 1968)
4.2.   H. Raether: *Excitations of Plasmons and Interband Transitions* Springer Tracts in Modern Physics, Vol. 28 (Springer, Berlin, Heidelberg 1980)
4.3.   D.S. Chemla, D.A. Miller: J. Opt. Soc. Am **B2**, 1155 (1985)
4.4.   H. Haug, S. Schmitt-Rink: J. Opt. Soc. Am **B2**, 1135 (1985)
4.5.   W.R. Su, J.R. Schrieffer, A.J. Heeger: Phys. Rev **B22**, 2099 (1980)
4.6.   D. Baeriswyl, G. Harbeke, H. Kiess, E. Meier, W. Meyer: Physica **117B & 118B**, 617 (1983)
4.7.   D. Baeriswyl: In *Theoretical Aspects of Bandstructure and Electronic Properties of Pseudo-One-Dimensional Solids*, ed. by R.H. Kamimura (Reidel, Dordrecht 1985)
4.8.   A.A. Ovchinnikov, I.I. Ukrainskii, G.V. Kventsel: (Sov. Phys.—Usp. **15**, 575 (1973)
4.9.   P. Horsch: Phys. Rev. **B24**, 7351 (1981)
4.10.   S. Kivelson, D.E. Heine: Phys. Rev. **B26**, 4278 (1982)
4.11.   D. Baeriswyl, K. Maki: Phys. Rev. **B31**, 6633 (1985)
4.12.   S. Kivelson, W.R. Su, J.R. Schrieffer, A.J. Heeger: Phys. Rev. Lett. 58, 1899 (1987)
4.13.   P.M. Grant, I.P. Batra: J. de Phys. **44**, C3-437 (1983)
4.14.   P.M. Grant, I.P. Batra: Sol. State Commun. **29**, 225 (1979)
4.15.   J. Ashkenazi, E. Ehrenfreund, Z. Vardeny, O. Brafman: Mol. Cryst. Liq. Cryst. **117**, 193 (1985)
4.16.   P. Vogl, D.K. Campbell, O.F. Sanhey: Synth. Met. **28**, D 513 (1989)
4.17.   J.L. Brédas, B. Thémans, J.G. Fripiat, J.M. André: Phys. Rev. **B29**, 6761 (1984)
4.18.   W.R. Salaneck: In *Handbook of Conducting Polymers*, ed. by T.A. Skotheim (Dekker, New York 1986) p. 1337
4.19.   W.K. Ford, C.B. Duke, W.R. Salaneck: J. Chem. Phys. **77**, 5030, 1982
4.20.   M.L. Elert, C.T. White, J.W. Mintmire: Mol. Cryst. Liq. Cryst. **125**, 329 (1985)
4.21.   M. Kertesz, J. Koller, A. Azman: J. Chem. Phys. **67**, 1180 (1977)
4.22.   M. Kertesz: Chem. Phys. **44**, 348 (1979)
4.23.   S. Suhai: Phys. Rev. **B27**, 3506 (1983)
4.24.   T. Woerner, A.G. MacDiarmid, A. Feldblum, A.J. Heeger: J. Polym. Sci., Polym. Lett. Ed. **22**, 119 (1984)
4.25.   G. Leising, R. Uitz, B. Ankele, W. Ottinger, F. Stelzer: Mol. Cryst. Liq. Cryst. **117**, 327 (1985)
4.26.   G. Leising, Phys. Rev. **B38**, 10313 (1988)
4.27.   J. Fink, G. Leising, Phys. Rev. **B34**, 5320 (1986)
4.28.   T. Mitani, S. Suga, Y. Tokewa, H. Koyama, I. Nakada, T. Koda: Int. J. of Quant. Chem. **18**, 655 (1980)
4.29.   J.J. Ritsko: Phys. Rev. **B26**, 2192 (1982)
4.30.   See ref. [4.27]
4.31.   C.R. Fincher jr., M. Ozaki, M. Tanaka, D. Peebles, L. Lauchlan, A.J. Heeger, A.G. MacDiarmid: Phys. Rev. **B20**, 1589 (1979)
4.32.   H. Fujimoto, K. Kamiya, J. Tanaka, M. Tanaka: Synth. Met. **10**, 367 (1985)
4.33.   J. Fink, B. Scheerer, W. Wernet, M. Monkenbusch, G. Wegner, H.-J. Freund, H. Gonska: Phys. Rev. **B34**, 1101 (1986)
4.34.   J. Fink, N. Nuecker, B. Scheerer, H. Neugebauer: Synth. Met. **18**, 163 (1987)
4.35.   H. Fark, J. Fink, B. Scheerer, M. Stamm, B. Tieke: Synth. Met. **17**, 583 (1987)
4.36.   E.G. Mele, M. Rice: Solid St. Commun. **34**, 339 (1980)
4.37.   G. Zannoni, G. Zerbi: J. of Mol. Structure **100**, 485 (1983)
4.38.   G. Zannoni, G. Gerbi: Solid St. Commun. **47**, 213 (1983)
4.39.   G. Zerbi, G. Zannoni, M. Gussoni, C. Castiglioni: Mol. Cryst. Liq. Cryst. **117**, 287 (1985)
4.40.   D. Rakovic, I. Bozovic, S.A. Stepanyan, L.A. Gribov: Phys. Rev. **B28**, 1997 (1983)
4.41.   D. Rakovic, I. Bozovic, S.A. Stepanyan, V.A. Dementiev: Phys. Rev. **B29**, 3412 (1984)
4.42.   D. Rakovic, I. Bozovic, S.A. Stepanyan, L.A. Gribov: Solid. St. Commun. **43**, 127 (1982)
4.43.   B. Horovitz: Sol. State Commun.: **41**, 729 (1989)
4.44.   B. Horovitz, Z. Vardeny, E. Ehrenfreund, O. Brafman: Synth. Met. **9**, 215 (1984)
4.45.   B. Horovitz, Z. Vardeny, E. Ehrenfreund, O. Brafman: J. Phys. C, Sol. State Phys. **19**, 7291 (1986)
4.46.   C. Castiglioni, J.T. Lopez Navarrete, G. Zerbi, M. Gussoni: Sol. State Commun. **65**, 625 (1988)
4.47.   M. Gussoni, C. Castiglioni, G. Zerbi: Synth. Met. **28**, D 375 (1989)

4.48. G. Zerbi, C. Castiglioni, J.T. Lopez Navarrete, T. Bogang, M. Gussoni: Synth. Met. **28**, D 359 (1989)
4.49. E. Mulazzi, G.P. Brivio, R. Tiziani: Mol. Cryst. Liq. Cryst. **117**, 343 (1985)
4.50. Z. Vardeny, O. Brafman, E. Ehrenfreund: Mol. Cryst. Liq. Cryst. **117**, 355 (1989)
4.51. E. Ehrenfreund, Z. Vardeny, O. Brafman, B. Horovitz: Mol. Cryst. Liq. Cryst. **117**, 367 (1989)
4.52. S. Haason, M. Galtier, J.L.-Sanvajol, J.B. Lère Porte, A. Banniol, B. Moukala: Synth. Met. **28**, C 317 (1989)
4.53. E.G. Mele: Solid St. Commun. **44**, 827 (1982)
4.54. H. Kuzmany, P. Knoll: Mol. Cryst. Liq. Cryst. **117**, 385 (1985)
4.55. E. Mulazzi, G.P. Bivio, S. Lefrant, E. Faülques, E. Perrin: Synth. Met. **17**, 325 (1987)
4.56. E. Mulazzi, G.P. Bivio, R. Tiziani: Mol. Cryst. Liq. Cryst. **117**, 343 (1985)
4.57. I. Harada, Y. Furukawa, M. Tsumi, H. Shirakawa, S. Ikeda: J. Chem. Phys. **73**, 4746 (1980)
4.58. E. Ehrenfreund, Z. Vardeny, O. Brafman, B. Horovitz: Phys. Rev. **B36**, 1535 (1987)
4.59. Cao Yong, Qian Renyan: Solid St. Commun. **54**, 211 (1985)
4.60. E.F. Steigmeier, H. Auderset, W. Kobel, D. Baeriswyl: Synth. Met. **17**, 219 (1987)
4.61. See e.g. L. Salem: *The Molecular Orbital Theory of Conjugated Systems.* (Benjamin, Reading MA, 1965)
4.62. J.L. Brédas, R.R. Chance, R. Silbey: Phys. Rev. **B26**, 5843 (1982)
4.63. K. Fesser, A.R. Bishop, D.K. Campbell: Phys. Rev. **B27**, 4804 (1983)
4.64. P. Vogl, D.K. Campbell, O.F. Sanhey: Synth. Met. **28**, D 513 (1989)
4.65. D. Baeriswyl, K. Maki: Synth. Met. **28**, D 507 (1989)
4.66. D. Baeriswyl, D.K. Campbell, S. Mazumdar: Phys. Rev. Lett. **56**, 1509 (1986)
4.67. Z.G. Soos, L.R. Ducasse: J. Chem. Phys. **78**, 4092 (1983)
4.68. P. Tavan, K. Schulten: J. Chem. Phys. **85**, 6602 (1986)
4.69. P. Tavan, K. Schulten: Phys. Rev. **B36**, 4337 (1987)
4.70. D.K. Campbell, D. Baeriswyl, S. Mazumdar: Synth. Met. **17**, 197 (1987)
4.71. U. Sum, K. Fesser, H. Büttner: Solid St. Commun. **61**, 607 (1987)
4.72. W.P. Su, J.R. Schrieffer: Proc. Nat. Acad. Sci. USA **77**, 5626 (1980)
4.73. G.P. Agrawal, C. Cojan, C. Flytzanis: Phys. Rev. **B17**, 776 (1987)
4.74. W. Wu: Phys. Rev. Lett. **61**, 1119 (1988)
4.75. W. Wu, S. Kivelson: Synth. Met. **28**, D 575 (1989)
4.76. L. Sebastian, G. Weiser: Phys. Rev. Lett. **46**, 1157 (1981)
4.77. B.I. Greene, J. Orenstein, B.R. Millard, L.R. Williams: Phys. Rev. Lett. **58**, 2750 (1987)
4.78. S. Schmitt-Rink, D.S. Chemla, H. Haug: Phys. Rev. **B37**, 941 (1988)
4.79. M. Sinclair, D. Moses, D. McBranch, A.J. Heeger, J. Yu, W.P. Su: Synth. Met. **28**, D 655 (1989)
4.80. B. Horovitz: Solid St. Commun. **41**, 729 (1982)
4.81. B. Horovitz: Phys. Rev. Lett. **47**, 1491 (1981)
4.82. S. Etemad, G.L. Baker, C.B. Roxlo, B.R. Weinberger, J. Orenstein: Mol. Cryst. Liq. Cryst. **117**, 275 (1985)
4.83. B.R. Weinberger, C.B. Roxlo, S. Etemad, G.L. Baker, J. Orenstein: Phys. Rev. Lett. **53**, 86 (1984)
4.84. J. Orenstein, G.L. Baker, Z. Vardeny: J. de Phys. **44**, C3-407 (1983)
4.85. Z. Vardeny, J. Orenstein, G.L. Baker: J. de Phys. **44**, C3-325 (1983)
4.86. G.B. Blanchet, C.R. Fincher, T.C. Chung, A.J. Heeger: Phys. Rev. Lett. **50**, 1938 (1983)
4.87. E.J. Mele: Sol. State Commun. **49**, 555 (1984)
4.88. For a review see R.H. Friend, D.D.C. Bradley, P. Townsend: J. of Phys. **D20**, 1367 (1983)
4.89. A.J. Heeger, S. Kivelson, J.R. Schrieffer, W.P. Su: Rev. Mod. Phys. **60**, 781 (1988)
4.90. Z. Vardeny, E. Ehrenfreund, O. Brafman: Mol. Cryst. Liq. Cryst. **117**, 245 (1985)
4.91. J. Orenstein, Z. Vardeny, J.L. Baker, G. Eagle, S. Etemad: Phys. Rev. **B30**, 786 (1984)
4.92. D.K. Campbell, D. Baeriswyl, S. Mazumdar: Synth. Met. **17**, 197 (1987)
4.93. J.D. Flood, A.J. Heeger: Phys. Rev. **B28**, 2356 (1983)
4.94. F. Moraes, Y.W. Park, A.J. Heeger: Synth. Met. **13**, 113 (1986)
4.95. C.G. Levey, D.V. Lang, S. Etemad, G.L. Baker, J. Orenstein: Synth. Met. **17**, 569 (1987)
4.96. I. Rothberg, T.M. Jedju, S. Etemad, G.L. Baker: Phys. Rev. Lett. **57**, 3229 (1986)
4.97. C.V. Shank, R. Yen, R.L. Fork, J. Orensstein, G.L. Baker: Phys. Rev. Lett. **49**, 1660 (1982)
4.98. A.R. Bishop, D.K. Campbell, P.S. Lomdahl, B. Horovitz, S.R. Phillpot: Phys. Rev. Lett. **52**, 671 (1984)
4.99. B. Horovitz, S.R. Phillpot, A.R. Bishop, Phys. Rev. Lett. **60**, 2210 (1988)
4.100. S.R. Phillpot, B. Horovitz, A.R. Bishop: Synth. Met. **28**, D 419 (1989)

4.101. B.E. Kohler: In 'Electr. Properties of Polym. and rel. Comp.', Springer Series in Solid State Sciences, Vol. 63 (1985)
4.102. Z.G. Soos, R.L. Ducasse: J. Chem. Phys. 78, 4092 (1983)
4.103. W.P. Su: Phys. Rev. B34, 2988 (1986)
4.104. G.T. Boyd: J. Opt. Soc. Am. 6, 689 (1989)
4.105. M. Thakur, Y. Shani, G.C. Chi, K. O'Brien: Synth. Met. 28, D 695 (1989)
4.106. F. Kajzar, S. Etemad, G.L. Baker, J. Messier: Synth. Met. 17, D 563 (1987)
4.107. J.L. Brédas, J.C. Scott, K. Yakushi, G.B. Street: Phys. Rev. B30, 1023 (1984)
4.108. J.H. Kaufman, N. Colaneri, J.C. Scott, G.B. Street: Phys. Rev. Lett. 53, 1005 (1984)
4.109. K. Kaneto, Y. Kohno, K. Yoshino: Mol. Cryst. Liq. Cryst. 118, 217 (1985)
4.110. T.-C. Chung, J.H. Kaufman, A.J. Heeger, F. Wudl: Mol. Cryst. Liq. Cryst. 118, 205 (1985)
4.111. G. Harbeke, D. Baeriswyl, H. Kiess, W. Kobel: Physica Scripta T13, 302 (1986)
4.112. G. Harbeke, E. Meier, W. Kobel, M. Egli, H. Kiess, E. Tosatti: Solid St. Commun. 55, 419 (1985)
4.113. M. Schärli, H. Kiess, G. Harbeke, W. Berlinger, K.W. Blazey, K.A. Müller: Synth. Met. 22, 317 (1988)
4.114. W. Hayes, F.L. Pratt, K.S. Wong, K. Kaneto, K. Yoshino: J. Phys. C18, L 555 (1985)
4.115. Shu Hotta, W. Shimotsuma, M. Taketani: Synth. Met. 10, 85 (1984/85)
4.116. Cao Yong, Qian Renyan: Solid St. Commun. 54, 211 (1985)
4.117. Z. Vardeny, E. Ehrenfreund, O. Brafman, A.J. Heeger, F. Wudl: Synth. Met. 18, 183 (1987)
4.118. Z. Vardeny, E. Ehrenfreund, O. Brafman, M. Nowak, H. Schaffer, A.J. Heeger, F. Wudl: Phys. Rev. Lett. 56, 671 (1986)
4.119. M. Capizzi, G.A. Thomas, F. DeRosa, R.N. Bhatt, T.M. Rice: Phys. Rev. Lett. 44, 1019 (1980)
4.120. E. Gerlach, P. Grosse: Festkoerperprobleme XVII, 175 (1977)
4.121. J.C.W. Chien, J.M. Warakomski, F.E. Karasz, W.L. Chia, C.F. Lillya: Phys. Rev. B28, 6937 (1983)
4.122. S. Kivelson, A.J. Heeger: Synth. Met. 17, 183 (1987)
4.123. E.J. Mele, M.J. Rice: Phys. Rev. B23, 5397 (1981)
4.124. R.H. Baughman, N.S. Murthy, G.S. Miller: J. Chem. Phys. 79, 515 (1983)
4.125. L.W. Shacklette, J.E. Toth: Phys. Rev. B32, 5892 (1985)
4.126. S. Jeyadev, E.M. Conwell: Phys. Rev. B33, 2530 (1989)
4.127. M. Tanaka, H. Fujimoto, K. Kamiya, J. Tanaka: Chem. Lett. 1983, 393 (1983)
4.128. J. Tanaka, M. Tanaka, H. Fujimoto, K. Kamiya: Mol. Cryst. Liq. Cryst. 117, 259 (1985)
4.129. T.C. Chung, F. Moraes, J.D. Flood, A.J. Heeger: Phys. Rev. B29, 2341 (1984)
4.130. J. Tanaka, K. Kamiya, M. Shimizu, M. Tanaka: Synth. Met. 13, 177 (1986)
4.131. S. Hasegawa, M. Oku, M. Shimizu, J. Tanaka: To be published in Synth. Met.
4.132. S. Hasegawa, K. Kamiya, J. Tanaka, M. Tanaka: Synth. Met. 14, 97 (1986)
4.133. K. Yakushi, L.J. Lauchlan, T.C. Clarke, G.B. Street: J. Chem. Phys. 79, 4774 (1983)
4.134. S. Hasegawa, K. Kamiya, J. Tanaka, M. Tanaka: Synth. Met. 18, 225 (1987)
4.135. F. Genoud, M. Guglielmi, M. Nechtschein, E. Genies, M. Salomon: Phys. Rev. Lett. 55, 118 (1985)
4.136. H. Kiess, M. Schärli, G. Harbeke: Macromol. Chem., Macromol. Symp. 24, 91 (1989)

5. Magnetic Properties of Conjugated Polymers

Pawan K. Kahol, W.G. Clark, and Michael Mehring

Magnetic resonance spectroscopy thrives upon the abundance of nuclear and electronic spins in matter. Conjugated polymers in particular usually contain the following nuclear spins: ^{1}H (100% natural abundance), ^{13}C (1.1%), ^{14}N (100%), ^{15}N(0.37%) and ^{2}H (0.01%). In special cases it is useful to isotopically enrich ^{13}C, ^{2}H and ^{15}N. The nuclear spins carry magnetic moments that can be detected in NMR (nuclear magnetic resonance) experiments, and which provide information on local fields at the nuclear site. These local fields depend on the electronic structure and dynamics of the polymer. Similar arguments hold for electronic spins, which are usually detected in an ESR (electron spin resonance) experiment. It is therefore not surprising that a wealth of important information can be obtained from such magnetic resonance (MR) or spin resonance experiments.

Whereas nuclear spins are highly abundant in condensed matter, electron spins are usually not. In conducting polymers they arise from either conjugational defects [5.1–3] and/or oxidation or reduction (loosely termed "doping") by chemical or electrochemical means. The renewed interest in conjugated polymers stems basically from two aspects: (i) the large conductivity variation which can be achieved upon doping and (ii) the physical picture which emerged from the treatment of the different types of defects as solitons, polarons and bipolarons [5.2–10]. Some of these 'objects' carry electron spin (soliton, polaron) while others do not (bipolarons). Magnetic resonance spectroscopy was therefore among the most important techniques used to unravel the nature of these defects.

Our main intention in this chapter is not to present a thorough review of the current literature on this subject, which is growing exponentially, but rather to focus on what we understand to be the most fundamental aspects of magnetic properties of conjugated polymers. This chapter should serve as a guide to the enormous variety of magnetic resonance experiments that have been performed on conducting polymers. Magnetic resonance spectroscopy, although widely used by chemists as an analytical tool in liquids and more rarely by physicists in condensed matter, is not as directly interpretable in terms of the structural and dynamical parameters of the system under investigation as, e.g., optical spectroscopy, diffraction methods, conductivity measurements etc. We therefore decided to give a brief account of magnetic properties and resonance techniques (Sect. 5.1).

The rest of this chapter has been arranged according to the following subjects: (i) structural and lattice dynamical aspects of the polymers (Sect. 5.2), (ii) structural and dynamical aspects of the conjugational defects (Sect. 5.3) and (iii) doping-induced characteristics relating to structure and dynamics (Sect. 5.4).

Section 5.5 gives a brief account of magnetic properties of polydiacetylenes and their elementary excitations. Section 5.6 summarises the results of some other conjugated polymers which have not yet been that thoroughly studied by magnetic resonance.

The major part of the investigation is concerned with polyacetylene, because it is the simplest of the conjugated polymers and shows most of the characteristic features. It is therefore quite natural that most of the magnetic resonance investigations covered in this chapter are also devoted to polyacetylene, which serves as a prototype conjugated polymer, even though it is very special among the conjugated polymers because of its degenerate ground state (in the *trans* isomer). However, the special properties of all other non-degenerate conjugated polymers such as polarons and bipolarons are also discussed in detail. A particularly thorough account of solitons in polyacetylene can be found in [5.12].

5.1 General Aspects of Magnetic Properties and Resonance Techniques

5.1.1 Susceptibility

The magnetic susceptibility per unit volume for isotropic materials without magnetic ordering is defined by [5.13–16]

$$\mu_0 M = \chi \cdot B, \tag{5.1.1}$$

where B denotes the macroscopic magnetic field intensity, M is the magnetic moment per unit volume and $\mu_0 = 4\pi \cdot 10^{-7} (H/m)$. The magnetic susceptibility is therefore a measure of the magnetization obtained in a unit field. If we neglect the orbital contributions, it measures the excess number of spins pointing parallel to the field over those pointing opposite to the field. Frequently, a more meaningful susceptibility referred to per gram (χ_m) or per mole (χ_M) of the substance is defined as

$$\chi_M = M' \chi_m = M' \chi / \rho \tag{5.1.2}$$

where ρ is the density and M' is the molecular weight. For diamagnetic materials, χ is usually almost independent of temperature and is negative. In ideal paramagnetic materials where spin–spin interactions are negligible χ is positive and follows the Curie law [5.13–16]

$$\chi_c = \mu_0 N I (I + 1) g^2 \mu_B^2 / 3 k_B T. \tag{5.1.3}$$

In (5.1.3) I is the spin quantum number, N is the number of spins per unit volume, μ_B is the Bohr magneton and g is the Lande factor ($g = 2.0023$ for free electrons). This relationship also holds for slightly delocalized electronic spins as long as their wavefunction does not overlap with neighbouring spins. On the other hand, when the electronic spins are delocalized and form energy bands with a finite density of states at the Fermi level, as in a metal, only those electrons within a range $k_B T$ of the Fermi level contribute to the susceptibility. In this case and for a free electron gas [5.13–16], one has the Pauli susceptibility

$$\chi_p = \tfrac{1}{4}\mu_0(g\mu_B)^2 N(\varepsilon_F), \tag{5.1.4}$$

where $N(\varepsilon_F)$ is the density of states (including both spin directions) at the Fermi energy ε_F. For a free-electron gas in three dimensions $N(\varepsilon_F) = 3/(2\varepsilon_F)$, whereas for a one-dimensional metal $N(\varepsilon_F) = 1/(2\varepsilon_F)$. Note that the Pauli susceptibility (5.1.4) is temperature independent, which is a typical signature of the metallic state.

In a simple one-dimensional tight-binding band structure, where the energy dispersion relation is given by

$$E(k) = -4t \sin^2(ka/2) \tag{5.1.5}$$

and where t is the transfer integral responsible for the electron transport along the chain, the density of states at the Fermi level can be expressed as

$$N(\varepsilon_F) = [\pi t \sin(\rho\pi/2)]^{-1}. \tag{5.1.6}$$

Here, ρ is the band-filling parameter. With one electron per site (atomic orbital) we have $\rho = 1$. This picture is based on a single-electron model. If electron–electron correlation are taken into account, as discussed in Sect. 5.3 and Chap. 8, the susceptibility is enhanced. Starting from the Hubbard model one can write within unrestricted Hartree-Fock approximation

$$\chi_s = \frac{\chi_p}{1 - \alpha} \tag{5.1.7}$$

with

$$\alpha = \tfrac{1}{2}UN(\varepsilon_F) \tag{5.1.8}$$

where U is the on-site Coulomb energy.

Since there is strong evidence for electron–electron correlations in conducting polymers (Sect. 5.3.1), it is necessary to include these correlations in the discussion of the susceptibility. As is evident from (5.1.7), the neglect of correlations results in an overestimation of the density of states $N(\varepsilon_F)$.

There are two categories of methods that are commonly used to measure χ: Static measurements (Faraday balance, SQUID magnetometer, etc.) and magnetic resonance, which is discussed in more detail below. Static methods measure the *total* susceptibility, and require additional measurements or analysis to separate χ into the diamagnetic (or core) and paramagnetic (orbital and spin) contributions. To obtain the spin part alone requires the knowledge of χ_{dia}, usually obtained from Pascal's constants, and χ_{orb}, which is usually unknown and difficult to estimate. Furthermore, the spin part is for all species of spins, and does not permit differentiation into the contribution from each type of spin.

The sensitivity of static methods is often sufficient to measure χ for electrons. Since, however, the nuclear spin susceptibility is normally several orders of magnitude smaller than the electronic susceptibility, it is ordinarily not practicable to use static methods to determine the nuclear contribution to χ.

In contrast to static measurements, magnetic resonance offers many advantages. It can detect fewer spins, it separates out the spin component naturally and it can usually be made specific to a single spin species by setting the spectrometer frequency and the applied magnetic field. The spin component of χ is simpler to interpret than the other ones. It can often be separated into χ_c and χ_p on the basis of its temperature dependence. This separation is somewhat more difficult in narrow bandwidth conductors, where χ_p does show a weak dependence.

The basis of magnetic resonance spectroscopy is the measurement of the energy difference between the spin states by applying on oscillating electromagnetic field perpendicular to the static field B_0 at either fixed field and variable frequency or fixed frequency and variable field, so as to satisfy the condition $\Delta E = h\gamma B_0 = h\nu = \hbar\omega_0$ where γ is the electronic ($\hbar\gamma_e = g\mu_B$) or nuclear (γ_n) gyromagnetic ratio. The other constants have their usual meaning [5.13–16]. This field induces transitions between the spin states corresponding to either absorption or emission of energy. If we denote $\chi''(\omega)$ as the absorption part of the spectrum, the total intensity I of the resonance signal is

$$I = \int \chi''(\omega)d\omega \tag{5.1.9}$$

which, with the help of Kramers–Kronig relations, can be shown to yield

$$I = \pi\omega_0\chi_0/2 \tag{5.1.10}$$

where χ_0 is the static spin susceptibility. That is, χ_0 can be obtained from the integrated intensity of the resonance curve. *Schumacher* and *Slichter* [5.15] have utilized this technique to calibrate the electronic spin susceptibility by performing the resonance experiment at the same radio frequency for a electronic spins (small B_0) and nuclear spins (large B_0). In metallic samples one has to be aware of the skin depth problem which leads to Dysonian lineshapes (see Sect. 5.4.3), and the analysis is more complicated [5.16].

5.1.2 Lineshapes, Linewidths and Lineshifts

Magnetic resonance experiments (NMR: nuclear magnetic resonance; ESR: electron spin resonance) are usually performed by applying a strong static magnetic field B_0 which splits the zero field levels of a spin I into $2I + 1$ levels which are separated by the Larmor frequency $\omega_0 = \gamma B_0$ [5.13–16]. Additional interactions result in further level shifts and splittings. Transitions between these levels are excited by applying a time varying field $B_1(t) = B_1 \cos \omega t$ perpendicular to B_0 at frequency ω near the Larmor frequency ω_L to a coil or cavity. In the standard CW (continuous wave) technique, the magnetic field B_0 is swept through the resonance in either a rapid passage, which yields the absorption signal directly, or by modulating the B_0 field and using a lock-in technique which records the derivative of the absorption signal.

Pulse techniques use either a single pulse, which is repeated regularly, or a well-defined sequence of pulses that create a spin echo [5.17]. A number of multi-pulse methods have been designed to serve special purposes which allow the separation of different interactions and different relaxation mechanisms [5.18,19]. This topic will be discussed in more detail in Sects. 5.1.3 and 5.1.5. The response of a pulse experiment is always a time domain signal, which after Fourier transformation leads again to a frequency spectrum. The spectrum may consist of a number of single lines corresponding to the level splittings due to (a) dipole–dipole, quadrupole, chemical shift, Knight shift or scalar coupling interactions, as in the NMR case or (b) due to hyperfine or fine structure interactions, g shifts etc. as in the ESR case. In solids, broad lines usually appear because of either strong relaxation mechanisms or anisotropic interactions.

Steady state magnetic resonance experiments generate a response that is a combination of absorption (in phase quadrature relative to the excitation) and dispersion (in phase with the excitation) which can be expressed in complex form as [5.13–16]

$$F(\omega) = F'(\omega) + iF''(\omega) \tag{5.1.11}$$

where the absorption ($F''(\omega)$) and the dispersion ($F'(\omega)$ lineshapes are connected by the Kramers–Kronig relation. We shall therefore discuss only the absorption lineshapes in the following.

Three different types of lineshapes, namely (i) homogeneous, (ii) inhomogeneous and (iii) heterogeneous will be briefly introduced here. Experimentally a strong radio frequency (rf) or microwave (mw) field is applied at a particular frequency of the line. Homogeneous lines become saturated over the full spectral width. They are usually governed by a spin–spin relaxation process (relaxation time T_2) and have Lorentzian shape. If dipole–dipole interactions dominate the lineshape again a homogeneous line results but usually of a Gaussian character due to the central limit theorem.

If a saturation field is applied to an inhomogeneous line, a "spectral hole" is burnt in the spectrum at the frequency of the applied rf or mw field. An inhomogeneous line is therefore caused by a distribution of local fields, where every spin "sees" a slightly different field and can therefore be saturated individually. Magnetic field inhomogeneity, chemical shift and Knight shift anisotropies and anisotropic hyperfine interactions are among the common causes of inhomogeneous lines [5.14–16].

If the spins experiencing different local fields are also coupled *to one another or to a third spin*, "spectral diffusion" can fill the hole in the spectrum. These effects can become even more pronounced in heterogeneous lines, which are mixtures of cases (i) and (ii). The coupling among different spins in a hetero-geneous line is partially quenched, but still effective in causing rapid spectral diffusion and connecting different parts of the spectrum. This appears to occur in some quadrupolar- and hyperfine-broadened spectra.

A rigorous lineshape calculation is both difficult and involved, but useful information can still be obtained from simple relations involving the moments and the linewidth. Two commonly used linewidth parameters are δ_0, the half width at half height of the absorption signal, and Δ_{pp}, the peak-to-peak separation of the first derivative spectrum. For Gaussian and Lorentzian lines, their correspondence is

$$\Delta_{pp} = [4/3]^{1/2}\delta_0 \quad \text{Lorentzian} \tag{5.1.12a}$$

$$\Delta_{pp} = [2/\ln 2]^{1/2}\delta_0 \quad \text{Gaussian.} \tag{5.1.12b}$$

The half linewidth δ_0 of a Lorentzian lineshape is often related to the spin–spin relaxation time T_2 by $\delta_0 = T_2^{-1}$ in rad s^{-1}, whereas the Gaussian lineshape is fully described by its second moment M_2, which leads to the relation

$$\delta_0 = [2\ln 2 M_2]^{1/2}. \tag{5.1.13}$$

Moments of the lineshapes are in general defined as

$$M_n = \int d\omega \omega^n F''(\omega); \quad n = 0, 1, 2. \tag{5.1.14}$$

The advantage of using moments over the lineshape calculations lies in the fact that the low-order moments can often be calculated rigorously. If one is dealing with a lineshape intermediate between a Gaussian and a Lorentzian, a simple formula relates the second (M_2) and the fourth (M_4) moments to the half width as [5.18]

$$\delta_0 = \left[\frac{\pi}{2} M_2/(\mu - 1.87)\right]^{1/2} \quad \mu \geq 3 \tag{5.1.15}$$

where $\mu = M_4/M_2^2$. For a Gaussian, $\mu = 3$.

Gaussian lineshapes usually occur when a large number of sources, such as other spins, contribute significantly to the local field of a given spin. Examples are dipole–dipole interactions in dense spin systems and hyperfine interactions. If the motion of the spins begins to dominate, M_4 increases, and the lineshape tends toward a Lorentzian form.

Another important lineshape is the Pake pattern, which results from dipolar coupled pairs of spins with a random orientation of their internuclear vector with respect to B_0.

For two spins S_j and S_k a doublet of lines occurs with separation [5.14,15]

$$\Delta\omega_{jk} = [3S(S+1)]^{1/2}\gamma^2\hbar r_{jk}^{-3}[3\cos^2\vartheta_{jk} - 1] \tag{5.1.16}$$

that depends upon the separation r_{jk} and the angle ϑ_{jk} of the interspin-vector with respect to the B_0. Equation (5.1.16) is valid only if $\omega_0 = \gamma_s B_0$ is much larger than $\Delta\omega_{jk}$.

For a large number of spins coupled by dipole–dipole interactions, the second moment M_2 is

$$M_{2j} = \sum_k \Delta\omega_{jk}^2 \tag{5.1.17}$$

which corresponds to linebroadening. Another interesting case of linebroadening results from a distribution of hyperfine interactions. Consider the simple example of an electron spin $S = 1/2$ coupled to a number N of protons ($I = 1/2$). This is a typical situation of a soliton or polaron (spin operator S) coupled to protons (spin operators I_j). For the Hamiltonian

$$H_{IS} = \sum_j A_{zzj} I_{zj} S_z - \omega_{0s} S_z - \omega_{0I} I_z, \tag{5.1.18}$$

where A_{zzj} relates to the principal value of the hyperfine tensor as

$$A_{zzj} = A_{11j}\sin^2\beta\cos^2\alpha + A_{22j}\sin^2\alpha\sin^2\beta + A_{33j}\cos^2\beta \tag{5.1.19}$$

and where α, β are the Euler angles of the magnitic field B_0 with respect to the principal axis system, the second moment, for $\omega_{0I} \gg A$, can be readily expressed as [5.20]

$$M_2 = \tfrac{1}{20}[A_{11}^2 + A_{22}^2 + A_{33}^2] + \tfrac{2}{3}[A_{11}A_{22} + A_{22}A_{33} + A_{33}A_{11}]\sum_j \rho_j^2 \tag{5.1.20}$$

where ρ_j is the electron spin density at the nuclear position j and the normalization condition $\sum_j \rho_j = 1$ is obeyed. An application of this equation to electron delocalization of solitons and polarons is discussed in Sect. 5.3.1

So far we have neglected the motion of the spins. Two different types of motion may be distinguished: (a) lattice (translational) motion and (b) spin-flip (orientational) motion. Both are often called spin diffusion in the literature. When the spin moves through the lattice it visits different places with different interaction energies. Its local field therefore fluctuates in time, i.e. spectral components are "exchanged". This exchange of spectral components can be brought about also by scalar or exchange coupling between a large number of spins that do not undergo lattice motion. We call the latter process spin-flip motion. Both exchange processes narrow the line to give a reduced linewidth

$$\delta \approx \frac{M_2}{\omega_e} \tag{5.1.21}$$

where M_2 is the second moment of the hyperfine interaction as discussed before and ω_e is the exchange frequency, which may also be expressed in terms of a correlation time $\tau = \omega_e^{-1}$.

Another important parameter in magnetic resonance spectroscopy is the *lineshift*. Besides the chemical shift which is caused by the closed electron shell around the nucleus, the Knight shift caused by paramagnetic conduction electrons is of particular importance in organic conductors. It is caused by the isotropic hyperfine interaction $\alpha_i = [A_{11} + A_{22} + A_{33}]/3$ of the conduction electrons with the nuclear spins and is given by [5.15,16,21–23]

$$\frac{\Delta\omega}{\omega_0} \equiv K = \frac{\chi_s a_i}{\hbar\gamma_e\gamma_n} \tag{5.1.22}$$

where χ_s is the electron spin susceptibility per conduction electron and the other parameters have their usual meaning as discussed before. Whereas these shifts are large in ordinary metals, they are rather small in organic conductors [5.21–23].

5.1.3 Spin Relaxation ($T_1^{\cdot}, T_2, T_{1\rho}$)

The location of a spectral line observed with magnetic resonance is determined by the static level splittings of the spin. Spin relaxation, on the other hand, is driven by time variations of the local field seen by the spin. There are basically two different types of spin relaxation processes, as described below.

(i) Longitudinal (T_1) relaxation

The static field B_0 is applied along the z direction of the spin coordinate frame. It establishes a Boltzmann distribution among the spin levels. If this Boltzmann distribution is perturbed somehow, it will relax back to the equilibrium state (longitudinal with B_0) with the characteristic time constant T_1. This process requires an exchange of energy between the spin and its surroundings at the frequency ω_0 corresponding to the transitions that occur between the spin states. It can happen directly, or indirectly through a Raman and/or Orbach process. The spin lattice relaxation rate $1/T_1$ always involves the random modulation of some spin interaction expressed by its second moment $\langle\Delta\omega_{x,y}^2\rangle$ and the spectral density function $J(\omega)$ of the fluctuation. In the case of the direct process, the spectral density at the particular splitting frequency $J(\omega_0)$ has to be considered for a two-level system or at the sum and difference frequencies for a multi-level system. The basic expression for $1/T_1$ in its simplest form is [5.14–18]

$$\frac{1}{T_1} = [\langle\Delta\omega_x^2\rangle + \langle\Delta\omega_y^2\rangle]J(\omega_0) \tag{5.1.23}$$

where ω_0 corresponds to the Lamor frequency of the electron spins if hyperfine interaction is the dominant source of fluctuating local fields.

The spectral density $J(\omega_0)$ is the Fourier transform of the correlation function

$$g(t) = \langle S(t)S(0)\rangle \tag{5.1.24}$$

where $S(t)$ is a dimensionless random parameter which reflects the dynamics of the system. In solids there are usually different modes with different correlation functions and it is often useful to distinguish them by their different q vectors:

$$g(q,t) = \langle S_q(t)S_{-q}(0)\rangle. \tag{5.1.25}$$

Since the spin resonance is a local property we arrive at

$$J(\omega) = \mathrm{Re}\left\{ \int_0^\infty dt g(t)e^{-i\omega t}\right\} \tag{5.1.26}$$

with

$$g(t) = \frac{a}{\pi}\int_0^{\pi/a} dq g(q,t)e^{-iqa} \tag{5.1.27}$$

where (5.1.27) is given for pedagogical reasons only for a one-dimensional Brillouin zone; it can be extended easily to any higher dimensional Brillouin zone.

A commonly used correlation function that corresponds to a constant transition probability per unit time for the loss of coherence in the fluctuating variable is

$$g(t) = \exp(-t/\tau) \tag{5.1.28a}$$

and leads to

$$J(\omega) = \frac{\tau}{1+\omega^2\tau^2}. \tag{5.1.28b}$$

It should be pointed out, however, that such a simple relation does not hold in highly anisotropic solids. The correlation function for quasi-one-dimensional motion can be expressed as [5.24a]

$$g(t) = e^{-t/\tau'}e^{-t(2/\tau_\parallel)}I_0(2t/\tau_\parallel) \tag{5.1.29}$$

where τ' and $\tau_\parallel$ are the correlation times for motion perpendicular and parallel to the unique axis. $I_0(x)$ is the modified Bessel function. The spectral density is readily obtained as [5.24a]

$$J(\omega) = (\tau_\parallel \tau')^{1/2}\left(\frac{[\Omega^2+1)(\Omega^2\varepsilon^2+(4+\varepsilon)^2)]^{1/2}+4+\varepsilon-\Omega^2\varepsilon}{2(\Omega^2+1)[\Omega^2\varepsilon^2+(4+\varepsilon)^2]}\right)^{1/2} \tag{5.1.30}$$

with $\varepsilon = \tau_\parallel/\tau'$ and $\Omega = \omega\tau'$.

In the limit $\omega\tau' \gg 1 \gg \omega\tau_\parallel$ the well-known expression [5.24a–26]

$$J(\omega) = 1/2\tau_\parallel(2\omega\tau_\parallel)^{-1/2} \tag{5.1.31}$$

results, with the characteristic $\omega^{-1/2}$ frequency dependence. For a treatment of relaxation processes using scaling arguments the reader is referred to reference [5.24b].

Measurements of T_1 are most directly performed by applying a series of pulses at the transition frequency in order to saturate the signal, i.e. the population difference is made zero. The magnetization recovery in the z direction $M_z(t)$

is subsequently monitored by a $\pi/2$ pulse excitation and usually obeys $M_z(t) = M_0[1 - \exp(-t/T_1)]$. Besides this saturation recovery technique, an initial π pulse can be applied in order to invert the z magnetization, after which a recovery towards Boltzmann equilibrium is observed as $M_z(t) = M_0[1 - 2\exp(-t/T_1)]$; this is called the inversion recovery technique. A third technique utilizes the saturation behaviour of a CW line which saturates with increasing rf (mw) field B_1 as $(1 + \gamma^2 B_1^2 T_1 T_2)^{-1}$. However, T_1 values determined from this behaviour are somewhat unreliable, since either the condition $T_1 = T_2$ must be fulfilled, or T_2 must be known in advance. Moreover, this simple saturation behaviour only holds for single-spin interactions and is in most practical cases not exactly applicable. It should therefore be used with great care.

(ii) Transverse relaxation (T_2)

Transverse relaxation (T_2) refers to irreversible loss of spin-phase memory for motion in the plane perpendicular to B_0. It is caused by random modulation of the level splittings with an average square deviation $\langle \Delta\omega_z^2 \rangle$ and, when the motional correlation time $\tau \ll T_2$, it is given by

$$\frac{1}{T_2} = \langle \Delta\omega_z^2 \rangle J(0) + \frac{1}{2T_1}, \qquad (5.1.32)$$

that is, T_2^{-1} is equal to $1/(2T_1)$ plus an additional contribution from the spectral density at zero frequency.

For "fast motion" or small B_0, i.e. when $\omega_0\tau \ll 1, J(0) \cong J(\omega_0)$ (5.1.28b), which leads to $T_1 \cong T_2$ provided $\langle \Delta\omega_z^2 \rangle \cong \langle \Delta\omega_x^2 \rangle \cong \langle \Delta\omega_y^2 \rangle$. In the limit of "slow motion" usually $J(0) \gg J(\omega_0)$), leading to $T_2 \ll T_1$.

Experimentally, these different regimes can easily be distinguished by measuring T_1 and T_2 separately. Reliable T_2 data can be obtained by using a Hahn spin echo sequence [5.17], where two rf (mw) pulses separated by time t_1 are applied at the resonance frequency. A spin echo occurs at time $t_E = 2t_1$ provided the line is inhomogeneously broadened. The spin echo decays with $\exp(-t_E/T_2)$ in general, unless the spins are diffusing rapidly or some highly anisotropic relaxation mechanisms are involved. If the line is homogeneous, T_2 can be obtained directly from the linewidth $\delta_0 = T_2^{-1}$ as was indicated in Sect. 5.1.2.

We now discuss $T_{1\rho}$, spin lattice relaxation time in the rotating frame. It is based upon "locking" the magnetization to a resonant or near-resonant field B_{1x} in the $x - y$ plane. The frequency that is then important in the spectral density is $\omega_1 = \gamma B_{1x}$, i.e. relaxation occurs along the rf field B_{1x} in addition to that at ω_0. Its rate is [5.18, 19]

$$\frac{1}{T_{1\rho}} = \langle \Delta\omega_z^2 \rangle J(\omega_1) + \frac{1}{2T_1}. \qquad (5.1.33)$$

The importance of $T_{1\rho}$ is that it can be used to explore a different region of $J(\omega)$, and leave a different time scale for the decay of the correlation function responsible for the nuclear relaxation.

A different kind of experiment that also permits extension of the frequency range of $J(\omega)$ is via magnetic field cycling. In this method, the amplitude of M_z is measured always at the same frequency ω_0 and field B_0, but B_0 is rapidly cycled to a different value at which spin-lattice relaxation occurs. After some relaxation has occurred, the field is rapidly cycled back to B_0 and the change in M_z recorded. By varying the time spent at the relaxation field, T_1 at that field is obtained. Measurements of this kind have been used to investigate $J(\omega)$ over an extremely wide range of ω in a conducting polymer [5.25].

By measuring the different relaxation rates and changing B_0 (fixed or by cycling) an enormous range of correlation times spanning $1 \leq \tau \leq 10^{-15}$ s can be investigated by NMR and ESR methods. Application of this method to conjugated polymers is discussed in the following sections. One should be aware, of course, that spin relaxation gives only the correlation function $g(t)$, from which it may be difficult to distinguish among different microscopic models for the dynamics responsible for the fluctuations.

In metals, there is an important relation between the Knight shift K and T_1, called the Korringa relation [5.23]:

$$K^2 T_1 T C_0 S_K = 1 \tag{5.1.34}$$

with

$$C_0 = \frac{4\pi k_B}{\hbar} \left(\frac{\gamma_n}{\gamma_e} \right)^2 .$$

$S_K \leq 1$ (to be discussed below) and T is the temperature. In metals K is practically temperature independent resulting in $T_1 T = \text{const}$, often called the Korringa law. In low-dimensional organic conductors, however, the situation is quite different. Here the one-dimensional nature of the transport and possible $2k_F$ fluctuations have to be taken into account [5.23, 25]. This can be done by the scaling factor S_K in (5.1.34), which in a three-dimensional free electron model is 1, but in 1D conductors should be replaced by [5.23]

$$S_K = \frac{1}{2} \left(\frac{\tau_\perp}{\tau_s} \right)^{1/2} \left[\frac{3}{5} \varepsilon j(\omega_n) + \left(1 + \frac{7}{5} \varepsilon \right) j(\omega_e) \right] \kappa_0(\alpha) + \frac{1}{2}(1 + 2\varepsilon)\kappa_{2K_F}(\alpha) \tag{5.1.35}$$

where τ_s is the 1D phonon-scattering time and $\varepsilon = d^2/a^2$ the ratio between the anisotropic (d: dipolar) and the isotropic hyperfine interaction a. The spectral density $j(\omega)$ at the nuclear Larmor frequency (ω_n) or the electron Larmor frequency (ω_e) can be obtained from

$$j(\omega) = \left[\frac{(1 + \omega^2 \tau_\perp^2)^{1/2} + 1}{2(1 + \omega^2 \tau_\perp^2)} \right]^{1/2} . \tag{5.1.36}$$

Electron–electron correlation is also taken into account in S_K through the parameters

$$\kappa_0(\alpha) = (1 - \alpha)^{1/2} \quad \text{and} \quad \kappa_{2K_F}(\alpha) = \frac{(1 - \alpha)^2}{[1 - \alpha F(2k_F)]^2}$$

where $\alpha = \frac{1}{2} U N(E_F)$ as discussed in Sect. 5.1.2,

$$F(2k_F) = \frac{1}{2} \ln(4.56 T_F / T) \tag{5.1.37}$$

is the Lindhard function and T_F is the Fermi temperature. The scaling factor S_K has been shown to reach rather large values ($S_K \cong 40$–80) in highly one-dimensional organic conductors [5.23].

5.1.4 Double Resonance Techniques

Double resonance implies that a multi-level system is excited by two different rf (mw) fields at different transition frequencies but applied at the same time, or at least within the coherence time T_2. The virtue of this technique is that the observation can be made on a stronger transition with good signal-to-noise ratio while irradiating a much weaker transition with small level spacings and, therefore, small population differences. Moreover, the observation of a double resonance effect proves the proximity of certain spins [5.15, 18].

In order to perform double resonance experiments, one needs a minimum of three levels and at least two allowed transitions. A more realistic case is, however, the four-level system, shown in Fig. 5.1 which appears when two different spins I and S are coupled to each other. This could be either an electron spin (S) and a nuclear spin (I) coupled by hyperfine interactions, or two nuclear spins (e.g. ^{13}C (S) and ^1H I)) coupled by dipole–dipole interaction or scalar interaction. In order to keep the treatment as comprehensible as possible, we follow the lines of *Slichter* [5.15]. As a specific example we first treat the case of ENDOR (electron nuclear double resonance) which we discuss by using the simplified Hamiltonian (high field approximation) given in units of $\hbar$ according to (5.1.38).

$$H = \omega_e S_z + A_{zz} I_z S_z - \omega_n I_z \tag{5.1.38}$$

where ω_e and ω_n are the electron and nuclear Larmor frequencies and A_{zz} is the z component of the hyperfine tensor. In the case of scalar coupling $a I \cdot S$, the Hamiltonian (5.1.38) can still be used, with $a = A_{zz}$ in the high field approximation. In passing we note that a is usually negative for protons in sp^2 radicals. The energy levels are

$$E(m_S m_I) = \omega_e m_S + A_{zz} m_S m_I - \omega_n m_I \tag{5.1.39}$$

for any Zeeman quantum number m_S and m_I. Here we discuss only the case $m_S = \pm 1/2$ and $m_I = \pm 1/2$, whose level diagram is shown in Fig. 5.1. There are two allowed electron spin transitions ($\Delta m_S = \pm 1$) at frequencies

$$\omega_{e\pm} = |\omega_e \pm A_{zz}/2| \tag{5.1.40}$$

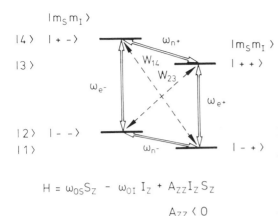

$$H = \omega_{0S} S_z - \omega_{0I} I_z + A_{zz} I_z S_z$$

$$A_{zz} < 0$$

Fig. 5.1. Four-level energy scheme corresponding to a nucleus of spin $I = 1/2$ coupled to an electron of spin $S = 1/2$ by the hyperfine interaction A_{zz}. The possible electron ($\omega_{e\pm}$) and nuclear spin ($\omega_{n\pm}$) transitions along with the relaxation rates W_{14}, W_{23} for the forbidden transitions are indicated

and two allowed nuclear spin transitions at

$$\omega_{n\pm} = |\omega_n \pm A_{zz}/2|. \tag{5.1.41}$$

The line splitting is the same for both transitions, i.e. A_{zz} can in principle be determined from either spectrum. However, if more than one nuclear spin is coupled to an electron spin, the situation changes. For N equivalent nuclear spins ($I = 1/2$) coupled to an electron, there are at least $(N + 1)$ ESR lines, but only two NMR (ENDOR) lines, whereas N nonequivalent nuclear spins ($S = 1/2$) result in 2^N ESR lines, but only $2N$ NMR (ENDOR) lines. Thus, for a fairly small number N, the ESR spectrum may become unresolved, whereas fewer lines in the NMR spectrum may lead to a well-resolved ENDOR spectrum [5.15, 27].

In the ENDOR technique one first saturates the ESR transition (e.g. $|2\rangle \leftrightarrow |4\rangle$ in Fig. 5.1) by applying a strong mw field at the particular transition. This nearly equalizes the populations p_2 and p_4. By applying a strong rf field to either nuclear transition ($|1\rangle \leftrightarrow |2\rangle$ or $|3\rangle \leftrightarrow |4\rangle$) at frequency $\omega_{n\pm}$ a partial repopulation of the levels $|2\rangle$ and $|4\rangle$ occurs, which leads to a change in the ESR signal. In this way the NMR, or rather the ENDOR transitions, at frequency $\omega_{n\pm}$ are indirectly observed when sweeping through the ENDOR lines [5.15, 27]. This discussion of the ENDOR effect is, however, oversimplified. Firstly, the truncated Hamiltonian used in (5.1.38) is not too realistic if the hyperfine interaction is comparable to ω_n, the Larmor frequency of the nuclei. In this case, off-diagonal elements must be taken into account and the eigenfrequencies (5.1.40, 41) have to be modified. Moreover, the intensity of the ENDOR lines does depend on these off-diagonal elements and therefore on frequency. These effects can be treated, however, in second-order perturbation theory [5.28]. Secondly, relaxation processes modify the spin dynamical processes during irradiation which again results in frequency-dependent line distortions [5.28]. A thorough investigation of these ENDOR effects and in particular a detailed discussion with respect to conjugated polymers, can be found in a series of review articles by *Dalton* and co-workers [5.29, 30] and *Thomann* [5.31]. In NMR pulsed version of double resonance techniques, SEDOR (spin echo double resonance) experiments [5.32, 33] have been used for a long time. More recently, these and other pulse techniques have been applied to the ENDOR situation [5.34, 35] with the benefit of increased sensitivity and better resolution. A more detailed account of the application of these techniques to polyacetylene can be found in [5.36].

Dynamic nuclear polarization (DNP) experiments on the other hand allow one to observe the NMR transition directly after having "pumped" at the ESR transition. It should be noted, however, that in DNP experiments the bulk nuclei of the samples are observed rather than the nuclei directly coupled to the electron spin. Two different regimes can be distinguished, namely (i) the Overhauser effect and (ii) the solid state effect.

(i) Overhauser Effect [5.15, 27]

Consider a situation where the electron spins are moving rapidly and all static hyperfine interactions are averaged to zero because of the motion. The ESR line in this case is homogeneous with broadening $\delta_0 = 1/T_2$, as discussed in Sect. 5.1.3. Again there is no hyperfine interaction to be observed at the nuclear transition. Saturation of the electron spin transitions ($1 \leftrightarrow 3$ and $2 \leftrightarrow 4$) leads, in combination with a flip-flop relaxation process (W_{23}, W_{32}), to a nuclear spin polarization which can be as large as γ_e/γ_n if no "leakage" effects are operative. When observing the nuclear spin polarization while varying the mw frequency at the ESR transition a symmetrical line around ω_e, typical for the Overhauser effect, is observed.

(ii) Solid-State Effect [5.15, 16]

In order to observe the solid-state effect, the spins must be fixed in space or at least are allowed to move only slowly compared with the hyperfine interaction, i.e. the ESR shows as broadening or splitting due to the hyperfine interaction. If a strong mw field is now applied at frequency $\omega_e \pm \omega_n$ rapid "forbidden" electron (flip)–nuclear (flop) transitions between levels $2 \leftrightarrow 3$ and $1 \leftrightarrow 4$ are induced. With the help of relaxation processes occurring between levels $2 \leftrightarrow 4$ and $1 \leftrightarrow 3$ a polarization of the nuclear sublevels develops which leads to greatly enhanced NMR signals. The characteristic feature of the solid-state effect is that an asymmetric pattern appears with a maximum at $\omega_e - \omega_n$ and a minimum at $\omega_e + \omega_n$ when observing the nuclear polarization while sweeping the mw frequency, in contrast to the Overhauser effect where the maximum polarization appears at ω_e.

In this way, DNP experiments allow one to distinguish between fixed and mobile electron spins. An application of this technique to polyacetylene and polyparaphenylene is discussed in Sect. 5.3.2.

The double resonance techniques discussed here can also be applied to pure nuclear spin systems, although there the effect is not that dramatic. Other schemes, to be discussed next, have therefore been developed for nuclear spins.

5.1.5 High-Resolution NMR

High-resolution NMR is a standard technique applied routinely to liquids by chemists. In ordinary liquids, the rotational rate of the molecules is much faster than the largest anisotropy of the spin interactions, resulting in an isotropic line spectrum with all anisotropies averaged to zero. In solids, conversely, all anisotropies which are allowed by local symmetry are expected to appear in the spectrum. The NMR and ESR spectra of solids are therefore usually broad and structureless, with certain exceptions. Here we discuss briefly the techniques which have been developed to overcome this obstacle [5.18, 19, 37–39].

Anisotropies in the spin interactions are in most cases described by second rank tensors. Since strong magnetic fields B_0, assumed to be parallel to the z direction, are applied, only the secular parts of the spin interactions play a dominant role in the spectrum. Correspondingly the typical orientational dependence of the spin interactions can be expressed as $P_2 (\cos \vartheta) = (3 \cos^2 \vartheta - 1)/2$, where ϑ is the angle between B_0 and the principal axis of the interaction. There are basically two different types of technique which utilize this orientational dependence, namely (i) magic angle sample spinning (MAS) and (ii) multiple-pulse experiments [5.18, 19, 37–39].

(i) Magic Angle Sample Spinning (MAS) [5.18, 37]

The sample is placed in a rotor, whose axis makes an angle $\vartheta_m = 54° \, 44^|$ (magic angle) with the magnetic field direction. Due to the fact that $P_2 (\cos \vartheta_m) = 0$, all anisotropies vanish under the condition that the rotor frequency ω_r is much larger than the magnitude of the spin interaction (in rad s^{-1}). In this case, only the isotropic part governs the spectrum, leading to a line spectrum. If ω_r is not large enough, sidebands appear which can in fact be utilized to obtain information about the tensor elements. This technique is extensively applied to ^{13}C nuclei and other rare nuclei in combination with cross-polarization experiments.

(ii) Multiple-Pulse Experiments [5.18, 19, 39]

Abundant nuclei such as ^1H have strong dipole–dipole interactions, characterized by their second moment M_2. Because $M_2^{1/2} \gg \omega_r$ for technically feasible rotational rates in real space, the dipole–dipole interaction must be averaged in spin space. This is performed by applying a sequence of properly spaced and phased pulses, in order to force a cyclic motion of the spins. This motion is designed to give a vanishing dipole–dipole interaction of the spins on average, leaving chemical shift interaction operative, although scaled by a scaling factor $S < 1$.

A different scheme for obtaining high resolution spectra in solids which works for rare spins is discussed below.

(iii) Proton-Enhanced Nuclear Induction Spectroscopy [5.18, 38]

Consider a rare nucleus (e.g. ^{13}C, 1.1% abundant) surrounded by abundant nuclei (e.g. ^1H). In order to obtain a detectable signal of the rare nuclei (spin S) they must be polarized with the help of the abundant nuclei (spin I). This is achieved in a sequence such as the one shown in Fig. 5.2. I spins are first spin-locked to obtain a low spin temperature and than brought into contact with the S spins by applying a rf field at ω_S. After the polarization is completed (contact time typical 1 ms–10 ms) the rf field is terminated and an S-spin free-induction decay (FID) results, which is observed at ω_S while still irradiating at ω_1 in order to decouple the I spins. Fourier transform of the FID leads to the S-spin spectrum. This technique is often combined with MAS.

We summarize by stating that a rich variety of magnetic resonance experiments is available and these have only been partially utilized in the investigation of conjugated polymers. Each technique has its specific advantages and drawbacks. Examples involving the application of these techniques to conjugated polymers will be discussed in the following sections.

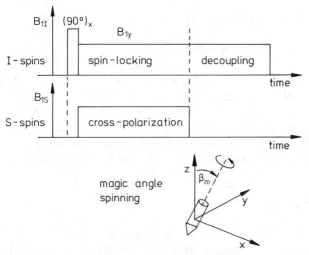

Fig. 5.2. Schematic sequence for cross-polarization (CP) magic angle spinning (MAS) NMR experiments. While the sample is rotated rapidly about an axis which is tilted by the magic angle ($\beta_m = 54°44'$) with repect to the static magnetic field (z-axis) a combined sequence of radio frequency pulses is applied to the ^{13}C spins (S-spins) and the nearby proton spin (I-spins). For details see [5.18]

5.2 Structure and Lattice Dynamics of Conjugated Polymers in the Non-Conducting Phase

5.2.1 Lattice Structure Determination from Dipole–Dipole Interactions

The structural arrangement of conjugated chains and their packing in the lattice implies that the proton nuclear resonance should be dominated by dipole–dipole interactions ($\propto r_{jk}^{-3}$, where r_{jk} is the separation between spins j and k) thereby leading to a broad, structureless and symmetric spectrum [5.14]. The geometric arrangement of the nuclear spins, as reflected in the lower order moments M_2 and M_4, has therefore been investigated using both cw and pulse techniques.

For $(CH)_x$, various structures have been proposed in the literature [5.40–50] and Table 5.1 lists the second moments calculated for some of the proposed structures of $trans$-$(CH)_x$. Experimental determination of M_2 by various authors [5.51–53] shows a large variation of M_2 values, ranging from 5–19 G^2 for $(CH)_x$. It is clear that discrimination between various structures requires accurate moments and good knowledge of the vibrational/librational motions of the chains. With regard to the first point, the moments evaluated from the cw spectrum using

$$M_{2n} = \int \omega^{2n} F(\omega) d\omega / \int F(\omega) d\omega \tag{5.2.1}$$

Table 5.1. Second moments M_2 calculated from various proposed structures for $trans$-$(CH)_x$ [5.57]. The moments for the structures (I–XV) proposed by *Enkelmann* et al. [5.49] have been calculated using $r_{C-C} = 146$ pm, $r_{C=C} = 135$ pm and a C=C–C angle of 120°, whereas those for the other structures involve $r_{C-C} = 144$ pm, $r_{C=C} = 136$ pm and an C=C–C angle of 122

Proposed structure	Calculated $M_2[G^2]$
Fincher et al. [5.43]	8.58
Baughman et al. [5.41]	8.86
Baughman et al. [5.42]	8.94
Shimamura et al. [5.44]	10.60
Enkelmann et al. [5.49]	
Structures I, II, XV	10.5 ± 0.1
Structure III	10.0
Structure IV	13.6
Structure V	12.9
Structure VI	9.5
Structures VII, IX	9.8
Structure VIII	36.9
Structures X, XIII	10.9
Structure XI	10.2
Structure XII	27.5
Structure XIV	11.9

suffer from noise and baseline artefacts of the experimentally obtained lineshape function $F(\omega)$. On the other hand, the moments from the FID, following an intense rf pulse, cannot be determined without mimicking the short-time behaviour (corresponding to the signal lost due to the spectrometer recovery). The short-time behaviour is especially important, since the FID can be written as

$$G(t) = 1 - M_2 t^2/2! + M_4 t^4/4! - \cdots \qquad (5.2.2)$$

where the moments are $2n$-order derivatives of $G(t)$ at $t = 0$. The use of poly-nomials or simple functionals which ignore the long-time features of the FID is highly unreliable to recover the lost signal. When the observed signal has the form of finite single-valued oscillatory functions, the FID may be represented as

$$G(t) = g(\alpha t)\exp(-\beta^2 t^2/2), \qquad (5.2.3)$$

where $g(\alpha t)$ should be chosen to yield a shape as close to $G(t)$ as possible. Depending on the ratio of the second and first nodes of the experimental signal, $g(\alpha t)$ can be chosen from the following functions: $J_0(\alpha t)$ with node ratio 2.29, $J_1(\alpha t)/(\alpha t)$ with 1.86, $J_2(\alpha t)/(\alpha t)^2$ with 1.64, $J_3(\alpha t)/(\alpha t)^3$ with 1.53, $\sin(\alpha t)/\alpha t$ with 2.0 or a Pake function (inverse Fourier transform of the Pake spectrum; see Sect. 5.2.2 and [5.14]) with node ratio 2.61.

The spectrometer recovery problem can, however, be eliminated altogether by using a magic echo sequence which yields zero-time resolution [5.54, 55]; the moments can then be estimated from the latter half of the echo by fitting it to, say, (5.2.3). The pulse sequence for the magic echo is shown in Fig. 5.3 and the echo occurs at $t = t_B/2$ where t_B is the duration of the pulse burst. Figure 5.4 shows the latter half of the echo for *trans*-$(CH)_x$ at 84 K and a fit to (5.2.3) with $g(\alpha t) = J_2(\alpha t)/(\alpha t)^2$; the values of the deduced second and fourth moments are $13.0 \pm 0.3\,G^2$ and $465 \pm 10\,G^4$, respectively. The broken line corresponds to (5.2.2) up to the second order in time, and the dash-dot line up to the fourth. The differences in obtaining moments from polynomial fits and functional fits can be demonstrated. The FID at $T = 200$ K [5.53] when fitted to (5.2.3) with $g(\alpha t) = J_1(\alpha t)/\alpha t$ gives a fit virtually indistinguishable from the

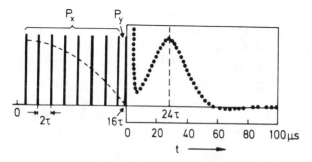

Fig. 5.3. Pulse sequence employed for the magic echo. A series of eight $\pi/2$ pulses in x direction spaced by 2τ is followed by a $\pi/2$ pulse in y direction of the rotating frame. The magic echo (data points from $(CH)_x$ experiment) is formed at 24τ

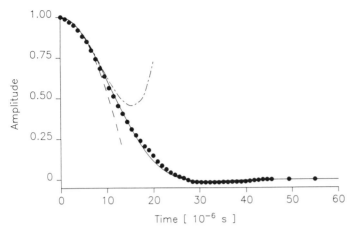

Fig. 5.4. Second half of the magic echo in *trans*-(CH)$_x$ at 84 K (filled circles); the continuous line is a fit to the data using (5.2.3) with $g(\alpha t) = J_2(\alpha t)/(\alpha t)^2$. The short-time behaviour of the magic echo obtained from the determined M_2 and M_4 is also shown to second order (broken line) and to fourth order (dashed-dot line) in time (c.f. (5.2.2))

experimental results but a value for M_2 ($7.5 \pm 0.2\,\text{G}^2$) much smaller than the value $11.8\,\text{G}^2$ obtained from the polynomial fit [5.53].

The second and fourth moments [5.56, 57], as obtained from the magic echo, are given in Table 5.2. Comparison of M_2 at 300 K with the proposed structures requires caution since (i) even pure *trans*-(CH)$_x$ contains some *cis* isomer or *cis* remnants [5.57], (ii) there are amorphous regions in the samples, (iii) a few percent($\approx 6\%$) of protons may have their origin in the catalyst residues [5.58] and (iv) there may be a contribution due to the vibrational/librational motions of the chains. These contributions usually lower the measured moments, although their precise magnitudes may not be easy to ascertain. For our good quality sample, if these contributions are within 10% of the measured values, the analysis seems to favour the structure due to *Shimamura* et al. [5.44] and a few others proposed from packing calculations [5.49]. The temperature dependence of M_2 does not exhibit any transition around 150 K as found by *Ikehata* et al. [5.52], as is also the case with other results [5.51, 53]. Consistently higher values of M_2, compared to previous work [5.51–53], are believed to be

Table 5.2. Values of the second and fourth moments at four different temperatures in *trans*-(CH)$_x$ as obtained from magic echo experiments by using a polynomial fit to the data (see text) [5.56]

	84 K	140 K	200 K	300 K
$M_2[\text{G}^2]$	13.0 ± 0.3	12.3 ± 0.3	11.8 ± 0.3	10.8 ± 0.3
$M_4[\text{G}^4]$	465 ± 10	396 ± 10	360 ± 10	314 ± 10

due to the advantage of the magic echo sequence compared with the other experimental techniques.

The temperature dependence of M_2 can be analysed in terms of the thermal expansion of the lattice constants along the crystallographic a and b axes. The calculations performed by effecting shrinkage to a and b by the same amount δ (which is the fractional contraction in a and b relative to their value at 300 K) yield a linear dependence

$$[M_2(\delta)]^{1/2} = [M_2(\delta = 0)]^{1/2} - m\delta \qquad (5.2.4)$$

with $m = 0.07$ and M_2 in G^2. The values of δ which correspond to the experimental M_2 values are plotted versus temperature in Fig. 5.5. It might be more informative, however, to express the results in terms of an expansion factor $\Delta(T)$ rather than a shrinking factor. Using the relationships $\delta(T) = [a(T) - a(300\,K)/a(300\,K)$ and $\Delta(T) = [a(T) - a(0)]/a(0)$, where $a(T)$ is the lattice constant a at temperature T, one obtains $\Delta(T)$, which is also plotted in Fig. 5.5. Note that both Δ and δ are assumed to be identical along the a and b axes. Similar magic echo experiments were also performed for polyparaphenylene (PPP) [5.56, 59] as discussed next.

The temperature dependence of the second moment is shown in Fig. 5.6 [5.56, 59]. A temperature-independent behaviour for $T < 230\,K$ and $T > 270\,K$

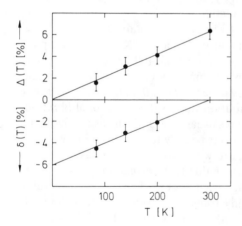

Fig. 5.5. Shrinkage (δ) and expansion (Δ) factors in *trans*-(CH)$_x$ as a function of temperature; the former is relative to the lattice parameters at 300 K whereas the latter is referred to absolute zero

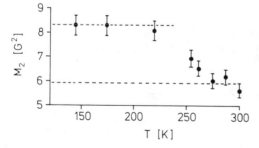

Fig. 5.6. Variation of the second moment M_2 as a function of temperature for polyparaphenylene using the magic echo sequence. The broken lines indicate the mean values of M_2 at low and high temperature [5.593]

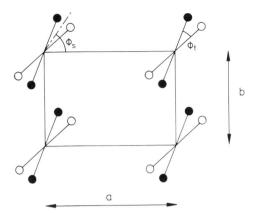

Fig. 5.7. Projection of the crystal structure of polyparaphenylene in the (a, b) plane

with a slow cross-over in the range 230–270 K is observed in both M_2 and M_4. To understand this behaviour, let us look at the crystal structure [5.60]: Space group $P2_1/a$, $a = 7.81$ Å, $b = 5.53$ Å, $c = 4.245$ Å and two chains per unit cell. There is uncertainty in the values of the monoclinic angle β and the setting angle ϕ_s of the chains; β can, e.g., be in the range $90°–100°$ and ϕ_s around $55°$. Projection of the crystal structure in the ab plane is shown in Fig. 5.7. The calculated second moment varies between $14.5\,G^2$ and $15.3\,G^2$ when a planar structure for the chains is assumed and β and ϕ_s are varied in the ranges $90°–104°$ and $53°–63°$, respectively. Since these values are large compared to the experimental values and because the M_2 for a single *planar* chain is about $11.8\,G^2$, the consecutive phenylene rings, as in p-terphenylene [5.61], must be twisted by an angle ϕ_t. The calculations including this orientational effect show drastic changes in M_2. For example, choosing $\beta = 90°$, $\phi_s = 55°$ and $\phi_t = 26°$, M_2 decreases to $9.8\,G^2$. It is to be noted that (i) M_2 is not very sensitive to the angles β and ϕ_s and (ii) M_2 cannot be lowered beyond $8\,G^2$ (for this $\phi_t = 40°$) by changing ϕ_t. Agreement between the experimental and calculated second moments is obtained for the following values: $\beta = 95 \pm 5°$, $\phi_s = 55 \pm 2°$ and $\phi_t = 30 \pm 4°$ in the low temperature phase.

It is clear that to explain the observed temperature dependence, at least two additional processes must be considered. The first is the partial averaging of the dipole–dipole interactions through thermally activated motion of the consecutive rings between two configurations and the second is the possible expansion of the lattice with temperature. As to the first process, the calculations involving flip–flop motion of the rings indicate reduction of M_2 from $9.8\,G^2$ to $6.4\,G^2$ which is quite near the M_2 value at room temperature. An approximate estimate of the activation energy for the collective motion of the rings can be made from the Arrhenius law

$$\tau(T) = \tau_0 \exp(E/k_B T). \tag{5.2.5}$$

One expects a change in M_2 at a temperature where the jump rate of the rings is of the order of the dipole–dipole interaction, i.e. $M_2^{1/2}\tau(T) \approx 1$. Taking $M_2 = 10\,G^2$, $T = 240\,K$ and $\tau_0 = 10^{-13}\,s$, the activation energy is about 32 kJ/mole [5.59]. This value should be contrasted with 2.5 kJ/mole obtained for p-terphenylene [5.61]. The chain length and intermolecular packing in PPP seem to play an important role in the phenyl ring flipping energy.

The role of lattice expansion in proton second moments is rather difficult to ascertain. The contribution of one proton due to the next three neighbours on a single chain is already about $6\,G^2$, i.e. minor changes in the twist angle can have significant effects on the value of the moments.

Polyparaphenylene sulphide (PPS) is a conjugated polymer in which the phenylene rings are essentially perpendicular to each other. The presence of S–C bonds imparts flexibility to the rings which can rotate about their long axes. The second moment at low temperatures is $\approx 7\,G^2$ [5.62] compared to $\approx 8.4\,G^2$ for PPP [5.56, 59]. Calculation of the rigid lattice M_2 from the known lattice structure [5.63] yields values much smaller than the observed ones, implying closer chain packing than proposed in [5.63]. The temperature variation of the second moment shows at least two activation energies, $\approx 6\,kJ/mole$ and $\approx 14\,kJ/mole$. There could in fact be a distribution of activation energies depending upon the rotation of the phenyl rings in different polymer morphologies.

5.2.2 Bond Length Determination from Dipole–Dipole Interactions

It is well known that an assembly of isotropically distributed spin-1/2 pairs (also single spins-1) gives rise to characteristic Pake spectrum whose singularities (corresponding to $\theta = 90°$ in (5.1.16) or shoulders ($\theta = 0°$ in (5.1.16)) directly yield an estimate of the internuclear distance between the spins. For an accurate determination of the internuclear distance, chemical shift and heteronuclear interactions must be suppressed as in transient nutation experiments [5.64] or delineated from the dipole–dipole interactions as in two-dimensional NMR experiments [5.65]. Bond length determinations for both cis- and trans-$(CH)_x$ have been performed on the same samples of $(CH)_x$ [5.65].

To obtain a dilute distribution of bonded pairs, whose bond lengths were to be determined, a mixture of 4% doubly ^{13}C-enriched acetylene (99% ^{13}C) and 96% natural abundant acetylene was polymerized to yield cis samples of $(CH)_x$. The trans samples were then prepared by thermal isomerization.

The experimental set-up for the transient nutation NMR is shown in Fig. 5.8 and is described in detail elsewhere [5.66, 67]. We briefly remark that the usual cross-polarization experiment for ^{13}C without magic angle spinning is first applied. Instead of observing just the ^{13}C FID after the CP pulse, a repeated cycle of pulses (e.g. 90°) is applied to the ^{13}C spins and the transient nutation is observed during the "windows". The corresponding spectrum is then obtained by Fourier transformation. Note that all linear spin interactions are averaged

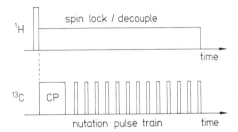

Fig. 5.8. Pulse scheme for the nutation experiment. The rf field on the protons is on continuously to suppress the ^{13}C–^{1}H interactions. The application of a nutation pulse train after the cross-polarization sequence minimizes the chemical shift effects. The nutation signal is captured between the rf windows [5.64, 66]

to zero, but the dipole–dipole interactions between ^{13}C spins are retained, though scaled by a factor 0.5.

The results of this experiment at 77 K in both the *cis* and *trans* isomers of $(CH)_x$ and their simulations, with bond lengths as parameters, are shown in Fig. 5.9 alongwith the schematic *cis* and *trans* structures. The existence of only one pair of singularities in *cis*-$(CH)_x$ (Fig. 5.9a) has been argued to be due to the Ziegler–Natta reaction which leaves the original pairs' doubly bonded. The spectrum simulation shown by a dotted line corresponds then to $r_{c=c} = 1.37$ Å (marked with asterisks in Fig. 5.9a). Two pairs of Pake singularities in *trans*-$(CH)_x$ have been interpreted as being due to C–C and C=C bonds and the spectrum simulations gives their values as 1.44 Å and 1.36 Å, respectively; these values were found to be independent of temperature in the range 4.2–300 K. It is to be noted that these bond lengths are the averages of a narrow distribution arising from a distribution of conjugated lengths [5.68, 69].

The values of the single and double bond lengths as deduced from the 2D experiment [5.65] (to be discussed in Sect. 5.2.3) are 1.45 ± 0.01 Å and 1.38 ± 0.01 Å respectively, and are in agreement with the results of [5.64].

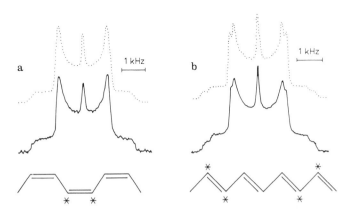

Fig. 5.9 a, b. Experimental and simulated (dotted line) nutation spectra at 77 K in (a) *cis*-$(CH)_x$ and (b) *trans*-$(CH)_x$ for the samples weakly enriched in ^{13}C pairs. The central peak corresponds to isolated ^{13}C spins. The deduced bond lengths are $r_{c=c} = 1.37$ Å for *cis*-$(CH)_x$ and $r_{c-c} = 1.44$ Å and $r_{c=c} = 1.36$ Å for *trans*-$(CH)_x$ [5.64]

5.2.3 Chemical Shift Tensor

Chemical shift tensor σ is a measure of the local anisotropic fields as seen by the nuclei. Although the quantum mechanical calculation of the chemical shift tensor is rather involved and reliable theoretical values are difficult to obtain for large molecules, it is nevertheless an experimentally accessible quantity and helps to differentiate between different types of conjugated structures. The measurement of σ provides information about the local electronic configuration and hence about the chemical structure. In the following we restrict ourselves to a phenomenological discussion of the experimentally determined values. There is some confusion in the literature concerning the sign of the chemical shift. We adopt here the notation where the observed frequency ω of a spectral line (for fixed magnetic field B_0) can be expressed as

$$\omega = \omega_0(1 - \sigma_{zz}) = \omega_0(1 + \delta), \tag{5.2.6}$$

with σ_{zz} being the component of the chemical shift tensor which is parallel to B_0 and where ω_0 is the Larmor frequency. Positive σ_{zz} therefore leads to a reduction in frequency (shielding) and $\delta = -\sigma_{zz}$. Note that the δ scale is commonly used in the chemistry literature. The chemical shift tensor (only the symmetric part) is represented by three principal values σ_{11}, σ_{22} and σ_{33} along the three principal axes. If the σ tensor is isotropic or its anisotropy is averaged by motion (including magic angle spinning) only the isotropic part $\sigma_i = (\sigma_{11} + \sigma_{22} + \sigma_{33})/3$ survives.

The isotropic part σ_i of the chemical shift tensor in $(CH)_x$ was first reported by *Maricq* et al. [5.70] who found values of -129 ppm and -139 ppm (with respect to tetramethylsilane) for cis and trans isomers, respectively. The various values existing in the literature fall in the range $-(126–129)$ ppm for *cis*-$(CH)_x$ and $-(138–139)$ ppm for *trans*-$(CH)_x$ [5.71–77]. The three principal components of σ have also been determined from the simulation of the CP powder spectra and/or the spinning side-bands analysis of the cross-polarization magic angle spinning (CPMAS) spectra [5.71, 75, 76]. The tensor components σ_{11}, σ_{22} and σ_{33}, obtained by various authors, show deviations for *cis*-$(CH)_x$ around mean values of -222 ± 6, -138 ± 1 and -21 ± 3 ppm, respectively. The corresponding mean values for the trans isomer are: -225 ± 9, -143 ± 3 and -42 ± 9 ppm, respectively. Comparison with the ^{13}C chemical shift tensors of other conjugated molecular solids [e.g., benzene and dihydromuconic acid (ADIMU)] is facilitated by Fig. 5.10. Consistent with these data, σ_{33} can be conjectured to be perpendicular to the plane of C–C=C whereas σ_{22} is nearly parallel to the C=C bond as in ADIMU; σ_{11} would then be orthogonal to σ_{22} and σ_{33}. Detailed theoretical investigations of the chemical shift tensor in polyacetylene and other polyenes were performed by *Ando* et al. (5.78–80].

Further evidence that σ_{33} is indeed perpendicular to the plane of C–C=C bonds comes from 2D CP experiments [4.57, 66]. These experiments are designed to separate different contributions to the spectrum, such as chemical

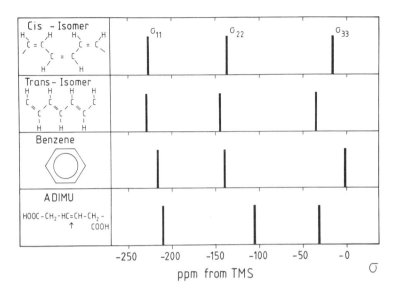

Fig. 5.10. Comparison of several ^{13}C chemical shift tensors of alternating double bonds [5.75]

shift interactions (causing a frequency shift ω_1) and dipole–dipole interactions (causing a splitting ω_2). The different interactions can be observed separately or in a two-dimensional plot with frequency axes ω_1 and ω_2. An early experiment of this sort was performed in 1982 in order to relate the chemical shift tensor to the CH bond [5.56], where

$$\omega_1 = \omega_{11} \cos^2 \alpha \sin^2 \beta + \omega_{22} \sin^2 \alpha \sin^2 \beta + \omega_{33} \cos^2 \beta \tag{5.2.7}$$

$$\omega_2 = \pm 3\gamma_I \gamma_S \hbar r^{-3} P_2(\cos \theta)/2 \tag{5.2.8}$$

$$\cos \theta = \sin \psi \sin \beta \cos (\alpha - \phi) + \cos \psi \cos \beta. \tag{5.2.9}$$

In these equations (α, β) are the Euler angles relating the chemical shift principal axes system (PAS) to the laboratory system, whereas (ψ, ϕ) are the polar angles of the carbon–proton internuclear vector in the chemical shift PAS. Simulation of the detected signal, after powder averaging over (α, β), was found to give good agreement with the experimental data with the following parameters [5.56]: $\sigma_{11} = -234$ ppm, $\sigma_{22} = -146$ ppm, $\sigma_{33} = -34$ ppm, $\phi = 23° \pm 2°$ and $\psi = 90° \pm 6°$. This gives an angle of $\approx 40°$ between the C–C bond and the σ_{11} axis, if the angle between the single and double bonds is assumed to be $120°$. The chemical shift PAS with respect to the –CH bond is sketched in Fig. 5.11.

A similar 2D NMR experiment was performed more recently in which the chemical shift tensor was related to the ^{13}C–^{13}C bond [5.65]. The resulting 2D spectrum exhibits ^{13}C chemical shift frequencies along the ω_1 axis and ^{13}C–^{13}C dipolar frequencies along the ω_2 axis. Identificaton of the various

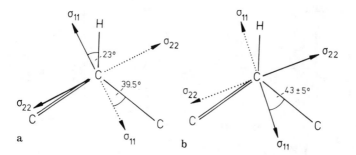

Fig. 5.11 a, b. Orientation of the chemical shift tensor in *trans*-(CH)$_x$ with respect to (**a**) the –CH bond [5.56] and (**b**) the C–C bond [5.65]

features of this spectrum shows that a cross section at $\sigma = \sigma_{33}$ yields two pairs of Pake singularities with their separation given via (5.2.8). The bond lengths for C–C and C=C determined from these singularities are 1.45 ± 0.01 Å and 1.38 ± 0.01 Å, respectively. The angle between the C–C bond and the σ_{11} axis obtained by this method is $\phi_{c-c} = 43 \pm 5°$ in agreement with [5.56]. The chemical shift determined in this way is shown in Fig. 5.11b.

The information about the chemical shift tensor, particularly its isotropic component, has been used to study the *cis-trans* thermal isomerization in (CH)$_x$ [5.72, 73]. Figure 5.12 shows such a CPMAS set of spectra from *Terao* et al. [5.76] indicating clearly *cis* (Fig. 5.12a) to *trans* (Fig. 5.12d) isomerization.

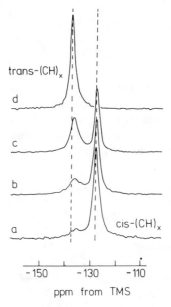

Fig. 5.12. ^{13}C cross-polarization magic angle spinning (CPMAS) spectra showing the *cis-trans* thermal isomerization in (CH)$_x$. The spectra *a* and *d* correspond to nearly pure *cis*- and pure *trans*-(CH)$_x$, whereas *b* and *c* correspond to a thermal treatment at 393 K in vacuum for 10 and 40 minutes, respectively [5.76]

Upfield shifts in Fig. 5.12b, c for the trans peak with respect to the position of the pure *trans* peak indicate that the structure of partially isomerized $(CH)_x$ is different that for pristine *trans*-$(CH)_x$. The *trans* peaks for the partially isomerized sample are broad relative to the pure *trans* peak, indicating a disordered structure or inhomogeneously broadened lines. This behaviour is rationalized by asserting that the isomerization proceeds inhomogeneously from the surface of the fibrils. A linear relationship between the decreasing linewidth of the *trans* peak and decreasing of the *cis* content (as determined from the Fourier transform infra red (FTIR) spectra) is found by *Gibson* et al. [5.81]. Furthermore, in contrast to earlier results [5.73], a linear 1:1 correlation of the isomeric compositions as determined by NMR and FTIR methods is established. Although Fig. 5.12 shows the general behaviour of *cis-trans* isomerization, it does not imply that these are the optimal conditions; detailed ESR linewidth investigations as a function of heating time [5.82] suggest a heating time of about five minutes in the range 453–493 K.

The instability of *trans*-$(CH)_x$ when exposed to air or oxygen is known to result from the formation of defect structures. A highly oxygen-contaminated sample shows in addition to the ordinary ^{13}C chemical shift tensor, another set of lines with [5.83] $\sigma_{11} = -242 \pm 7$ ppm, $\sigma_2 = -174 \pm 16$ ppm and $\sigma_{33} = -132 \pm 13$ ppm. Comparison of this tensor with that of model compounds allows one to draw some conclusions concerning the chemical structure of the defect [5.83]. Contrary to the previously proposed defect structures [5.84, 85], the defect seems to have the chemical structure of a carboxylic group or that of a structure involving a double bond between carbon and oxygen atoms. Such a defect also explains the brittleness of oxygen-contaminated $(CH)_x$.

In polyparaphenylene (PPP) two different kinds of carbons (*A* and *B*) can be distinguished, i.e. those which are bonded to protons (*A*) and those which take part in the phenyl–phenyl bond(*B*). They give rise to distinguishable ^{13}C chemical shift spectra, which are similar to those of the oligomers [5.86]. A complete analysis of the ^{13}C chemical shift tensor was also performed for the same class of materials [5.56, 87]. By using the spinning side-band analysis of the CPMAS spectra the following values were obtained: For tensor *A* $\sigma_{11} = -213 \pm 3$ ppm, $\sigma_{22} = -145 \pm 2$ ppm and $\sigma_{33} = -19 \pm 1$ ppm for tensor *B* $\sigma_{11} = -229 \pm 3$ ppm, $\sigma_{22} = -161 \pm 4$ ppm and $\sigma_{33} = -17 \pm 1$ ppm [5.56, 87]. The direction of the principal axes in the molecular frame is the same as discussed in the case of polyacetylene, i.e. σ_{33} is perpendicular to the molecular plane. Several oligomers were also investigated, as shown in Fig. 5.13 [5.87]. There seems to be a linear relationship between the chain length n (number of phenyl rings) and σ_{33}–σ_{11} for the bonding carbons (*B*) (Fig. 5.14). The value for PPP is shown by the unfilled circle and seems to suggest a chain length of 8 rings. The PPP chain most likely consists of more rings and the linear relationship might not hold for longer chains. Moreover, there is substantial ring flip motion as discussed in Sect. 5.2.2, which tends to partially average the chemical shift tensor. No firm conclusion about the effective chain length can therefore be drawn from this kind of analysis.

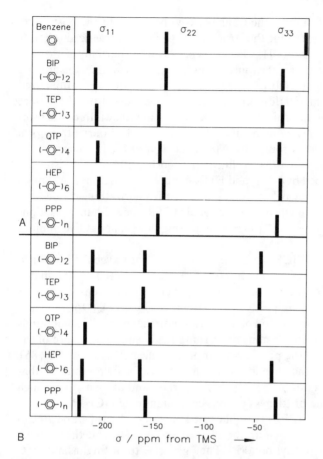

Fig. 5.13. Components of the chemical shift tensor for polyparaphenylene and its oligomers for proton-bonded (A type) and proton-nonbonded (B type) ^{13}C spins; the tensor components for benzene are also shown for direct comparison [5.87]

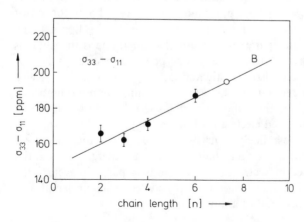

Fig. 5.14. ^{13}C chemical shift asymmetry parameter ($\sigma_{33} - \sigma_{11}$) as a function of the chain length n for the oligomers of polyparaphenylene. Open circle corresponds to the value for polyparaphenylene

Isotropic ^{13}C chemical shift values were also obtained for other conjugated polymers. In polypyrrole (PPy) the measured values [5.88, 89] are $\sigma_A = -105.2$ ppm and $\sigma_B = -125.2$ ppm. Similar values have also been reported for polythiophene [5.88, 90–92], polyaniline [5.88] and polyparaphenylene sulphide [5.93, 94].

5.3 Spin Dynamics of Conjugational Defects in the Non-Conducting Phase

It has been known for over two decades that conjugated polymers contain conjugational defects carrying electronic spins [5.95]. The recent synthesis of freestanding films of $(CH)_x$ [5.96], which show conducting behaviour on doping, has led to renewed interest in both the pristine and doped forms of these polymers.

In this section we discuss magnetic studies of the undoped polymers in the non-conducting form. This topic is important in part because it deals with the pure state to which the dopant is added. It is of special importance in the case of *trans*-polyacetylene, because in this material it is only the uncharged form of the soliton defect that has an unpaired spin and can, therefore, be studied through its magnetic effects. The main questions that have occupied investigators have been concerned with the structure of the defect, its interactions with the surroundings and other defects, and the microscopic nature of its motion. As far as the motion is concerned, special emphasis has been given to the measurements that would elucidate its expected one-dimensional (1D) nature. Here, we review how ESR and ENDOR lineshapes, dynamic nuclear polarization, nuclear spin-lattice relaxation and electron spin relaxation have been used to explore this subject.

5.3.1 ESR and ENDOR Lineshapes

ESR studies on pristine *trans*-$(CH)_x$ were the first to suggest [5.97–99] that the ESR signal arises from a mobile π-electron bond-alternating defect ($g \approx 2.0023$). The existence of π electrons leads to anisotropies both in the g value ($\Delta g = -0.0017$ in ideal π electron radicals) and the hyperfine interaction; these anisotropies can be ascertained using, e.g. oriented samples. The *cis* isomer also exhibits a weak ESR signal whose origin is believed to be in the "locking in" of some *trans* isomers between *cis*-$(CH)_x$ regions. PPP, synthesized by two commonly employed techniques [5.100, 101], exhibits ESR signal due to π electrons but yields different magnetic behaviour [5.102]. The g value for PPy (≈ 2.0025) is typical of π electrons [5.103] and two different types of defects, with nearly the same g value, have been seen via ESR.

In a standard ESR experiment, all types of defects contribute to the ESR signal according to their concentration and thus a defect of small concentration cannot usually be distinguished in the presence of other defects even if they differ substantially in their linewidth. In these situations, time-resolved ESR [5.56, 104, 105] can be employed. From such experiments two types of defects were observed in cis-$(CH)_x$, one delocalized over several lattice sites and the other localized to a single –CH unit; the concentration of the latter defects was estimated to be about 1/100 of the former. In trans-$(CH)_x$, on the other hand, delocalized "pinned" and mobile defects were observed; the distinction between these two disappears below 50 K. Besides the uncontrolled presence of oxygen which leads to pinning of the defects [5.106, 107], presence of remnant cis linkages and the electronic rearrangements between the cis–transoid and trans–cisoid configurations has also been suggested as leading to different kinds of pinning defects [5.57].

The principal components of the hyperfine tensor, due to a π radical coupled to a proton, can approximately be expressed as $A_{11} = -(1 - \alpha)a$, $A_{22} = -a$ and $A_{33} = -(1 + \alpha)a$, where α accounts for the anisotropy and a is the isotropic part of the hyperfine coupling. For π electrons, α is usually between 0.5–0.6 and a can be in the range 20–30 G. Very often, the characteristic values of the malonic acid radical $A_{11} = -11\,\text{G}$, $A_{22} = -23\,\text{G}$ and $A_{33} = -35\,\text{G}$ are chosen [5.20]. Since the ESR lines in conjugated polymers are narrow, partial averaging of the hyperfine tensor can take place either dynamically (motionally or via multi-spin exchange) or through spin delocalization, or both. As both ESR and ENDOR experiments measure the hyperfine interactions, the question of spin delocalization can be answered if the dynamical effects related with motion can be minimized, e.g. by working at low temperatures. In this case the Hamiltonian

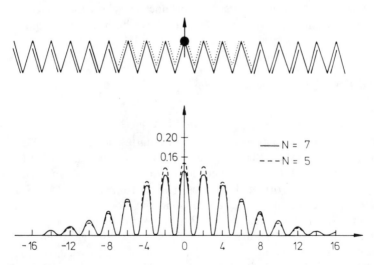

Fig. 5.15. Delocalized shape of the soliton in trans-$(CH)_x$ as predicted theoretically by Su et al. [5.2]

of the system can be written as

$$\mathscr{H} = \omega_e S_z - \omega_n I_z + \sum_j I_j A_j S, \tag{5.3.1}$$

where ω_e and ω_n are the electron and nuclear Larmor frequencies. The ESR spectra can be simulated with a particular distribution of hyperfine tensors A_j ($\propto \rho_j$, the spin density at site j). The theory of *Su* et al. (SSH) [5.2] predicts $\rho_j = (1/l) \operatorname{sech}^2(j/l) \cos^2(j\pi/2)$ for the soliton defect in *trans*-(CH)$_x$ (Fig. 5.15); here l is the half width of the delocalized soliton and $j = 0, \pm 2, \pm 4, \ldots$, where $j = 0$ is the centre of the soliton.

Weinberger et al. [5.99], using the SSH spin distribution for $l = 7$ and with only the isotropic part of the hyperfine tensor, could simulate well the ESR spectrum in *trans*-(CD)$_x$. *Kuroda* et al. [5.108] later included the hyperfine anisotropy whilst *Sasai* and *Fukotome* [5.104] calculated the ESR spectrum utilizing Pariser–Parr–Pople Hamiltonian. This has been considered as evidence for the SSH spin density of the defect. Interestingly enough, not only the SSH model for ρ_j but also others [5.20] give similar quality results for the ESR

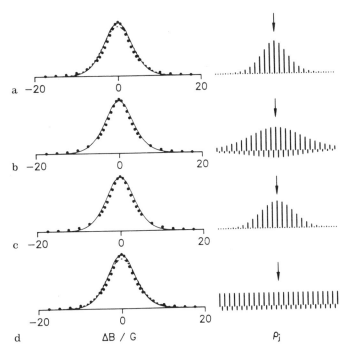

Fig. 5.16 a–d. Experimental (at $T = 4.2$ K) and calculated ESR powder spectra, using different spin density distributions, for *trans*-(CH)$_x$: **a** SSH with $j_{max} = 50$ and $l = 8$ (full curve) and $l = 7$ (broken curve); **b** ASD (i.e. SSH with finite spin density on odd sites) with $l = 15$, $j_{max} = 50$ and $ug = 0.3$; **c** the Gaussian distribution with $l = 8$ and $j_{max} = 50$ and **d** the rectangular distribution with $ug = 0.3$ and $l = j_{max} = 24$ (full curve) and $l = j_{max} = 20$ (broken curve). For each case, the form of the spin density function is also shown. The arrows denote the central $j = 0$ position of the distribution function

line. Figure 5.16 exemplifies this point, showing clearly that the SSH alternating spin density (SSH form but with the inclusion of negative spin densities on odd sites), Gaussian or even the rectangular spin density distribution give similar results with reasonable physical parameters. This is understandable in the light of the discussion in Sect. 5.1.4. It was shown there that the second moment M_2 suffices to describe the linewidth of the ESR line, which is Gaussian (central limit theorem) when a large number of protons are coupled to a single electron spin. In defining the spin densities, the following notation is used [5.20]: with g and u as the amplitudes of ρ_j at, respectively, even (gerade) and odd (ungerade) sites j, $g - u = 1$, $ug = u/g$, $\sum \rho_j = 1$ and $j = 0, \pm 1, \pm 2....$ and l is the half-width of the spin distribution function. A modified SSH spin-density distribution can now be expressed in the form [5.20]

$$\rho_j = (1/l)\text{sech}^2(j/l)[g\cos^2(j\pi/2) - u\sin^2(j\pi/2)] \tag{5.3.2}$$

or for an arbitrary but properly normalized distribution function $f(x)$

$$\rho_j = f(j/l)[g\cos^2(j\pi/2) - u\sin^2(j\pi/2)]. \tag{5.3.3}$$

As many different functional forms fit the ESR spectrum, the conclusion is straightforward: ESR spectrum by itself does not unambiguously define the spin density distribution of the conjugational defect.

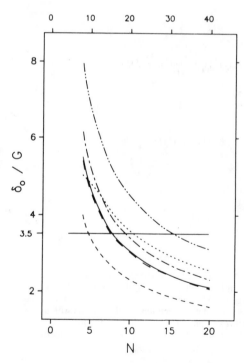

Fig. 5.17. The half-width at half maximum height of the ESR spectrum as a function of l (the half-width of the soliton) calculated using M_2 and M_4 for various forms of ρ_j: SSH (continuous line), Lorentz (small broken lines), Gauss (long broken lines), ASD with $ug = 0.1$ (dash-dotted lines), ASD with $ug = 0.3$ (dash-dot-dot lines) and rectangular with $ug = 0.3$ (dotted line). The scale at the top corresponds to the rectangular distribution and the straight line at $\delta_0 = 3.5\,\text{G}$ corresponds to a typical value for trans-$(CH)_x$ at 4.2 K

Since the width of the ESR spectrum is intimately related to the width of ρ_j, a plot showing the calculated half-width at half-height of the ESR spectrum versus l is shown in Fig. 5.17 for various functional forms of ρ_j. It is again clear that the width l can only be inferred from the ESR spectrum if the precise shape of ρ_j is known. The reason for this is simple. In the ESR spectrum which, for a simple case of $\omega_{01} \gg A_{zz}$, is the Fourier transform of

$$G(\alpha, \beta, t) = \prod_j \cos(A_{zzj}t/2), \tag{5.3.4}$$

each spectral line involves multiple convolutions of N spectral lines alongwith their dependence on all the hyperfine couplings through their orientations (α, β) with respect to the principal axis system [5.20].

The situation is however different in the case of ENDOR where the spectrum results from simple summation rather than multiple product of the various spectral lines, as in the ESR case. Any form of ρ_j must therefore reproduce simultaneously the ENDOR spectrum. Detailed ENDOR investigations of $(CH)_x$ (and its composites) have been made by *Dalton* and co-workers [5.110–115] and by *Kuroda* et al. [5.108, 116, 117]. At very low temperatures, the observation of similar ENDOR spectra from *cis*- and *trans*-$(CH)_x$ imply the same nature of the defects [5.113]. The spectra consist of two well-defined peaks on either side of the free proton Larmor frequency and a central peak, possibly due to distant ENDOR, exceptions being the results of [5.111] at 10 K, of [5.118] at 4 K and more recently of [5.119] at 4.2 K. The interpretation of *Dalton* is in terms of two hyperfine tensors with spin densities 0.06 and -0.02 per carbon site implying thereby a rectangular form of ρ_j with delocalization length 50 [5.112–114]. Since ENDOR reflects only the static properties, it is argued that the SSH soliton diffuses, in a time less than nuclear T_1, over ≈ 50 –CH units giving thereby an average spin density which is measured by ENDOR [5.112, 120]. The explanation for the negative spin densities on odd sites is due to the electron–electron Coulomb interactions [5.120]. The main reason for the above interpretation is the absence of a dominant peak, due to the weak hyperfine couplings, near the free proton Larmor frequency (implicit in SSH ρ_j with or without $ug = 0$) which is usually observed in π-radical systems. The cut-off of the weak hyperfine couplings of the SSH ρ_j in the above way is contested by *Kuroda* et al. [5.115, 117] who argue that SSH ρ_j, truncated for small values but including the negative spin densities, gives an equally adequate description of the ENDOR results. More recent pulsed-ENDOR investigations on Feast type (Durham route) polyacetylene have demonstrated that the double peak pattern can be accounted for by spin dynamical effects rather than vanishing hyperfine interactions [5.121]. This is substantiated by SEDOR (spin echo double resonance) experiments, which clearly identify the weak hyperfine interactions in the middle of the spectrum [5.122]. Figure 5.18 shows the Davies-type pulsed-ENDOR spectrum of Feast-type *trans*-$(CH)_x$ at 20 K according to [5.121–123] together with the proposed spin density. Characteristics of the spin-density distribution are reflected in the wings of the ENDOR spectrum, whereas the hole in the

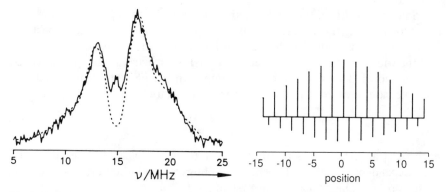

Fig. 5.18. Pulsed ENDOR spectrum at 20 K of Feast-type *trans*-(CH)$_x$ (left) according to [5.121, 122]. Right: Proposed spin-density distribution including alternating positive and negative spin-density with half-width $l = 11$ and negative/positive spin-density with half-width $l = 11$ and negative/positive spin-density ratio $u/g = 0.43$. The theoretical calculation of the ENDOR spectrum based on the proposed spin-density distribution is shown as a dashed line. The peak in the middle of the spectrum can be explained by distant hyperfine interactions due to neighbouring chains

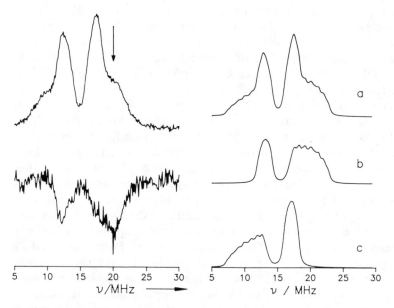

Fig. 5.19. Pulsed triple experiment on Feast-type *trans*-(CH)$_x$ at 20 K according to [5.123]. Left: The arrow in the pulsed ENDOR spectrum marks the position of triple irradiation. The corresponding difference triple spectrum is shown below. It reflects the correlation between positive and negative spin densities. Right: Simulated ENDOR spectra for the $+1/2$ and $-1/2$ sublevels, respectively, of the electron spin. The observed ENDOR spectrum is the sum of these sub-level spectra (lower part)

middle of the Davies-type ENDOR spectrum is due to spin alignment effects. These spin alignment effects are also present in CW ENDOR and must not be interpreted as spectral features. Independent triple-resonance experiments were performed [5.123] which revealed the ratio of negative to positive spin density $ug = 0.43$. Figure 5.19 shows the result of such a triple experiment where rf irradiation is performed at two independent ENDOR frequencies in addition to the mw irradiation. The important result of such an experiment is its possibility to determine the relative sign of hyperfine interactions. In the triple spectrum shown in Fig. 5.19 two spectral features are visible, one above the nuclear Larmor frequency and one below. This proves that the relative sign of the hyperfine interaction is negative. From numerous experiments of this type the negative/positive ratio of hyperfine interactions in $trans$-$(CH)_x$ was determined with an average value of $ug = 0.43$ [5.123]. A value of 0.44 has recently been obtained from CW ENDOR and ENDOR-induced ESR experiments on stretched samples of cis-rich $(CH)_x$ [5.124]. These values for ug are all rather large. Note that in the SSH model, $ug = 0$ since electron–electron correlations are neglected. In an antiferromagnetic chain with strong electron–electron correlations, $ug = 1$. A value of $ug = 0.43$ observed in polyacetylene therefore suggests a Hubbard $U = 7\,eV$ if the transfer integral $t = 2.5\,eV$ [5.125]. Similar negative spin densities appear also in Hartree–Fock calculations [5.126], INDO calculations [5.127] and calculations based on the density functional method [5.128]. A detailed discussion of electron–electron correlation is given in Chap. 1 (Sect. 1.4).

With the value for ug fixed by triple resonance experiments, it is possible to fit the ENDOR spectra with e.g. a spin-density distribution such as the one in (5.3.1) by varying only one parameter, namely the length l. The best fit was obtained for $l = 11$ [5.121]. The half width of the soliton seems to be slightly more extended than proposed by the SSH model. It should be noted that there are a number of ingredients for fitting the ENDOR spectra: (i) proper hyperfine tensor values A_{11}, A_{22}, A_{33} have to be chosen, (ii) transition matrix elements up to second order have to be taken into account and (iii) the full orientational dependence of all anisotropic parameters must be used and a powder average must be performed. Moreover, different spin alignment effects have to be considered. ENDOR spectra based on these prerequisites were calculated in [5.121, 122] and were found to give reasonable fits with the observed spectral features.

The low temperature ENDOR spectrum of PPP is very similar to that of $trans$-$(CH)_x$ shown in Fig. 5.18. The spectrum spans a frequency range of $\approx 10\,MHz$. This indicates spin delocalization over at least 4 rings. It is therefore suggested that the electron spin resides on a polaron-like defect. ENDOR results on a stretched film of polyphenylene-vinylene also show a fairly delocalized π-electron defect [5.129].

Finally, the question of spin delocalization has been addressed through the measurements of hyperfine couplings via electron spin echo envelope modulation patterns [5.130, 131]. Here the electron spin echo amplitude is modulated when the pulse spacing τ between the two microwave pulses exciting the spin echo is varied. The Fourier transform of this transient phenomenon results in a spectrum which reflects the hyperfine interactions. Because of the strong non-linear excitation, "multi-quantum" transitions are also excited. These appear at multiples of the nuclear Larmor frequency ω_{0I}. The number of multiple quantum lines depends on both the strength of excitation and the number of nuclei coupled to the electron spin. Firm conclusions can, however, only be drawn when the excitation mechanism is properly taken into account. This point has also been raised [5.132] with respect to the interpretation of the proton ENDOR spectrum in ^{13}C-enriched samples of $(CH)_x$ [5.112, 131].

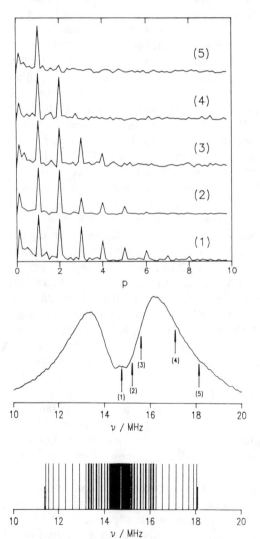

Fig. 5.20. Multi-quantum ENDOR spectra of Feast-type *trans*-$(CH)_x$ at 20 K according to [5.133]. Left: Different sets of multi-quantum ENDOR spectra for different frequency positions. Right: ENDOR spectrum (top) with arrows pointing at the different frequency positions where multi-quantum spectra were taken. Below, the spectral density of hyperfine lines corresponding to the spin-density distribution shown in Fig. 5.18 is represented as a stick spectrum. Note the decrease of spectral density when moving away from the centre

An alternative way of observing multi-quantum spectra of ENDOR sub-levels is by direct excitation of the ENDOR sublevels. Figure 5.20 shows spectra of this sort [5.133]. The higher the number of visible multi-quantum spectra at a particular ENDOR frequency, the higher the spectral density of hyperfine lines there. If the spin-density distribution function is monotonically decaying, as shown in Fig. 5.18, a large number of spectral lines is expected in the centre of the ENDOR spectrum, i.e. a large number of multi-quantum lines is also visible. In the wings of the ENDOR spectrum, however, which corresponds to the centre of the soliton, the spectral lines are spaced further apart, resulting in fewer multi-quantum lines [5.133].

5.3.2 Dynamic Nuclear Polarization

Dynamic nuclear polarization (DNP) experiments have been used to study the motion and trapping of the conjugational defect in $(CH)_x$. It is expected that the DNP due to a stationary defect will display a solid-state effect (SSE) and that of a rapidly moving one will show an Overhauser effect (OE). In this case, rapid motion means that the resultant time varying hyperfine interaction has components to its power spectrum at the ESR frequency ω_e. The two effects can be distinguished experimentally from the fact that the OE is symmetric about the ESR absorption, whereas the SSE is antisymmetric.

Figure 5.21a, shows the DNP of protons as a function of frequency for various mixtures of cis- and trans-$(CH)_x$ at 5.5 K and 300 K [5.134]. The trans-rich material shows a symmetric DNP line at 300 K and an antisymmetric one at 5.5 K. These results were interpreted as evidence for rapid motion (OE) at high temperature and a trapped conjugational defect at low temperature. Varying the relative content of the trans isomer in $(CH)_x$ (Fig. 5.21b) yielded a mixture of both the SSE and OE [5.134, 135].

Such measurements on trans-$(CH)_x$ were extended by Clark et al. [5.136] to the entire temperature range 1.5–300 K. By decomposing the result into a symmetric (OE) and antisymmetric (SSE) part at each temperature they obtained separately the SSE and OE as a function of temperature shown in Fig. 5.21c [5.137]. They were able to gain further quantitative insight into the dynamical aspects of these features through a model for the combined OE and SSE that occurs for an arbitrary translational motion of a dilute electron spin [5.143]. The motion is expressed in terms of the autocorrelation function of the electron–nuclear interactions with a characteristic correlation time τ. Two forms of the correlation function have been considered, namely exponential decay $\exp(-t/\tau)$ and the decay corresponding to the one-dimensional diffusion of the localized defect. Their results for the SSE and OE enhancements are shown in Fig. 5.22. In this plot, ω_n is the NMR frequency and Δ is the linewidth of the Lorentzian ESR line. Because of the general applicability of these highly informative results, we discuss them further here. For $\Delta\tau \gg 1$, a pure SSE and for $\Delta\tau \ll 1$ a pure OE is predicted. Substantial attenuation of the SSE begins to occur around

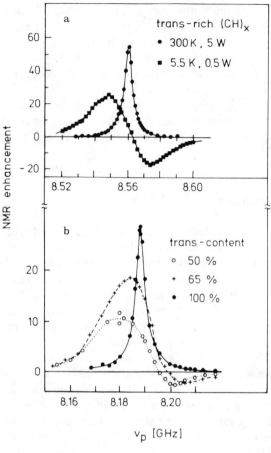

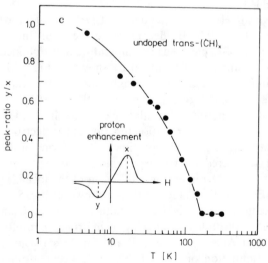

Fig. 5.21 a–c. Enhancement of the proton NMR amplitude $(P/P_0 - 1)$ versus proton frequency v_p as obtained by dynamic nuclear polarization experiments where P and P_0 denote, respectively, the NMR signal amplitudes with and without microwave pumping: **a** in *trans*-rich $(CH)_x$ at 300 K (filled circles) and 5.5 K (filled squares) [5.134]; **b** in $(CH)_x$ with 50% (open circles), 65% (pulses) and 100% (filled circles) trans content at ambient temperature [5.135] and **c** peak ratio (y/x) for the nuclear enhancement as a function of temperature (filled circles) with a solid line guide to the eye [5.136]. $y/x = 1$ signifies a pure solid-state effect and $y/x = 0$, pure Overhauser effect

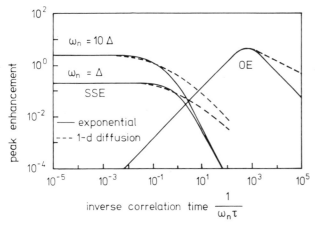

Fig. 5.22. Theoretical values for the peak solid-state and Overhauser effects versus the inverses correlation time $(1/\omega_n\tau)$ of the electron motion. Two correlation functions for the modulations of the electron–nuclear hyperfine interactions are considered: exponential and that due to the 1D diffusive motion of the electron [5.137]

$\Delta\tau \cong 1$ and continues as $\Delta\tau$ decreases (i.e. when the correlation time of the motion decreases). The behaviour of the SSE when its attenuation starts depends strongly on the dimensionality of motion. The OE is maximum when $\omega_e\tau \cong 1$. When $\omega_e\tau < 1$, the OE depends linearly on $1/\tau$ and is insensitive to the dimensionality of the motion; the dimensionality shows up only when $\omega_e\tau > 1$. Finally, due to the large difference between ω_n and ω_e, the SSE disappears before the OE maximizes itself leaving a dip in the total enhancement.

On the basis of the above model, a single correlation function (or a single correlation time) for all the electronic defects is in disagreement with the DNP results on trans-$(CH)_x$. First, the dip in the enhancement curve, as mentioned above, is not observed in the experimental results (Fig. 5.21c). Second, as the SSE decreases the peak separation of the SSE curve should increase, which was not found. This implies that the solitons are trapped at low T, mobile at high T and over the range $10\,K \leq T \leq 150\,K$, trapped solitons gradually become mobile. The proton enhancement can thus be written as the sum of SSE and OE contributions, weighted according to the concentrations of trapped and diffusive spins. Using the model for the density of states of trapped and diffusive spins shown in Fig. 5.21c, Clark et al. [5.136, 137] could simulate well the observed temperature dependence of the OE and SSE. In particular, they find the largest diffusion rate of the mobile solitons to be $1.3 \times 10^{13}\,s^{-1}$. The temperature dependence of the diffusion rate is found to vary with temperature as $T^{1/2}$, in contrast to T^2 dependence deduced from NMR T_1^{-1} results [5.137]. The analysis of the DNP just described is based upon the assumption that the effect of electron–electron interactions on the DNP is negligible. Recently, it has been suggested that the observed effects are due to a temperature-dependent exchange interaction among the conjugational defects and that the defect is immobile at *all* temperatures [5.115]. For this to apply, however, it is necessary that $J \geqq \omega_e$. In this case, however, the ESR line would be exchange narrowed and no ENDOR spectra such as the ones described in the previous section could have been observed.

PPP at room temperature, contrary to $(CH)_x$, exhibits only SSE [5.138] indicating that the conjugational defects are localized at all temperatures. The same is true of the polymers obtained from PPP by a linkage of the aromatic rings with segments involving unsaturated bonds or heteroatoms, e.g. in poly-paraazophenylene [5.139].

5.3.3 Nuclear Spin Lattice Relaxation

Nuclear spin relaxation can be a powerful probe of electron spin motion because it is sensitive to the magnetic field fluctuations of a moving electron. This approach has been applied intensively to *trans*-$(CH)_x$, but with conflicting interpretations of the experimental results. The first such measurements were those of *Nechtschein* et al. [5.107], who found a proton spin-lattice relaxation rate $T_1^{-1} \alpha \omega^{-1/2}$ over the temperature range 4.2–295 K, where ω is the proton Larmor

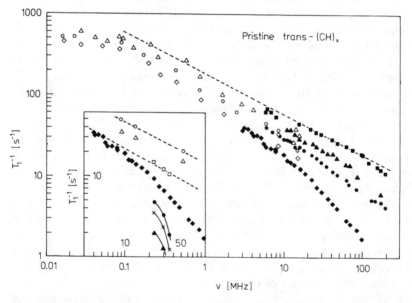

Fig. 5.23. Experimental results of the proton spin-lattice relaxation rate T_1^{-1} as a function of proton frequency v: at 295 K (filled squares) [5.107], at 77 K (filled triangles) [5.107], at 30 K (filled circles) [5.107], at 4.2 K (filled diamonds) [5.107], at 150 K (open triangles) [5.136], at 35 K (open circles) [5.136], at 7 K (open diamonds) [5.136]; in the insert: at 295 K (open circles) [5.143], at 77 K (open triangles) [5.143], at 4.2 K (filled diamonds) [5.107], at 4.5 K (open squares) [5.144], at 1.0 K (filled circles) [5.143], at 0.85 K (crosses) [5.143] and at 0.60 K (filled triangles) [5.144]. The broken lines show the $\omega^{-1/2}$ behaviour of the relaxation rates

frequency. There can be different interpretations of this behaviour: (i) one-dimensional diffusion of the conjugational defect, (ii) limited nuclear spin-flip diffusion to localized paramagnetic defects [5.140, 141] and (iii) modulations of the dipolar interactions due to thermal fluctuations in the orientation of the –CH bonds [5.142]. The last possibility can be ruled out since there is conclusive evidence of the overwhelming presence of hyperfine interactions. Similarly, possibility (ii) can be ruled out at least at temperatures where paramagnetic conjugational defects lead to proton signal enhancements of the pure OE type; at lower temperatures there may be some contribution due to this mechanism. The potential mechanism of proton relaxation may thus be due to one-dimensional diffusion of the defect as asserted by the authors themselves [5.107].

The results for T_1^{-1} versus ω from various authors [5.107, 136, 143, 144] are plotted in Fig. 5.23 (see figure caption for details). The main findings of the results are: (i) the range of the $\omega^{-1/2}$ behaviour depends on the temperature T, (ii) a more or less frequency independence of T_1^{-1} for ω approximately less than or equal to 80 kHz, (iii) depending upon T, T_1^{-1} deviates from $\omega^{-1/2}$ behaviour at higher ω and (iv) T_1^{-1} decreases with T.

Nechtschein's interpretation focused on the $\omega^{-1/2}$ region and makes use of the spectral density function $J(\omega) = (2D\omega/b^2)^{-1/2}$, that due to the one-dimensional diffusion of the conjugational defect, in the expression for T_1^{-1}

$$T_1^{-1} = k_B T \chi [3d^2 J(\omega_n)/5 + (a^2 + 7d^2/5)J(\omega_e)], \tag{5.3.5}$$

where χ is the normalized spin susceptibility, a and d are the isotropic and dipolar electron–proton hyperfine couplings, D is the diffusion coefficient and b is the elementary hopping unit equal to 2.46 Å in the case of trans-(CH)$_x$. Instead of using the truncated spectral density function in (5.3.5) one might as well use the full expression as given by (5.1.30). Equation (5.3.5) assumes no delocalization of the electronic spin. If the spin is delocalized over many sites (called a super site [5.145]) with half-width N, the diffusion rate in $J(\omega)$ involves $D/(Nb)^2$ instead of D/b^2, if the hopping involves jumps from one super site to another. The diffusion coefficient in this case is obtained from the diffusion rate by multiplying it with $(Nb)^2$ instead of b^2 as in the case of a non-delocalized spin. The situation for the single-site hopping of the delocalized spins is different and the appropriate expression, similar to the one derived in [5.145], must be used for $J(\omega)$. If super sites of width Nb are considered, the hyperfine interaction is also scaled, leading to reduced second moments a^2/N and d^2/N in (5.3.5). Together with D/N^2b^2 in $J(\omega)$ the width N drops out at least in the limit $\omega\tau_\parallel \ll 1$ where $\tau_\parallel$ is the single-site hopping time (see Sect. 5.1.3). As will be seen later, the difference between the single-site and super-site hopping cases clearly shows up in $J(\omega)$ at higher frequencies. In almost all the previous estimates of D except [5.24a], the implied soliton motion is via super-site hopping or single-site hopping of the non-delocalized spin. The diffusion coefficient at room temperature, deduced in [5.107] using (5.3.5) and $J(\omega)$ for a single site spin, is $D \approx 3 \times 10^{-2}$ cm^2/s; this value remains unchanged for a multi site spin with super-site hopping.

Since $T_1^{-1} \propto D^{-1/2}$ in this model, and since experimental T_1^{-1} decreases with increase of T, it is implied that the diffusion coefficients are larger at lower temperatures, in disagreement with the DNP results; this has been the case with the analysis of *Kume* et al. [5.143] which yields $D \propto T^{-1/2}$. To explain the T dependence of T_1^{-1} (and simultaneously the ESR linewidths), a statistical model was introduced in which χ was decomposed into localized-spin part (n_l) and diffusive-spin part (n_d). The concentration of the fixed spins according to this model is

$$C = \frac{n_l}{n_l + n_d} = \frac{p}{p + (1-p)\exp(-E/k_B T)}, \qquad (5.3.6)$$

where p is the concentration of the traps with trapping energy E. The motivation to introduce the concept of traps was provided by a large spread in the values of ESR linewidths from sample to sample and interpreted as being due to the presence of oxygen which pins the mobile defects. As far as the proton T_1^{-1} is concerned, fairly reproducible values were obtained on samples with different ESR linewidths thereby implying that the trapped or localized spins do not contribute to the relaxation rate. Assuming a distribution in the trapping energy (0–650 K) and arguing that the nuclear relaxation via localized spins was very inefficient, they rationalized the temperature dependence of T_1^{-1} and deduced $D \propto T^2$ behaviour.

There has been a few attempts to ascertain the role of spin-flip diffusion of the nuclear species to localized paramagnetic defects. Evidence [5.136] is claimed to be provided by smaller relaxation rates in *trans*-$(CD_{0.92}H_{0.08})_x$ compared to pristine *trans*-$(CH)_x$ although the former contained 2.6 times more unpaired spins; this was interpreted as being due to much smaller spin-flip diffusion rate for deuterons. *Ziliox* et al. [5.58], however, find about 6% extra protons attached to the residual catalysts which do not experience any direct interactions with the mobile electron spins. This therefore implies that the dynamics of the conjugational defect may not be unambiguous from T_1^{-1} results at low proton concentrations. A similar strategy was adopted by *Scott* and *Clarke* [5.146] who measured ^{13}C and ^1H relaxation rates on ^{13}C enriched/^1H diluted samples of *trans*-$(CH)_x$ as a function of frequency ω.

Although proton relaxation in pristine *trans*-$(CH)_x$ exhibits $\omega^{-1/2}$ behaviour, ^{13}C relaxation rates showed a frequency-independent behaviour for three samples with ^1H and ^{13}C concentrations of 100:98, 2:90 and 2:20. Their conclusion that the spin-flip diffusion process does affect the relaxation rates is also supported by other studies [5.147, 148].

The necessity of combining the one-dimensional diffusive motion of the conjugational defect and the phenomenon of nuclear spin-flip diffusion has been realised by *Kahol* et al. [5.24a] through a "confined soliton" model. The model is based on the assertion that mobile solitons, whose concentration varies with temperature, move one-dimensionally over a small part of the chain (of length l) and relax these nuclei directly; the remainder nuclei relax via nuclear spin-flip diffusion. A set of diffusion equations were solved for the magnetization

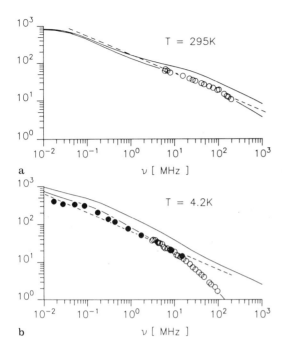

a

b

Fig. 5.24 a, b. Proton spin-lattice relaxation rate T_1^{-1} as a function of proton frequency at 295 K (**a**) and 4.2 K (**b**): experimental results are shown by circles (open [5.107] and filled [5.136]) whilst those in continuous lines are from the confined soliton model using $J(\omega)$ corresponding to single-site hopping (line passing through the experimental points) and super-site hopping of the delocalized soliton [5.24a]

which was then expanded through the "normal modes" analysis to yield an expression for T_1^{-1}. The results of this model for proton T_1^{-1} at 295 K and 4.2 K are shown in Fig. 5.24. The following information has gone into the numerical calculations: (a) the concept of trap sites not using (5.3.4) explicitly but treating C as a parameter, (b) complete hyperfine tensor, (c) neglect of the contribution due to pinned solitons, (d) spin-flip diffusion rate in the fast diffusion regime, (e) the interruption in the 1D motion via a phenomenological correlation function and (f) the spectral density function $J(\omega)$ appropriate to a confined soliton, whose spin density distribution extends over N carbon sites (rectangular shape with width $N = 15$). The form of $J(\omega)$, due to single-site hopping rather than super-site hopping [5.145] of the delocalized spin differs at higher frequencies as is shown in Fig. 5.25; the parameters of Fig. 5.24 have been used in this plot. The analysis yields for the diffusion coefficient a value $(3.0 \times 10^{-3} \text{ cm}^2/\text{s})$ which is about an order of magnitude smaller than that of [5.107]. The model has also been applied to show the frequency independence of ^{13}C relaxation rates [5.24a].

The relaxation rate at 4.5 K reported in [5.144] does show $\omega^{-1/2}$ behaviour but this is clearly not the case at lower temperatures, as has been claimed (Fig. 5.23 inset). At these low temperatures, thermally activated behaviour of the type $T_1^{-1} \propto \exp(-g_{\text{eff}}\mu_B B_0/k_B T)$, where g_{eff} is nearly twice the free-electron g value, is observed [5.136].

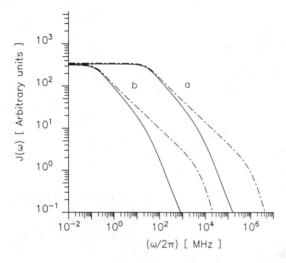

Fig. 5.25. Calculated spectral density function $J(\omega)$ for a confined and delocalized spin-1/2 due to single-site (continuous line) and super-suite (dash-dot line) hopping motion. The results (a) and (b) correspond to $T = 295$ K and $T = 4.2$ K, respectively; the same set of parameters as used in the analysis of the T_1^{-1} data of Fig. 5.22 is used in the calculations [5.24a]

The discussion of the cut-off behaviour at low frequencies and that of the diffusion coefficients is deferred to the next section. Additional references for the relaxation rates as a function of temperature or frequency are [5.144, 150].

PPP is a system that shows only the solid-state effect (SSE) [5.138] implying the "stationary" nature of the defects. Further evidence of the immobile electronic spins is provided by the $t^{1/2}$ dependence of the decay of the magnetization in a $T_{1\rho}$-type experiment [5.151]; this dependence results from the interactions of the fixed paramagnetic spins with the nuclear spins without spin-flip diffusion process [5.152]. The linear plot of the relaxation rate, deduced from this experiment, as a function of $1/T$ implies the activated nature of the interactions in the range 130–300 K. That the proton–proton dipolar interactions are modulated via thermal activation of the phenylene rings was seen to occur in Sect. 5.2.1. The same mechanism seems to dictate the relaxation behaviour in the above experiment [5.151]. This is further seen by the value of the activation energy (4 kJ/mole [5.151]) which differs, however, from the approximate analysis of the second moments (see Sect. 5.2.1). The spin-lattice relaxation rate in PPP [5.138] shows a relatively faster decrease with increase in temperature for $T > 200$ K compared to $T < 200$ K. Dipolar relaxation due to thermally modulated proton–proton interactions (BPP-type [5.153]) is

$$T_1^{-1} = C_i\left(\frac{\tau}{1 + \omega^2\tau^2} + \frac{4\tau}{1 + 4\omega^2\tau^2}\right), \tag{5.3.7}$$

where $C_i = 3\gamma^4\hbar^2\sum_i r_{ij}^{-6}/10$ and τ is given by (5.2.5). Since this equation predicts a minimum in T_1^{-1} when $\omega\tau \approx 1$ and strong temperature dependence around this minimum (not observed in the experimental rates on the low temperature side), activated paramagnetic relaxation of the form $\exp(E_P/k_B T)$ seems to dominate the other relaxation mechanism at low temperatures.

5.3.4 Electron Spin Relaxation

For the electron spin-lattice relaxation rate T_1^{-1} in *trans*-$(CH)_x$ it has been reported that $T_1^{-1} \sim \omega_e^{-1/2}$ (ω_e: ESR frequency), which was observed in the NMR case [5.154, 155]; the ESR linewidth also follows the same frequency dependence [5.156] though its interpretation is not as straightforward as that for the relaxation rate T_1^{-1}. This behaviour is shown in Fig. 5.26 for *trans*-$(CH)_x$ and *trans*-$(CD)_x$ at $T = 300$ K [5.154]; filled circles correspond to that for *trans*-$(CH)_x$ obtained by *Robinson* et al. [5.157]. Based upon all the available NMR and ESR information, these features indicate 1D motion of the conjugational defect.

As to the interactions responsible for relaxation, electron–nuclear hyperfine and electron–electron dipolar interactions seem to be the dominant ones. The evidence of these interactions is provided [5.154] by the linear dependence of T_1^{-1} on spin concentration of the defects (Fig. 5.27). That the limiting values of the intercepts (corresponding to $C \to 0$) for protonated and deuterated $(CH)_x$

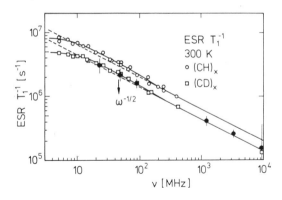

Fig. 5.26. ESR spin-lattice relaxation rate T_1^{-1} as a function of electron frequency $\nu (= 1/2\pi)$ at 300 K for *trans*-$(CH)_x$ (open circles) and *trans*-$(CD)_x$ (open squares) [5.154]. The filled circles are the results of *Robinson* et al. [5.157] The broken line indicates the $\omega^{-1/2}$ dependence of the relaxation rates

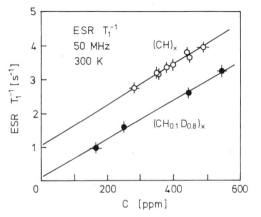

Fig. 5.27. Dependence of the ESR relaxation rate T_1^{-1} as a function of electron spin concentration at 300 K and at 50 MHz in *trans*-$(CH)_x$ (open circles) and *trans*-$(CH_{0.1}D_{0.9})_x$ (filled circles). Different spin concentrations were obtained using different isomerization conditions [5.154, 156]

are proportional to the number of protons or deuterons, they give an estimate of T_1^{-1} due to hyperfine interactions; the rest is then due to electron–electron dipolar and/or exchange interactions. Since $T_1^{-1} \propto \omega^{-1/2}$, dipolar and hyperfine contributions of T_1^{-1} can be represented as [5.162, 164]

$$T_1^{-1}(\text{dipolar}) = (1.0 \pm 0.2) \times 10^{14}\, C/T^{1/2}(\text{s}^{-1}) \tag{5.3.8}$$

$$T_1^{-1}(\text{hyp.}) = (1.6 \pm 0.3) \times 10^{10}/\omega^{1/2}(\text{s}^{-1}) \tag{5.3.9}$$

where C is the concentration of the defects per carbon atom. In the following, analytical expressions for the relaxation rates $T_1^{-1}(\text{dipolar})$ and $T_1^{-1}(\text{hyperfine})$, which involve the dipolar and hyperfine interactions modulated by the 1D motion of the solitons, have been calculated and directly compared with (5.3.10, 11) to obtain the diffusion rates D (in rad s^{-1}) or the diffusion coefficients D (in cm^2/s [5.154].) The value of D at 300 K as obtained from the dipolar part is $(2.3 \pm 1.0)10^{-2}$ cm^2/s whereas that obtained from the hyperfine part is $(0.7 \pm 0.3)10^{-2}$ cm^2/s. A rather similar interpretation to the frequency dependence of T_1^{-1}, but considering only the electron–electron dipolar interactions modulated by the 1D motion of the solitons (induced by soliton–phonon interactions), is given in [5.157]. The temperature dependence of T_1^{-1} is nicely reproduced [5.157] but the implied behaviour $D \propto T^{-\alpha}(\alpha \to 1/2$ to $5/2)$ is in contradiction to NMR and DNP results; various other mechanisms for the temperature dependence of the relaxation rates are also discussed in [5.111, 158, 159].

Other investigations of the electron dynamics have involved cw ESR lineshapes, ESR linewidths and electron spin echoes. The trapped/diffusing-spin model [5.106, 107, 160], discussed briefly in Sect. 5.33, when adapted for the ESR linewidth including the dipolar and hyperfine contributions of the trapped and diffusing spins, gives satisfactory results over the entire temperature range. The numerical expression for the ESR linewidth, as obtained in [5.107], shows that it arises mainly from the pinned defects. Various points supporting this model are: (a) frequency and temperature behaviour of the DNP analysis (Sect. 5.3.2), (b) two characteristic decay rates of the proton magnetization in oxygen-exposed materials [5.106], (c) nearly the same proton T_1^{-1} for samples giving widely different ESR linewidths [5.106], (d) time-resolved ESR which shows the presence of both the pinned and mobile defects [5.56, 104] and (d) a satisfactory account of the temperature dependence for both ESR linewidths and NMR T_1^{-1}.

Mizoguchi et al. [5.154] have deduced $D(T)$, in the spirit of the trapped/diffusing-spin model, by measuring the frequency dependence of the ESR linewidth for $T > 200$ K. The linewidth is decomposed into three parts: frequency-independent secular width due to motionally narrowed dipolar and hyperfine interactions, frequency-dependent non-secular width and the width due to trapping of the spins. Simulations of the frequency dependence of the non-secular part, as described in Sect. 5.3.3, yields $D(T)$ very close to those of [5.107].

Tang et al. [5.161], on the other hand, have analysed the ESR lineshapes in terms of hyperfine interactions modulated only by the 1D motion of the mobile solitons. It can be shown from the Kubo-Tomita formalism that, for the above situation, the ESR lineshape is given by

$$G(t) \propto \exp(-\langle \Delta\omega^2 \rangle (\pi/\alpha\gamma)^{1/2} t) \tag{5.3.10}$$

when $\alpha \gg \gamma \gg 1/t$, and

$$G(t) \propto \exp(-4\langle \Delta\omega^2 \rangle \alpha^{1/2} t^{3/2}/3) \tag{5.3.11}$$

When $\alpha \gg 1/t \gg \gamma$. The low temperature data has been fitted to a general expression [which has (5.3.10, 11) as limiting cases] to obtain the on-chain ($\alpha = D/N^2 b^2$) and off-chain (γ) diffusion rates. Their values for the diffusion coefficients are on the lower side of those obtained in [5.107]. In particular, the diffusion coefficient at room temperature is $\approx 4.5 \times 10^{-4} \, \text{cm}^2/\text{s}$ with an anisotropy in the diffusion coefficients of ≈ 1300.

In light of the time-resolved ESR experiments [5.56, 104] which show narrow and broad lines due to the presence of mobile and pinned solitons, the above analysis should be strictly applied to the motionally narrowed ESR lines only. The expression relating the linewidth ($1/T_2^*$) to α, γ and $\langle \Delta\omega^2 \rangle$ can be obtained from (5.3.10, 11) as

$$1/T_2^* = \langle \Delta\omega^2 \rangle (\pi/\alpha\gamma)^{1/2}; \qquad \alpha \gg \gamma \gg 1/T_2^* \tag{5.3.12}$$
$$1/T_2^* = (4\langle \Delta\omega^2 \rangle/3)^{2/3} \alpha^{1/3}; \qquad \alpha \gg 1/T_2^* \gg \gamma \tag{5.3.13}$$

For the 1D motion interrupted only at the boundaries of the mobility range ($\alpha N^2 b^2 = \gamma l^2$), (5.3.12) gives a value of $\approx 3.2 \times 10^{-3} \, \text{cm}^2/\text{s}$ using: $\langle \Delta\omega^2 \rangle = 23 \, \text{G}^2$, $1/T_2^* = 2.35 \times 10^6 \, \text{s}^{-1}$ [5.104] and $(l/b) = 150$ [5.24a]. This value of D is very close to that obtained in [5.24a] from the analysis of the proton T_1^{-1}. Note that this analysis is subject to the condition $\alpha \gg \gamma \gg 1/T_2^*$ and applies only when the lines are Lorentzian.

The dynamics of the conjugational defects has also been studied from analysis of the echo decays [5.162]. The time constant denoting the spin echo decay is T_{2m}, the phase memory time whose temperature dependence is shown in Fig. 5.28 for deuterated (CH)$_x$; the temperature dependence of $1/T_2^*$ and $1/T_{1e}$ is also shown alongside [5.161]. It is clear that the ESR line is inhomogeneously broadened below $\approx 170 \, \text{K}$. Since $1/T_{1e}$ is much smaller than $1/T_2^*$ and $1/T_{2m}$, the main contribution to the ESR line comes from the spin processes which cause transverse relaxation. Both the hyperfine and electron dipolar interactions play their roles: the former is evidenced by smaller phase memory times in trans-(CH)$_x$ and the latter by the composites of trans-(CH)$_x$ [5.111]. The dynamics can be investigated by simulating the echo decay behaviour employing specific models or equivalently modelling the phase memory times. Shiren et al. [5.162] have made use of a Gauss-Markov model assuming 1D motion of the defect along the chain and Lorentzian hopping between the chains. Their analysis seems self consistent in the sense that they determined the effective hyperfine coupling constant ($\approx 3 \, \text{G}$) from the echo decay with minimum T_{2m} and used this value to obtain the temperature dependence of the diffusion coefficients. The values deduced for D in this way are 2–3 orders of magnitude smaller than

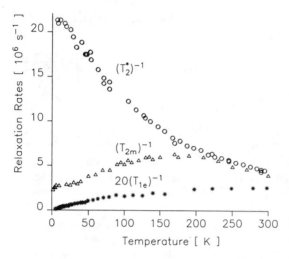

Fig. 5.28. Temperature dependence of the phase memory rate $(1/T_{2m})$, cw ESR half linewidth $(1/T_2^*)$ and spin-echo electron spin-lattice relaxation rate $(1/T_{1e})$ in *trans*-$(CD)_x$ [5.162]

that of [5.107]. The reason for this discrepancy is supplied by *Nechtschein* et al. [5.107] who argue that in the analysis of the dynamical effects, spin delocalization effects are not important and that a value of 23.4 G for the hyperfine constant removes the apparent contradiction between the spin echo and the proton T_1^{-1} diffusion coefficients.

A plot showing $D(T)$, deduced from NMR [5.58, 107, 160], ESR [5.154–156] DNP [5.136] and theoretical calculations [5.163, 164], is given in Fig. 5.29. Taking as reference the results of Nechtschein et al. [5.107], since they are compatible with both NMR and ESR data, a large variation is evident. Estimates of D from [5.24a, 154–156] are the closest to [5.107] mainly because the trapped/diffusing-spin model is invoked in these estimates. The values of D obtained in [5.161], not shown in Fig. 5.29, are off by up to two orders of magnitude, as are also the estimates from spin-echo experiments [5.162] taken as evidence against the trapped/diffusing-spin model. Since the narrow dynamic time-resolved ESR line yields the diffusion coefficient at room temperature within an order of magnitude of [5.107], it is implied that even at room temperature a finite number of pinned defects would contribute to the ESR linewidth, as is the case with the much smaller diffusion coefficient obtained by *Tang* et al. [5.161]. The theoretical analysis of *Maki* [5.164], according to which the coupling of the solitons to the acoustic phonons is responsible for the soliton diffusion, gives a diffusion coefficient at room temperature only one order of magnitude larger than that obtained in [5.107]; the temperature dependence of D is $D \propto T^{1/2}$ as shown by the broken straight line. A similar temperature dependence is found from analysis of the DNP results (dashdot-dot straight line).

Going back to the frequency dependence of the ESR T_1^{-1} and linewidths (and also proton T_1^{-1}), a cut-off at low frequencies (ν_c) is observed. Different mechanisms for this feature are possible, but its precise origin is still unclear. Direct electron–electron interactions, trapping on the chain, inter-chain transfer

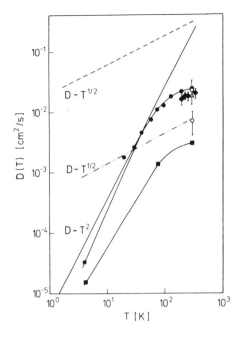

Fig. 5.29. Temperature dependence of the diffusion coefficient D as a function of temperature T from: NMR (filled circles) [5.107], NMR (open triangle) [5.143], NMR (filled squares) [5.24a], ESR (open circles) [5.154], ESR (filled diamonds) [5.156], DNP (dashed-dot-dot line) [5.137], **theory** (continuous straight line) [5.163] and **theory** (broken straight line) [5.164]

or the back and forth of the electron defect interrupted (e.g. by trapping for a finite time) by the boundaries of the mobility domain could lead to this feature. Whatever the mechanism, two cut-offs are expected, one due to $J(\omega_n)$ and the other to $\approx J(\omega_e)$. The analysis of the $T_{1\rho}$ experiment by *Holczer* et al. [5.103] gives the cut-off $v_c \approx 7\,\mathrm{MHz}$. The experiment of *Clark* et al. [5.136] gives v_c at $\approx 55\,\mathrm{MHz}$ whereas it is found at around $\approx 10\,\mathrm{MHz}$ in ESR studies [5.154, 156]. Since the cut-offs due to $J(\omega_n)$ and $J(\omega_e)$ differ by a factor of ≈ 658, the cut-off as implied by the above experiments must be due to $J(\omega_e)$. Analysis by *Kahol* et al. [5.24a] gives v_c at room temperature at $\approx 25\,\mathrm{MHz}$ due to $J(\omega_e)$ and there are indications of a "left-over" cut-off, due to $J(\omega_n)$, at much higher frequencies. It must be emphasized that none of the interpretations of the cut-off frequency can be taken seriously since at these low frequencies the high field approximation used in all expressions for the relaxation rates breaks down. Any interpretation requires dealing with the Hamiltonian valid at low fields where the dipolar or hyperfine interactions are comparable to the Zeeman field. The temperature dependence of v_c is also contradictory. v_c found in [5.154, 156] is frequency dependent, whereas that in [5.136] occurs at the same frequency.

5.3.5 Light-Induced ESR

This section originates from the prediction by Su and Schrieffer [5.7] that light-induced electron-hole pairs on an isolated chain of *trans*-$(CH)_x$ would become charged soliton–antisoliton pairs on a sub-picosecond time scale. The interest in light-induced ESR (LESR) thus stems from characterizing the nature of light-induced carriers with regard to their spin-charge properties. It is important to realise that LESR experiments do not probe the systems on the time scale in which a soliton pair relaxes.

The LESR experiments of *Flood* and *Heeger* [5.165] were the first to show that the ratio of the photo-produced neutral to charged solitons (called the branching ratio) in *trans*-$(CH)_x$ was of the order of 10^{-2}. That is, the quantum efficiency for the photo-production of spins is much smaller than that of the spinless charge carriers; this has since been shown theoretically [5.166]. LESR experiments carried out subsequently with higher sensitivity also show no signal corresponding to the light-induced spins on carefully prepared samples [5.167]. These experiments support the mechanism of [5.7] that an electron–hole pair on the same chain evolves rapidly into a charged soliton–antisoliton pair. Two other mechanisms have also been proposed. Initially, the excitation of electrons and holes on neighbouring chains would lead to polarons moving along and between chains. Either two polarons appearing on the same chain can combine to yield a pair of charged solitons or the polaron can convert an already existing neutral soliton to a charged soliton. The latter possibility suggested by *Orenstein* [5.168] was claimed to have been observed in his samples which showed a decrease in the ESR signal [5.168]. However, the effect is weak enough to be caused simply by a temperature increase combined with Curie's law. In a recent paper [5.169], *Levey* et al. seem to have solved the problem of sample heating. They observe an LESR signal whose dependence on excitation wavelength and temperature correlates with the photo-induced absorption band at 1.35 eV (due to uncharged species) and shows no correlations with the band at 0.45 eV (that due to charged species). Since charged species are expected to participate in one way or another (on the time scale of the LESR experiment) in diminishing or enhancing the dark ESR signal, lack of any correlations with the 0.45 eV band support the conclusion of direct photo-production of charged solitons. The LESR signal has therefore been interpreted as being due to photo-production and subsequent trapping of neutral solitons, the observations not predicted by existing soliton theories.

In polythiophene (PT), *Moraes* et al. [5.170] found an increase in the ESR intensity during light illumination. The light-induced line was shifted upfields by $\Delta g/g \approx 10^{-4}$ and had a linewidth of 5.4 G. Combined with the light-induced infrared absorption results, the carriers were concluded to be both charged and spin-carrying species. In a recent paper, however, *Vardeny* et al. [5.171] did observe a spin-1/2 (polaron) excitation profile peaked around 1.5 eV, but this was not observed on the same sample when annealed to improve the structural order. These excitations have been interpreted as due to deep defect states arising from structural imperfections. Thus, in conclusion, the light-induced carriers are spinless and have been identified with bipolarons. LESR experiments on BF_4^--doped PT are also explained via polarons serving as carriers and bipolarons [5.172].

Light-induced triplet spins, on the other hand, have been observed recently in polydiacetylenes [5.173, 174], details of which will be covered in Sect. 5.5.3.

5.4 Magnetic Properties of Conjugated Polymers in the Conducting Phase

5.4.1 Susceptibility

The spin susceptibility of pristine Shirakawa $trans$-$(CH)_x$ exhibits Curie law behaviour [5.175, 176] down to 1.5 K, as is expected from the relatively few $(10^{-4}-10^{-3}$ spins per carbon atom) conjugational defects created during the isomerization process; the spin susceptibility of cis-$(CH)_x$ is smaller by one to two orders of magnitude ($\approx 10^{-8}$ emu/mole carbon) [5.82, 97, 177–179]. The spin susceptibility of Durham $trans$-$(CH)_x$ deviates from the Curie behaviour below 30 K, believed to be due to the presence of "impurities" in the sample [5.180].

One of the first susceptibility measurements on doped $(CH)_x$ was performed on $(CH \cdot yAsF_5)_x$ over the concentration range $0 \leq y \leq 0.13$ and temperature range 77–295 K, by *Weinberger* et al. [5.179] using both the Faraday balance and Schumacher–Slichter methods. For lightly doped $(CH)_x$ anamolously small Curie susceptibility, indicating the non-magnetic nature of the localized states on doping, was observed, whereas for $y \geq 0.02$ only a temperature-independent χ_p was found. These results were in contradiction with those of *Tomkiewicz* et al. [5.177] who, from the ESR absorption lines, found only the Pauli susceptibility even at the lowest doping levels; subsequent reinvestigations by these authors gave essentially similar results [5.181]. On the other hand, *Ikehata* et al. [5.182], like *Weinberger* et al. [5.179], found extremely small χ_p until $y \approx 0.05$ and a step-like increase at $y \approx 0.07$. Since *Tomkiewicz* et al. [5.177, 181] find significant χ_p even at small doping levels, their doping procedure seems to create metallic islands. A cyclic doping procedure has been argued [5.183, 184] to yield uniform doping for the AsF_5 dopant in $trans$-$(CH)_x$ (yielding Lorentzian lines with Curie behaviour up to $y \approx 0.001$) but, in the case of cis-$(CH)_x$, even this elaborate procedure produces small metallic domains at small concentrations. The authors of [5.183, 184] have discussed these differences in terms of the dopant diffusion time and the charge transfer time. We would like to make two remarks here. Firstly, direct integration of the Dysonian lines, as in [5.177, 181], over-estimates considerably the number density of the spins [5.185]. Secondly, the conducting material can also, in some cases, affect the spin concentration estimates by lowering the cavity quality factor or/and distorting the microwave field. Comparison of the ESR intensity with an external reference, by substitution method, may not be quantitative; simultaneously recorded reference ESR spectrum must be used in such situations [5.186].

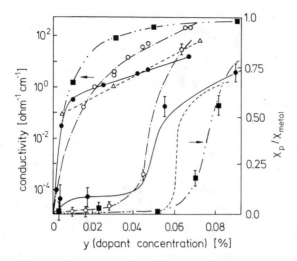

Fig. 5.30. Behaviour of the Pauli susceptibility χ_p/χ_{metal} (right hand scale) and dc conductivity (left hand scale) as a function of dopant concentration y in doped $(CH)_x$ for various dopants: iodine (continuous line and filled circles) [5.187], sodium (broken line and open triangles) [5.188], ClO_4^- (dash-dot line and open circles) [5.187] and AsF_5 (dash-dot-dot line and filled rectangles) [5.182, 189]

The general behaviour of Pauli susceptibility χ_p for various dopants [5.182, 187, 188] is shown in Fig. 5.30 along with the conductivity data; in the figure $\chi_{metal} = 3.9 \times 10^{-6}$ emu/mole C [5.187] and the conductivity data for $(CH \cdot yAsF_5)_x$ is taken from [5.189]. A rather universal behaviour, though not as dramatic as with the iodine dopant, is indicated, namely extremely small χ_p in the low ($y \approx 0$) and intermediate ($y < 0.05$) regions and sudden build-up around $y = 0.07$. Such plots are usually used to argue that the conductivity at low and intermediate doping levels is not correlated with χ_p and that spinless, charged species must be involved in the mechanism of conduction. We wish to point out, however, that the conductivity data are logarithmically plotted whereas the susceptibility data are linearly presented, suggestive of an apparently different dependence on the dopant concentration. Nevertheless, change in the basic character of the species contributing to χ_p is observed around $y \approx 0.07$. At very small values of y, χ_p is extremely small. The Curie spin concentration decreases [5.182] and the results may be rationalized with the soliton doping mechanism, i.e. the midgap states are ionized first leading to charged solitons rather than localized states at the edge of the valence band. The fact that the Curie spins do not decrease significantly until $y \approx 0.001$ whereas, according to the soliton doping mechanism, they are expected to decrease from around $y = 0.0004$ might be due to (i) the diffusion kinetics of the dopant and (ii) the co-existence of polarons and solitons. The latter mechanism is invoked explicitly in the interpretation of the results in Na [5.190] and iodine [5.186]-doped $(CH)_x$.

In the following we briefly review the experimental observations and current explanations of the susceptibility data in other doped conjugated polymers with non-degenerate ground state; the dopant-induced spins here may be visualized as polarons which do not exhibit spin-charge inversion property, typical of the solitons. The y dependence of the Curie spins in these polymers at low doping

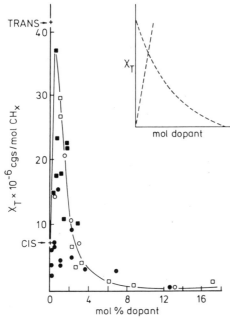

Fig. 5.31. Total magnetic susceptibility as a function of dopant concentration (mole percent) in 70% cis-(CH)$_x$ for the following dopants [5.119]: MoF$_6$ (open circles), WF$_6$ (filled squares), ReF$_6$ (open squares) and UF$_6$ (filled circles); +cis and +trans correspond to those for undoped 70% cis- and 100% trans-(CH)$_x$. The solid line is a guide for the eye

levels is thus expected to be different from that for (CH)$_x$. One of the first examples is that of (PPy·yAsF$_5$)$_x$ which showed an increase in the concentration of Curie spins until $y \approx 0.1$, from where the concentration decreased slowly [5.191]; note that the concentration of the Curie spins decreases with increase in y in Shirakawa (CH)$_x$ for the dopant AsF$_5$. Similar results were also found in (PPy·yFeCl$_3$)$_x$ [5.192]. The intensity of the ESR spectra for PPy doped with gaseous oxygen first increases and then continuously decreases [5.103]. The increase in the number of Curie spins as a function of dopant concentration is also shown by PT [5.193] and cis-(CH)$_x$ [5.119]. The total magnetic susceptibility as a function of y is shown in Fig. 5.31 for samples of 70% cis-(CH)$_x$ doped with hexafluorides of Mo, W, Re and U [5.119]. Similar to the behaviour of other polymers with non-degenerate ground state, the susceptibility increases with the increase in the dopant level, and shows a sharp maximum in the range $y = 0.08–0.15$. At higher doping levels, the susceptibility decreases. PT has been doped with iodine [5.194], BF$_4^-$ [5.193], ClO$_4^-$ [5.195] and SF$_6$ [5.196] via different doping procedures. It is intriguing that the absolute values of the spin susceptibility and the values of the dopant concentration where the susceptibility shows a maximum are quite dependent on the dopant type and doping procedure. Combined with the conducting results, it is fair to conclude that the nature of the dopant and doping procedure affects the details of the charge carrier generation.

Whereas the increase in the Curie spins as a function of y can be qualitatively explained as being due to the formation of polarons, the decrease in the susceptibility has been postulated by many authors to be due to the creation of spinless bipolarons. In situ electrochemical doping has been done in PPy and PT to find out correlations between the amount of doping and the number of induced spins [5.195, 197] and the data for PPy reveal such a direct correlation up to a certain doping level, i.e. the spin concentration increases with increasing dc voltage, reaches a maximum and then decreases. This behaviour is reversible and has been discussed in terms of polaron–bipolaron statistics [5.198, 199]. However, the in situ measurements of the susceptibility from the ESR spectra in PT (doped with ClO_4^-) show virtually the same number of Curie spins as in the pristine sample, and only a small value for the Pauli suceptibility [5.195], thus suggesting the overwhelming formation of non-magnetic bipolarons. These results also show the dependence of the charge carriers on the nature of the dopant ions. *Schärli* et al. [5.200] have recently reported detailed ESR investigations of tens of samples of poly-(3-methyl thiophene) (PMT) doped electrochemically with BF_4^-. Whilst we will give more details of their work in Sect. 5.4.2, we cite the following from their paper: "...the number of mobile spins increases linearly with the dopant concentration, indicating that highly mobile polarons are the dominant charge states". We remark, however, that this interpretation does not necessarily resolve the various discrepancies.

As the existence of polarons and/or bipolarons has implications in the interpretation of the conductivity data, it is important to obtain magnitudes of the Pauli susceptibility χ_p and hence the density of states at the Fermi level $N(\varepsilon_F)$ (5.1.4) for an understanding of the metallic behaviour of these polymers. Table 5.3 summarizes the density of states $N(\varepsilon_F)$ for a few of the heavily doped polymers [5.182, 187, 188, 191, 202, 204]. For example, in heavily doped (CH-y $AsF_5)_x$, $N(\varepsilon_F) = 0.10$ states/eV-C [5.182] which is near to 0.16–0.20 states/eV-C expected for $(CH)_x$) with zero band gap [5.201]. In PT doped with BF_4^-, $N(\varepsilon_F) \approx 0.2$ states/eV-ring [5.202]. In contrast, $N(\varepsilon_F) \approx 1.0$ states/eV-ring at $y = 0.24$ for PT chemically doped with AsF_5 [5.203].

Measured susceptibility in the above types of systems is mainly due to the sum of Curie and Pauli susceptibilities, and is analyzed through its variation as a function of temperature (Sect. 5.1.1). It has been observed that PPP doped with AsF_5 [5.205] or alkali metals [5.206] and PPy doped with ClO_4^- [5.207] exhibit a non-Curie-like behaviour. Figure 5.32 shows such a behaviour for three samples of AsF_5-doped PPP, labelled as B, C and D. The model chosen

Table 5.3. Density of states at the Fermi level, $N(\varepsilon_F)$, in heavily doped conducting polymers in units of states/eV/monomer unit

Polymer	AsF_6^-	I_3^-	ClO_4^-	BF_4^-	Na^+
Polyacetylene	0.10 [5.182]	0.09 [5.187]	0.12 [5.187]		0.07 [5.188]
PPP	0.28 [5.191]				
PT	1.0 [5.203]		0.0 [5.203]	0.2 [5.202]	
PPy	0.04 [5.204]			0.18 [5.202]	

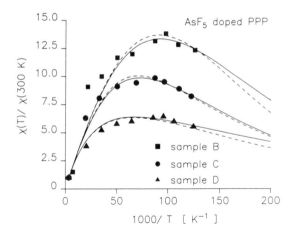

Fig. 5.32. Experimental and calculated $\chi(T)/\chi(300\,\text{K})$ versus inverse temperature for three samples (B, C and D [5.205]) of AsF$_5$-doped polyparaphenylene. The curves shown in continuous and broken lines are the results of theoretical calculations using two models for the distribution function $P(J)$ as described in the text. The spin concentration of sample B is twice as large as that for sample C and nearly 12 times the value for sample D

in [5.205, 206] is that of two exchange-coupled spins (polarons) exhibiting singlet and triplet states with their separation equal to $-2J$. Here the ground state is antiferromagnetic (singlet) if $J < 0$. The statistical nature of the distribution of the spins is incorporated via a parameter Z which defines the number of equivalent positions a second spin (comprising the pair) can assume relative to the first spin. This model yields reasonably good results up to the reported temperatures with an antiferromagnetic coupling of $\approx 25\,\text{K}$. These systems, however, are expected to have a random distribution of spins and therefore a distribution of exchange couplings. *Kahol* and *Mehring* [5.208] have analysed this data using an exchange-coupled pairs (ECP) model with a distribution function $P(J)$ for the J couplings. $\chi(T)$ for a single pair is

$$\chi = 2\,(g\mu_\text{B})^2(3 + \exp{(-2\beta J))^{-1}}/k_\text{B}T \tag{5.4.1}$$

whereas that for the entire system is

$$\chi(T) = \int_0^{J_0} \chi_\text{ECP}(T, J)P(J)dJ. \tag{5.4.2}$$

The results calculated from (5.4.1, 2) and using $P(J)$ for an isotropic distribution of spins are also plotted in Fig. 5.32. The results shown as a continuous line correspond to J couplings which decrease exponentially with separation between the spins (r) whereas those shown as a broken line correspond to J decreasing as the inverse cube of r. In this analysis, the peak position of $P(J)$ versus J gives a measure of the antiferromagnetic couplings, which is of the order of $10\,\text{K}$ [5.208]. The overall temperature behaviour in PPy doped with ClO_4^- is also described satisfactorily by the above exchange-coupled pairs model. It is thus seen that heavily doped PPP and PPy form some sort of antiferromagnetic polaron lattice, thereby providing strong evidence for electron–electron correlation, as discussed in Sect. 5.3.1 and treated theoretically in Chap. 1.

5.4.2 ESR Lineshapes and Linewidths

Analysis of the various ESR characteristics (e.g., lineshape, linewidth asymmetry ratio A/B, g value etc.) including their dependence on the microwave power, provides useful information about the nature, origin and dynamics of the paramagnetic conjugational defects. Different types of defects can be identified from the ESR spectrum via differences in the linewidths and/or g values provided their intensities are comparable. For example, the absorption ESR line at room temperature is Gaussian (with peak-to-peak linewidth $\Delta B_{pp} = 5$–10 G) in the *cis* isomer of $(CH)_x$ whereas it is Lorentzian with $\Delta B_{pp} < 1$ G in the *trans* isomer.

The presence of a narrow ($\Delta B_{pp} \approx 1$ G) and a broad line in both *cis*- and *trans*-$(CH)_x$, doped lightly and slowly with AsF_5 and SbF_5, has been reported below room temperatures [5.209]. The two lines in *cis*-$(CH)_x$ have been interpreted as being due to pinned solitons in the undoped regions and mobile solitons in doped regions. In *trans*-$(CH)_x$, on the other hand, the implication is either that of two distinct types of defects or of inhomogeneous doping [5.190]. Contrary to these features, the spectra corresponding to iodine and bromine dopants [5.209] show a single and relatively broad line. One possibility is that of homogeneous doping [5.209] yielding a single line, whereas the other may be that of inhomogeneous doping in which the resonance of the mobile spins is not observed, possibly due to large broadening of the ESR line through spin-orbit interactions. The ESR spectra of pristine PPy [5.103] also exhibits a superposition of a narrow Lorentzian line ($\Delta B_{pp} \approx 0.4$ G) and a broad line ($\Delta B_{pp} \approx 2.8$ G). When oxidized with gaseous oxygen the narrow line disappears around $y \approx 0.005$. With γ-ray-induced SF_6 doping of PT, the spectrum of the neutral state ($\Delta B_{PP} \sim 7$ G) becomes asymmetric due to the superposition of a narrow component ($\Delta B_p \sim 1$ G) [5.196].

Evolution of the ESR spectra as a function of the doping time has been reported in many systems [5.82, 210–213] in order to understand the nature of the spins induced upon doping and their possible involvement in the mechanism of conductivity. Information about the linewidths and lineshapes can also be utilized to investigate the process of *cis–trans* isomerization of $(CH)_x$ induced chemically [5.175], thermally [5.96, 180, 214, 215] or through the dopants [5.216]. We show below how the dopant-induced *cis–trans* isomerization can be followed via ESR. Figure 5.33 shows one of the earliest ESR spectra (Fig. 5.33a) on $[CH \cdot yNa)]_x$ [5.82] and one of the most recent (Fig. 5.33b) on $[CH \cdot yLi]_x$ [5.210], the former obtained through chemical doping and the latter through electro-chemical doping. The spectra (from top to bottom) correspond to progressive levels of doping; the symbols Q and I in Fig. 5.33b denote, respectively, the charges and instant currents necessary for the observation of the shown features at various doping levels y (or t_d, the doping times). The evolution of the spectral features is similar in both cases; narrow-line features are due to the *trans* isomer whereas broad-line features are due to the *cis* isomer. The absolute dopant level y where complete or nearly complete isomerization occurs depends on the technique employed, rate of doping, the diffusion (a): cis -(CH)$_x$, Na doping   (b): cis -(CH)$_x$, Li doping

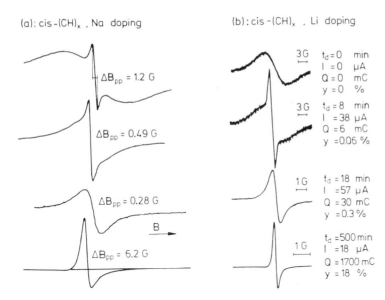

Fig. 5.33 a, b. Evolution of the ESR signal for various levels of doping in chemically (**a**) and electrochemically (**b**) doped (CH·yNa)$_x$ [5.82] and (CH·yLi)$_x$ [5.210], respectively. The arrow shows the increasing field direction. For electrochemically doped (CH·yLi)$_x$, the currents, charges, doping times and doping levels are indicated alongside the spectra. At high doping levels, the lineshape is Dysonian in contrast to the Gaussian lineshape for the pure *cis* material. Note the difference in the peak-to-peak linewidth as the spectra evolve towards Dysonian shapes

characteristics of the dopant species and the characteristic charge transfer rate. A discussion of these rates is given by *Chien* et al. [5.184] for iodine and AsF$_5$ dopants in [CH]$_x$. Possible evidence for the diffusion of the dopant species comes from, besides others [5.217–219], the work by *Bernier* and co-workers [5.210, 220] who, at a certain doping level in [CH·yLi]$_x$, first observe a superposition of a narrow Lorentzian and a broad Gaussian line and then, after a waiting time of 24 hours, lose the narrow signal altogether. This is thought to be due to charge transfer of the slowly diffusing charged dopants to the neutral solitons within the [CH]$_x$ fibrils.

The dependence of the peak-to-peak linewidth ΔB_{pp} on y is shown in Fig. 5.34c (filled circles) for [CH·yLi]$_x$ [5.221, 222]. The linewidth decreases from 7 G to 0.8 G at a doping level in the range 0.003–0.01, and then around $y = 0.05$ it decreases steplike to 0.4 G [5.221, 222]. These results have been interpreted as being due to complete isomerization upto $y \approx 0.01$, at $y \approx 0.05$ a metallic state is reached.

The ESR spectra in heavily doped samples (for example, the bottom panel of the spectra in Fig 5.33a, b) may exhibit asymmetric Dysonian lineshapes due to attenuation of the microwave field, if the sample thickness is much larger than the skin depth [5.223, 224]. The various relaxation times affecting the line-shape in heavily doped samples are T_2, T_δ and T_d which denote, respectively,

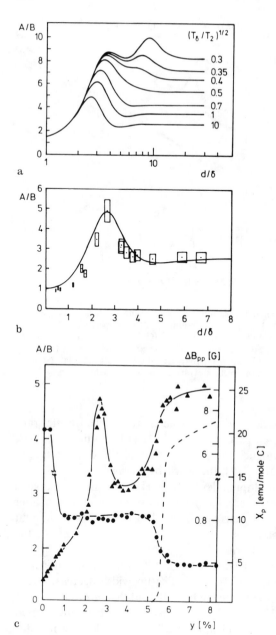

Fig. 5.34 a. Asymmetry ratio A/B of the Dysonian lineshapes as a function of d/δ for various values of $(T_\delta/T_2)^{1/2}$ as calculated from the theoretical expressions [5.223–226]; **b** A/B ratio versus d/δ for AsF$_5$-doped pellets of polyparaphenylene. The continuous-line curve corresponds to the calculations for $T_\delta/T_2 > 100$; **c** Evolution of the asymmetry ratio A/B versus dopant level y for a cis-$(CH)_x$ film doped electrochemically with Li. The dependence on y of the ESR peak-to-peak linewidth is also shown. The solid line in both these cases is a guide for the eye. To emphasize the transition to the "metallic" state, the behaviour of the Pauli susceptibility [5.188] in $(CH \cdot yNa)_x$ is also plotted as a dashed line (According to [5.221, 222]

the spin–spin relaxation time, the time for the particle to diffuse through the skin depth and the time for the particle to traverse the sample thickness. The expression for the lineshape involves, besides these relaxation times, the strengths of the microwave, static and resonance fields, electrical conductivity, magnetic susceptibility, skin depth δ and the thickness of the sample d. Exact expressions exist [5.223–225] for the case of (i) thin films ($d < \delta, T_d \ll T_\delta$) which yield symmetric lineshapes and (ii) thick films ($d > \delta$, $T_d \gg T_\delta, T_2$) which yield asymmetric lines. For thick films, exact expressions also exist [5.223–225] for (a) $T_\delta/T_2 \ll 1$ and $T_\delta/T_2 \to 0$ (narrow lines and highly conducting samples) and (b) $T_\delta/T_2 \gg 1$ (broad lines and slowly diffusing spins). For the latter case, $A/B = 2.7$ for a Lorentzian line and 2.0 for a Gaussian line provided the spins are distributed uniformly; if they are located only at the surface $A/B = 1.0$. Thus, the observation of asymmetric lines is not necessarily direct evidence of the diffusive motion of the spins. The behaviour of A/B versus d/δ, as calculated from the theoretical results of $Dyson$ [5.223], is shown in Fig. 5.34a [5.226]. Figure 5.34b shows the experimental results on AsF$_5$-doped PPP [5.226] along with the theoretical curve for slowly diffusing spins ($T_\delta/T_2 > 100$). Figure 5.34c shows the asymmetry behaviour of electrochemically doped thick films of (CH·yLi)$_x$ [5.221]. The behaviour is similar to that predicted by $Dyson$ (Fig. 5.34a), for $T_\delta/T_2 \gg 1$, up to $y \approx 0.05$ but then, instead of levelling out, A/B increases to a constant value of 5.5. This is interpreted as being due to a fundamental change in the mobility of the defects; these features are reversible. A/B generally increases with the dopant concentration in a non-trival way. The value of A/B for Na-doped (CH)$_x$ increases from ~1.3 (at $y \sim 0.01$) to ~8 (at $y \sim 0.08$). A larger value of A/B for Na-doped compared with that for Li-doped (CH)$_x$ also corre-lates with higher conductivity of Na-doped (CH)$_x$. Heavily K doped cis-(CH)$_x$ leads to higher conductivity on annealing and hence to increased A/B ratio [5.227]. The relationship between the conductivity and the A/B ratio is further evident when (CH)$_x$ is oxidised with different metal halides [5.228].

The linewidth at high dopant levels in (CH·yLi)$_x$ decreases or stays constant, whereas it increases in Na- [5.188, 82], iodine- [5.180, 229, 230] and bromine-doped [5.230, 231] (CH)$_x$. Spin-flip scattering due to larger spin-orbit couplings to sodium, iodine and bromine ions may be the likely explanation. The results of $Bartl$ et al. [5.229] who could observe ESR signals in (CH·yI)$_x$ up to $y \approx 0.30$, in contrast to $y = 0.10$ [5.229] and $y = 0.15$ [5.232], are similar to [5.188] on (CH·yNa)$_x$. Bromine-doped (CH)$_x$ shows a significant g shift [5.230, 231], not observed with iodine [5.229]. The observation that the linewidth for (CH·yBr)$_x$ at $y = 0.55$ is nearly equal to that for pristine cis-(CH)$_x$ is concluded [5.230] to imply that doping proceeds via substitution and addition reactions to C=C bonds; this breaks the chain conjugation and confines the solitons as in pristine cis-(CH)$_x$. Similar conclusions are also drawn from the ESR spectra on iodine-doped polyparaazophenylene (PPN) [5.139]. That the halogen dopants behave differently is also clear from iodine- [5.203] and BF$_4^-$- [5.193] doped PT; whereas in the former the linewidth increases continuously, it decreases at first in the latter from ≈ 9 G to ≈ 0.4 G and then increases slowly to 3 G at $y = 0.30$.

Quantitative analysis of the linewidth cannot be made for inhomogeneously doped ESR spectra. For homogeneously doped lines, the various important sources of line broadening/narrowing are: unresolved hyperfine interaction with or without motional narrowing, g-factor anisotropy or its inhomogeneous distribution, dipole–dipole and exchange interactions and spin-orbit interactions. ESR linewidth usually results from both the static and dynamic contributions. Dipole–dipole and exchange interactions may become very important when the concentration of the paramagnetic conjugational defects becomes large. Anisotropy of the g-factor may contribute statically about 5% to the total linewidth. Spin-orbit interactions contribute dynamically to the linewidth and the theory [5.233, 234] predicts a Z^α (where $\alpha = 4$) dependence of the linewidth, Z being the atomic number of the dopant. For $(CH)_x$ doped with alkali metals, qualitative agreement with the above-mentioned value of α is observed [5.211, 235]. Heavily doped PPP also shows some dependence on Z [5.206]. The theory of *Elliot* [5.235] predicts $(\Delta g)^2 = \tau \Delta B$ where τ is the electron–phonon scattering time. It is misleading to interpret Δg, however, as a real lineshift. Rather, it is the fluctuation in the g value when the electron orbit is changed due to a scattering event with correlation time τ. Since the wavefunction of the counter ions is only weakly mixed into the conduction electron band the effect of spin-orbit coupling due to counter ions is expected to be rather weak, unlike in metals. An example showing the effect of spin-orbit couplings is given in Fig. 5.35. At high temperatures, ΔB_{pp} increases almost linearly with increase in temperature [5.235], which can be attributed to a decrease in the correlation time τ. It is interpreted as being due to the phonon-activated inter-chain hopping which dominates the spin relaxation via the spin-orbit coupling. At low temperatures, where ΔB_{pp} increases with decrease in temperature, it is argued that the inter-chain hopping

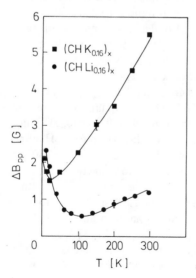

Fig. 5.35. Temperature dependence of the ESR peak-to-peak linewidth in *cis*-$(CH)_x$ doped heavily with Li and K. The solid lines are a guide for the eye. The linewidth is seen to increase linearly with temperature at high temperatures (According to [5.235])

rate becomes small and the relaxation via the spin-orbit coupling is taken over by other processes, e.g. exchange between spins and/or hyperfine interactions. PT doped with iodine shows a similar behaviour, but when doped with AsF_5 yields almost temperature-independent linewidth [5.203]. Due to much smaller spin-orbit couplings on AsF_5 ions, the linewidth in $(CH \cdot yAsF_5)_x$ shows no linear dependence on temperature; the linewidth decreases instead with increase in temperature [5.209]. Peak-to-peak linewidth of 6–12 G in K doped $(CH)_x$ increases to 40–60 G during annealing and the main contribution is likely to be the spin-orbit coupling due to the increased hybridization of potassium s and carbon p_z orbitals [5.227]. For PPP doped with AsF_5, ΔB_{pp} increases from ≈ 0.2 G at the lowest temperature to ≈ 0.7 G at room temperature; this is interpreted as being due to increased exchange interactions at low temperatures [5.226]. The presence of hyperfine interactions is evidenced from larger linewidths in PPP than in its deuterated counterpart [5.206]. It is thus clear that, due to the presence of many types of interactions in doped polymers, quantitative analysis of the linewidth is sample dependent.

Qualitative estimates of the static and dynamic contributions to the linewidth can, however, be made using the CW T_{2*} and spin echo results T_1 and T_{2m}. Phase memory times obtained from the spin echo experiments give a measure of the irreversible contributions to the ESR linewidth; the reversible part can be obtained by using, e.g., relation $\Delta B_{pp} = 2(T_{2*}^{-1} - T_{2m}^{-1})/\sqrt{3}$. Since the irreversible contributions are mainly driven by electron–electron interactions, the role of conductivity in heavily doped systems in changing the values of T_1 and T_{2m} at different temperatures must be properly understood for a correct interpretation of the ESR linewidths. T_{2m} seems to be directly proportional to conductivity as found in PPP [5.236, 237], whereas T_1 shows inverse dependence on conductivity, as found in iodine-doped $trans$-$(CH)_x$ [5.238].

Finally, we briefly summarize the findings and conclusions of *Schärli* et al. [5.200] from their ESR investigations of BF_4^--doped PMT. Having studied the dependence of the ESR lineshape on the microwave power, *Schärli* et al. decomposed the experimental spectrum into its Lorentzian and Gaussian components. The Gaussian component contains relatively few concentration-independent spins and exhibits T^{-1} behaviour. On the other hand, the number of spins obtained from the Lorentzian component increases linearly with the dopant level by about 1 spin per injected charge up to $y \approx 0.1$, implying the generation of polarons. Subsequent decrease in the number of spins has been proposed by *Schärli* as being due to the increased metallic character of the polarons rather than due to the formation of spinless bipolarons. The "linear" temperature dependence of ΔB_{pp} at the highest doping level (similar to the behaviour shown in Fig. 5.35) has been argued by them to be due to the "metallic" nature of PMT, as predicted theoretically by *Elliott* [5.233]. Since "PMT behaves as a metal", these authors have further suggested that macroscopic electrical conductivity as opposed to ESR signals is limited by certain types of hopping processes.

5.4.3 NMR Results

The dopant-induced *cis–trans* isomerization in $(CH)_x$ has been investigated from measurements of the proton second moments M_2 [5.51–53, 239]. The second moment for undoped *cis*-$(CH)_x$ is larger by a few Gauss than for undoped *trans*-$(CH)_x$ and therefore a decrease in M_2 is expected as a function of dopant concentration as *cis-trans* isomerization proceeds. Iodine doping of *cis*-$(CH)_x$ changes M_2 from $\approx 12\,G^2$ at $y = 0$ to $\approx 10.5\,G^2$ at $y = 0.005$, indicating a substantial amount of the *trans* isomer [5.51]; interestingly, electrochemical doping of *cis*-$(CH)_x$ with Li also indicates a similar dopant level for complete *cis–trans* isomerization [5.210]. On the contrary, much higher doping levels are indicated elsewhere [5.239] which may possibly be due to different doping procedures. The y dependence of M_2 is consistent with the mechanism of dopant intercalation between the b and c planes of $(CH)_x$ [5.51]; the mechanism of lattice swelling is implied when the moments become smaller than those for undoped *trans*-$(CH)_x$.

One of the more direct NMR methods to monitor dopant-induced *cis–trans* isomerization, dopant homogeneity and identification of the counter ions is via ^{13}C CPMAS technique whereby the evolution characteristics of the spectra with doping level are investigated. No significant change in the CPMAS spectra of $(CH \cdot yAsF_5)_x$ until $y = 0.07$ and a sudden shift (with line broadening) to low fields was first reported by *Peo* et al. [5.189] who attributed this to the appearance of Knight shift. The fact that the chemical shift spectrum yields microscopic information about the local electronic environment of the nuclear spins, absence of the features characteristic of the doped material was unexpected, particularly because the dopants transfer charges to/form the polymer chains. Reinvestigations by *Clarke* and Scott [5.240] led to the recognition of some subtle differences. It was found that the main peaks (at -127 ppm and ≈ -136 ppm due to the *cis* and *trans* isomers, respectively) lay on top of a relatively broad peak (downfield shifted to ≈ -150 ppm) possibly due to the presence of charged solitons [5.109]. Approximate estimates [5.241] for the downfield shift and the Knight shift in *trans*-$(CH \cdot 0.076AsF_5)_x$, -12 ppm and -2 ppm respectively, further support the viewpoint that the downfield shift has its origin in the removal of electrons from the polymer backbone. Comparison of the spectra for *cis*- and *trans*-$(CH)_x$, doped to an intermediate level $y = 0.03$, shows that *cis*-$(CH)_x$ gives two lines characteristic of undoped and heavily doped $(CH)_x$ regions, whereas the *trans*-sample shows a broad downfield shifted line and a broad line typical of the undoped *trans* material. These features suggest inhomogeneous doping and much higher dopant levels are necessary for complete *cis–trans* isomerization.

Besides these basic features, *Heinmaa* et al. [5.77] observed another signal centred around -50 ppm in $(CH \cdot yI)_x$ for $y > 0.03$. Their results, showing the evolution of the ^{13}C CPMAS spectra as a function of y, are given in Fig. 5.36 for both *cis*- and *trans*-$(CH)_x$. With progressive doping of *cis*-$(CH)_x$, the peak due to the *cis* component at ≈ -127 ppm disappears and a new peak ≈ -150 ppm builds up, broadens and increases in intensity. On the other hand, the parent

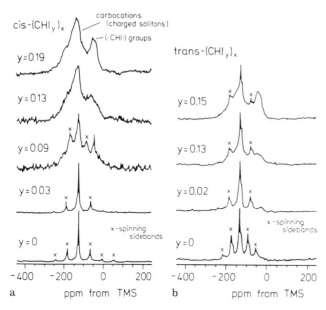

Fig. 5.36 a, b. Evolution of the cross-polarization magic angle spinning spectra of iodine-doped (**a**) *cis*- and (**b**) *trans*-(CH)$_x$ at various doping levels. **a** For $y = 0$, a weak and partially resolved line at -137 ppm in addition to the main peak at -127 ppm shows the presence of some *trans* material in the starting *cis*-(CH)$_x$. With progressing doping, the line around -127 ppm disappears and two new peaks ≈ -150 ppm and ≈ -50 ppm appear. The broad downfield shifted line at ≈ -150 ppm arises from the polycarbocations (e.g. charged solitons) whilst that at ≈ -50 ppm from the covalently bonded iodine; **b** The main peak at ≈ -136 ppm (corresponding to pure *trans*-(CH)$_x$ and the weak signal at ≈ -127 ppm persist right up to the highest doping level ($y = 0.15$); these lie on top of another broad peak which is not resolved even at $y = 0.15$. The occurrence of a peak at ≈ -50 ppm, similar to *cis*-rich (CH)$_x$, is also observed (According to [5.77])

peak at ≈ -137 ppm in *trans*-(CH)$_x$ persists to high doping levels (Fig. 5.36). Thus, similar to AsF$_5$, iodine leads to more homogeneous samples of *cis*-(CH)$_x$ than of *trans*-(CH)$_x$. Quantitative estimates of the *cis* and *trans* contents from the intensity of the peaks may not be correct in the experiments involving a cross-polarization step since the intensities may depend on the cross-polarization time [5.242]. Similar to the peak at ≈ -50 ppm in (CH·yI)$_x$, Br-doped (CH)$_x$ exhibits a peak around -60 ppm [5.241] and SbCl$_5$-doped (CH)$_x$ around -62 ppm [5.243]. This peak, in all these cases, has been assigned to carbons covalently bonded to I, Br or Cl. Contrary to the downfield shift in halogen-doped (CH)$_x$, K-doped (CH)$_x$ at $y = 0.09$ exhibits a line shifted highfields by ≈ -12 ppm due to the increased π-electron density at carbon atoms; an approximate estimate of this shift is ≈ -14 ppm [5.241].

In conjugated systems other than (CH)$_x$, where ^{13}C spins are either bonded to H atoms (*A* type) or non-bonded (*B* type) or bonded to other atoms, the study of the CPMAS spectra has provided useful information. The *A* and *B* chemical shifts in pristine PT, (thiophene.0.12BF$_4^-$)$_x$ and 2,5-dimethyl thiophene are similar (within $\pm 2\%$) [5.88, 91, 92]; this suggests that most of the charge resides on the sulphur atom if the charged species are spinless bipolarons. The Knight shift due to the polarons, if polarons and bipolarons co-existed, as argued for example in PPy [5.197], would be of the order of -2 ppm [5.88]. A separate set of measurements on neutral and ClO_4^--doped PT, however, show significant shift of the A and B lines towards low fields [5.90]. ^{13}C shifts in PPS doped heavily with SO_3^- [5.93] and TaF_6 [5.94] are also very similar to its undoped state. Electrochemically and chemically oxidized PPy (with BF_4^- [5.88], ClO_4^- [5.89] and I [5.89]) shows that the A-type line shifts to low fields depending upon the degree of oxidation whereas the position of the B line is not much altered. In PPP doped with AsF_5 [5.191], no shifts (± 1 ppm) in the A and B chemical shifts have been detected even at a doping level of $y = 0.17$ per phenylene ring. The results on PPP doped with $SbCl_5$ and IF_5, obtained from analysis of the ^{13}C powder chemical shift spectra, also show only very small shifts [5.244].

As to the dynamics, spin-lattice relaxation rates as a function of dopant level, frequency and temperature have been reported. In $trans$-$(CH \cdot 0.10 AsF_5)_x$, T_1^{-1} is smaller by a factor of 10 relative to undoped $trans$-$(CH)_x$ but exhibits $\omega^{-1/2}$ behaviour [5.135]. Similarly, in $trans$-$(CH)_x$ doped with about 15 mole percent Br, T_1^{-1} is reduced by a factor 2 and exhibits $\omega^{-1/2}$ behaviour [5.143]. Since Br shortens the conjugation length and thus confines the solitons over short segments (shown also by reduced Overhauser effect (OE) [5.231]), the observed frequency dependence would then imply 1D motion of the conduction electrons [5.135]. This interpretation is supported by the observation of Korringa relationship in Br-doped $(CH)_x$ [5.143]. The frequency dependence of proton NMR T_1^{-1} in PPS doped heavily with SO_3^- [5.245] is well described by the inverse square law. Non-exponential recovery of the proton magnetization in this system has been interpreted as being due to inhomogeneous doping even in heavily doped samples. Analysis similar to that of defects in $trans$-$(CH)_x$ yields here much smaller diffusion coefficients compared to those for undoped $trans$-$(CH)_x$.

That doping proceeds via the formation of charged solitons can be inferred from the y dependence of T_1^{-1} in iodine- and Na-doped $trans$-$(CH)_x$ [5.239]. The relaxation rate decreases monotonically, as can be explained by the decrease in number and mobility of the neutral solitons. T_1^{-1} in cis-$(CH)_x$, however, first increases due to the cis-$trans$ isomerization, and then decreases.

Thus, the various sources for the relaxation mechanism in doped conjugated polymers are: (1) conduction electrons, (ii) pinned solitons (neutral), polarons or paramagnetic defects, (iii) mobile solitons (neutral) and polarons, (iv) mobile charged solitons or bipolarons and (v) modulations of the interactions via molecular motions (BPP type). The T dependence of the relaxation rate in iodine doped $(CH)_x$ has been analysed [5.150] considering the following three contributions: (a) metal-like contribution dependent linearly on T and as $\omega^{-1/2}$ on frequency, (b) fixed paramagnetic centres and (c) charged solitons. The overall frequency behaviour of T_1^{-1} remains of the $\omega^{-1/2}$ type. Inverse square root dependence on frequency is also observed in ClO_4^--doped PPy [5.207] whereas OE is observed in BF_4^--doped PPy [5.246]. Since the ESR lines are narrow (≈ 0.2–0.3 G) the spins must be highly mobile [5.207] or exchange rapidly. In $(Py \cdot yBF_4^-)_x$ [5.246], the $^{19}F\ T_1^{-1}$ dependence on T shows a BPP-type behaviour whereas $^1H\ T_1^{-1}$ shows only a weak temperature dependence. The origin of $^{19}F\ T_1^{-1}$ is mainly due to the modulations of the ^{19}F–^{19}F dipolar interactions by the rotation of the BF_4^- ions; this is supported by a large increase in the value of the ^{19}F second moment (around 100 K) due to freezing of the ionic motion. The same mechanism, however, contributes little to $^1H\ T_1^{-1}$. This is seen by comparison with ClO_4^--doped PPy; the relaxation behaviour and the rates in both cases are similar and it is known that the ionic motion of the ClO_4^- ions affects T_1^{-1} only slightly due to the much smaller nuclear dipolar interaction. Thus the main contribution to ^1H relaxation rates seems to originate from the coupling of the protons with the electronic spins.

5.5 Magnetic Properties of Polydiacetylenes (PDA)

Polydiacetylenes (PDA) are of particular importance, since they are the only known polymers which can be obtained in single crystal form [5.247–256]. This guarantees large chain lengths and high purity materials which are not contaminated by catalysts and/or oxygen. Moreover the interpretation of spectra, optical and magnetic resonance spectra, becomes more precise for single crystals than for powders or amorphous materials. The spectral lines in single crystals are unusally sharp and well resolved and often show orientational dependences which yield further information about the microscopic structure on the molecular level. Due to limited space we can only dwell briefly in this section on some magnetic resonance results. Numerous original publications have appeared on the subject and we select only a few. Reviews by *Sixl* [5.257a] and *Schwoerer* and *Niederwald* [5.257b] on the solid-state polymerization reaction process in diacetylene crystals should be consulted for further details.

5.5.1 Structure

Almost perfect and large polymer single crystals can be obtained from monomers by solid-state polymerization. Other polymerization processes lead to less perfect crystals, so we shall therefore restrict ourselves to the solid-state process. In the first step monomer crystals are grown from a solution of diacetylenes of the chemical form [5.247–257]

$$R - C{\equiv}C\text{-}C{\equiv}C - R; \quad R_1 - C{\equiv}C\text{-}C{\equiv}C - R_2; \quad \begin{matrix} C{\equiv}C\text{-}C{\equiv}C \\ \diagdown R_3 \diagup \end{matrix}$$

282   P.K. Kahol et al.

Table 5.4. Chemical structure of diacetylenes

$$R-C{\equiv}C-C{\equiv}C-R, \quad R_1-C{\equiv}C-C{\equiv}C-R_2; \quad \underset{R_3}{\overset{C{\equiv}C-C{\equiv}C}{\smile}}$$

Nomenclature	Substituents
TS-6	R: $-CH_2OSO_2C_6H_4CH_3$
TS-12	R: $-(CH_2)_4OSO_2C_6H_4CH_3$
FBS-6	R: $-CH_2OSO_2C_6H_4F$
TCDU	R: $-(CH_2)_4OCONHC_6H_5$
TCDA	R_1: $-(CH_2)_9CH_3$; R_2: $-(CH_2)_8CO$
BPG	R_3: $-OCO(CH_2)_3OCO-$

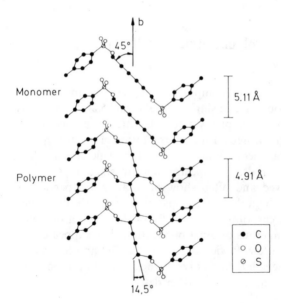

Fig. 5.37. Atomic positions in monomer and polymer TS-6 crystals deduced from X-ray data [5.251]

where R, R_1, R_2 and R_3 are different substituents. Table 5.4 lists a few diacetylene substituents and the currently used nomenclature. All these diacetylene monomers form single crystals, however some of them are especially suited for obtaining long chain polymer single crystals. The general structure of the repetitive unit in partially polymerized single crystals is shown in Fig. 5.37. Note the slight change in lattice constant upon polymerization. The side groups are moved slightly away from their position in the monomer crystal after polymerization, whereas the diacetylene unit is tilted drastically. Several experimental techniques have confirmed that extremely long chains of more than 1000 monomer units exist in partially polymerized crystals [5.257–262].

5.5.2 Solid-State Polymerization

It is evident from Fig. 5.37 that the solid-state polymerization process proceeds with a minimum of atomic displacements starting from the monomer crystal. This so-called "topochemical principle" demands that the position and orientation of reactive groups in the monomer crystal is favourable for linking neighbouring groups with a minimum of rearrangement of the molecules (principle of least motion). The optimal arrangement of the monomer molecules is obtained by an appropriate choice of the substituents R.

The polymerization proceeds via intermediate structures of the form

$$n[R-C\equiv C-C\equiv C-R] \xrightarrow{hv}.$$

Intensive optical and ESR investigations have revealed that the "acetylene" structure is dominant in long polymer chains, whereas the "butatriene" structure exists only for certain types of short chains [5.257, 263–267]. Sixl and co-workers [5.268–269] have developed a model based on MO calculations which can explain the transition from the butatriene structure to the acetylene structure at a chain length of about 5–6 monomer units.

The best studied PDA single crystal up to now is poly-TS-6. All atomic positions and the lattice parameters for the monomer and for the polymer crystal are precisely known from X-ray analysis. Solid state polymerization has been achieved in this system by UV-, X-ray- or γ-irradiation, and simply by heating. Photo-polymerization is usually performed at low temperatures (10 K) in order to avoid competative thermal polymerization. The mechanism proceeds from the monomer to an initial dimer, which is essential for the subsequent polymerization. Once the dimer formation is achieved, e.g. by photo-polymerization, the following trimer, tetramer etc. units can be generated either thermally or photo-chemically. The detailed mechanisms of this process were clarified by spectroscopic techniques [5.257]. Table 5.5 lists the so-far identified reaction

Table 5.5. Symbols, spin multiplicities, structures and notations for reaction intermediates which are schematically represented by the trimer unit [5.262]

Symbols spin S	Structure trimer unit	Notations chain length n
$-DR-$ diradicals $S = 0, S^* = 1$		butariene $n \geqq 2$
$-DC-$ dicarbenes $S = 0, S^* = 1, 2$		acetylene $n \geqq 7$
$-AC-$ asymmetric carbenes $S = 1$		acetylene $n \geqq 2$
$-SO-$ stable oligomer $S = 0$		acetylene $n \geqq 3$

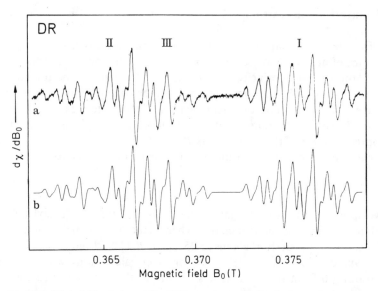

Fig. 5.38. High-field of the triplet ESR spectrum of diradical reaction intermediates at 77 K. *a* experimental spectra, *b* simulated spectra. The hyperfine structure in the simulated spectra results from two CH_2 groups [5.257a]

intermediates including their spin states [5.257, 262, 269]. Note that besides the stable oligomer (–SO–) all intermediate structures have triplet-spin states (S = 1) and in the case of dicarbenes (–DC–), also a quintet state (S = 2). ESR spectroscopy is has proven to be a favourable tool for the investigation of these reaction intermediates [5.262–269].

A typical triplet ESR spectrum of the diradical series is shown in Fig. 5.38 [5.257a]. Note that a triplet ESR spectrum consists in general of a pair of lines which appear symmetrical about the Larmor frequency (or field) and where the linesplitting corresponds to the fine structure (dipole–dipole splitting of the two S = 1/2 electrons). Figure 5.38 represents only the high-field part of the spectrum, i.e. it should consist only of a single line if only one species would be present. As is evident from Fig. 5.38, however, three different –DR– species with differing fine-structure splittings ($I > III > II$) are visible. The corresponding chain lengths are expected therefore to be in reverse order ($n_I < n_{III} < n_{II}$). In addition, Fig. 5.38 shows some hyperfine structure which derives from the coupling of the triplet spin with the surrounding proton spins of two CH_2 groups. The lower spectrum represents the corresponding simulated spectrum.

Whereas the spectral information reveals the structure of the reaction intermediate, the intensity of the ESR spectrum corresponds to the number of spins. The ESR intensity therefore allows one to investigate the reaction kinetics, as demonstrated in Fig. 5.39 [5.257a]. The different diradical species (I, II and III) are formed under UV-light irradiation. Their evolution with irradiation time can be followed by observing the ESR intensity of the corresponding triplet

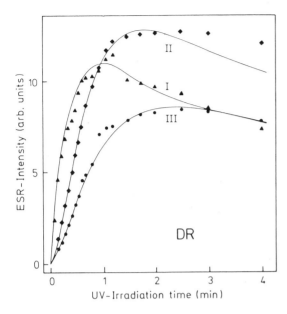

Fig. 5.39. Evolution of three different diradical species (I, II, III) upon *uv* irradiation. The ESR intensity of the corresponding lines in Fig. 5.38 serves as a measure for the concentration of the species [5.257]

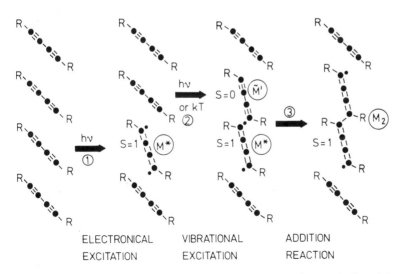

ELECTRONICAL EXCITATION   VIBRATIONAL EXCITATION   ADDITION REACTION

Fig. 5.40. Reaction scheme of the dimer initiation reaction: 1) Photo-excitation of the metastable monomer diradical M^*; 2) Displacement of the adjacent monomer molecule M'; 3) Formation of the diradical dimer molecule M_2 by 1, 4 addition [5.257]

states, as shown in Fig. 5.39. In a similar fashion a number of reaction intermediates (see Table 5.5) were observed and characterized. Their reaction steps
and kinetics were completely analyzed and are fairly well understood.

The following picture emerges from this analysis. The initial step is the
dimer formation which is usually achieved by UV-, γ- or X-ray irradiation at
low temperatures (10 K). This creates a long-lived metastable excited state of
the monomer (M^*) which represents the reaction centre of the initiation reaction.
It has been proposed that the monomer diradical $T_1(M)=DR_1$ with the structure
$R-\overset{\circ}{C}=C=C=\overset{\circ}{C}-R$ could play this role. An adjacent monomer can now be added
to form a dimer-diradical DR_2, most likely in its triplet state, which finally
relaxes to the ground state DR_2. The different steps of this mechanism were
investigated in detail. This detailed reaction scheme is represented in Fig. 5.40
[5.257]. Upon further thermal- or photo-excitation, trimers, tetramers etc., and
finally polymers are formed. This mechanism has been monitored in great detail
by ESR and ENDOR spectroscopy [5.262–269].

Among these studies, ESR and ENDOR of the quintet states has attracted
particular interest [5.268, 269]. The quintet states arise from the electron–electron
coupling of two triplet carbenes. The quintet state has an excitation energy of
$\Delta\varepsilon_{SQ}$ above the singlet ground state which depends on the "separation" of the
two carbenes. The fine structure observed in the ESR spectra of the quintet
states is again determined by the separation of the two carbenes [5.268, 269].

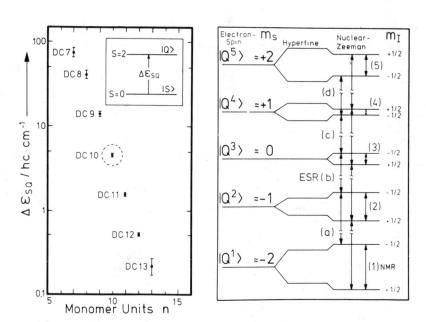

Fig. 5.41. Left: Experimentally determined values of the singlet-quintet splittings $\Delta\varepsilon_{SQ}$ versus the
number of monomer units ($n = 7, 8, \ldots, 13$) in the dicarbene radical [5.268]. Right: quintet state
splittings in an external field B_0 including hyperfine interaction with one proton [5.268d]. The
allowed ESR ($a\ldots d$) and ENDOR ($1, \ldots, 5$) transitions are indicated

A correlation between these two quantities is therefore expected. The number n of the monomer units in a dicarbene determines this separation. During the photo-polymerization process, different dicarbenes with different n can be produced and are readily distinguished by their fine structure constants from ESR spectra [5.268, 269]. The experimentally determined fine structure is therefore a direct measure of the number of monomer units n and is used to disentangle

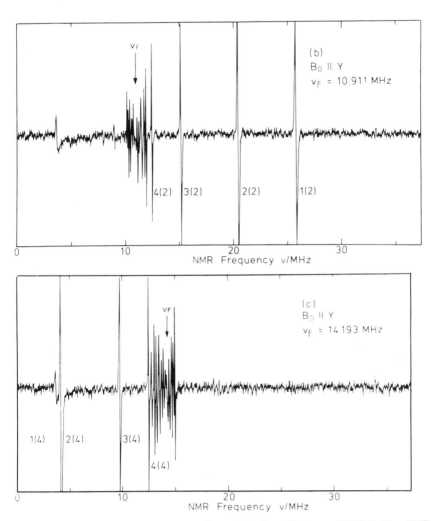

Fig. 5.42. Quintet ENDOR spectra according to *Hartl* and *Schwoerer* [5.268d]. Top: ENDOR spectrum, detected at the ESR transition b in Fig. 5.41. The Larmor frequency of "free" protons is indicated by v_F. The ENDOR lines are labelled $j(u)$, where $j = 1, \ldots, 22$ refers to the jth proton in the dicarbene and $u = 1, \ldots, 5$ indicates the quintet state $|Q\rangle$ (see Fig. 5.41). Strong hyperfine interactions are observed for protons 1, 2, 3 and 4 with decreasing strength
Bottom: ENDOR spectrum detected at the ESR transition c (see Fig. 5.41). The strongest hyperfine lines (below v_F) appear for protons 1, 2 and 3 in the quintet state $|Q^4\rangle$

the ESR spectra. On the other hand, the intensity of the ESR spectrum depends on the quintet energy $\Delta\varepsilon_{SQ}$ which appears as an activation energy of the corresponding ESR signal. From temperature-dependent measurements of the ESR intensity $\Delta\varepsilon_{SQ}$ has been determined [5.268, 269] as is plotted in Fig. 5.41 (left). If l_m is the length of the monomer unit, the carbene–carbene distance may be expressed as $D=n\,l_m$ for n monomer units. The experimental data for $\Delta\varepsilon_{SQ}$ shown in Fig. 5.41 follow the rule

$$\Delta\varepsilon_{SQ} = \Delta\varepsilon_{SQ}^0 \exp(-D/D_0)$$

with $D_0 = 5.4\,\text{Å}$ [5.268d].

In the following we will discuss the hyperfine coupling of the surrounding protons to the quintet state. We restrict ourselves to the quintet state with $\Delta\varepsilon_{SQ} = 4.5\,\text{cm}^{-1}$ which is encircled in Fig. 5.41 (left). In a large magnetic field B_0 the quintet spin $S=2$ splits into 5 states $|Q^u\rangle$ with $u=1,2,3,4,5$ as shown in Fig. 5.41 (right). Each of these states is split again by hyperfine coupling to a proton $(I=1/2)$, as indicated in Fig. 5.41 (right). In practice many protons couple to the quintet state and a large number of hyperfine lines is expected. These are usually not resolved in the ESR spectrum.

ENDOR spectroscopy has therefore been employed in order to resolve the hyperfine structure of this quintet state [5.268d]. Two examples for the ENDOR spectra observed in the quintet state with $\Delta\varepsilon_{SQ} = 4.5\,\text{cm}^{-1}$ are shown in Fig. 5.42. Different ESR transitions are labelled a–d in Fig. 5.41 (right). These ESR transitions are monitored while sweeping through the ENDOR spectral range (0–40 MHz). The variation of the ESR intensity is then plotted versus the applied radio-frequency field. Figure 5.42 displays those ENDOR spectra where the ESR lines (b) (left) and (c) (right) were used as "monitors". Different hyperfine lines labelled $j(u)$ are observed, where $j=1,\ldots,22$ indicates the position of the proton in the dicarbene radical, beginning with the outer ends and where u labels the quintet Zeeman state $|Q^u\rangle$. It is evident that the strongest hyperfine

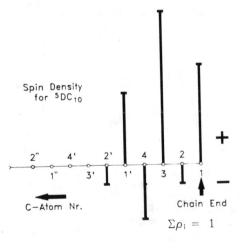

Spin Density for $^5DC_{10}$

C–Atom Nr.

Chain End

$\Sigma\rho_i = 1$

Fig. 5.43. Spin density distribution at the chain end of DC_{10} dicarbenes obtained from proton ENDOR spectra [5.268d] (courtesy of M. Schwoerer)

interaction (largest distance from ν_F) is observed for the outer protons, indicating the highest spin density at these positions. The actual spin density distribution as evaluated from proton ENDOR spectra for the corresponding carbon sites of the DC_{10} dicarbenes is shown in Fig. 5.43. Note the alternating sign change due to electron–electron correlations. For further details we refer to [5.268d]. Additional information was obtained from optical spectra and their quantum mechanical interpretation. As expected, the wavelength of the optical absorption edge decreases with increasing chain length. For further details of the optical spectra, the special literature should be consulted [5.257, and references therein].

5.5.3 Quasi-Particle Excitations

In the following we discuss the excitation of quasi-particles in long polymer chains of polydiacetylenes. A number of quasi-particles can be generated by optical excitation of inter-band transitions ($\pi\pi^*$). A list of possible quasi-particles is shown in Fig. 5.44.

They are believed to appear after electron–hole photo-excitations followed by lattice relaxation. Some of these excited states were observed by optical absorption and emission spectroscopy [5.270–273]. Time-resolved optical spectroscopy has proven to be particularly important in detecting these quasi-particle excitations.

Configuration   Quasiparticle

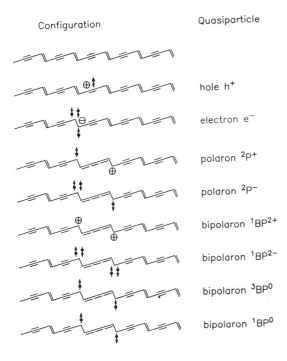

hole h^+

electron e^-

polaron $^2p^+$

polaron $^2p^-$

bipolaron $^1BP^{2+}$

bipolaron $^1BP^{2-}$

bipolaron $^3BP^0$

bipolaron $^1BP^0$

Fig. 5.44. Possible quasi-particle excitations of the polydiacetylene backbone. Radical electrons are labelled by dots together with their spin states [5.257a]

In the following we focus on a photo-excited triplet state in polydiacetylene which was first observed by *Robins* et al. [5.173]. The magnetic field dependence of the decay rates and fine-structure splitting which were investigated by Robins by the ODMR (optical detection of magnetic resonance) technique proved the existence of this triplet state [5.173]. Subsequent direct observation of the triplet state, whose ESR spectrum is shown in Fig. 5.45, by *Winter* et al. [5.174] led to the conclusion that a pair of soliton-like quasi-particles in a triplet state is generated after optical excitation at 337 nm. The ESR spectrum in Fig. 5.45 shows a pair of lines with opposite amplitudes which is the signature of a triplet state whose $m = 0$ level is less populated than the $m = \pm 1$ levels. The $\Delta m = +1$ transition therefore appears in "emission" whereas the $\Delta m = -1$ transition appears in "absorption" mode. The observed time evolution of these signals after laser pulse excitation leads to the conclusion that during the laser pulse a pair of triplets is formed on-chain with all triplet levels *equally* populated, i.e. in a totally symmetric singlet state as proposed by *Tavan* and *Schulten* [5.274].

Appreciable spin polarization only develops over the course of time and then decays again. The average extension of the triplet states is 3.4 Å, as obtained from the measured fine-structure parameter $D = -0.0654\,\text{cm}^{-1}$. The observation of initial equal population of all triplet levels demands the existence of a triplet pair, which has initially singlet symmetry (totally symmetric) [5.174]. Due to the subsequent selective decay of the $m = 0$ level, a spin polarization develops which leads to the observed spectrum (Fig. 5.45). The complete scenario is out-lined in Fig. 5.46. The photo-excited quasi-particles are allowed to separate on the chain and become individual objects. The delayed emission from the triplet state to the ground state was observed at 720 nm, which also supports the fission

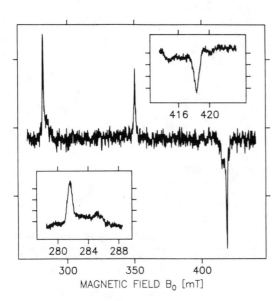

Fig. 5.45. Electron-spin echo spectrum of photo-excited triplet states in PTS, 15 µs after laser pulse excitation ($T = 10\,\text{K}$). Note the absorptive (low field) and emissive (high field) polari-zation of the triplet lines [5.174]

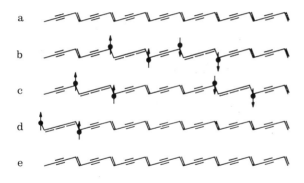

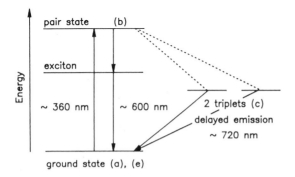

Fig. 5.46. Scenario of triplet soliton-pair formation on a poly-diacetylene chain after *uv* excitation [5.174]. Bottom: Energy level diagram together with relevant optical data according to [5.272]

into a triplet pair [5.174]. We remark that the existence of this triplet state in a conjugated polymer again demonstrates that strong electron–electron correlations are present in conjugated polymers [5.274]. In fact, in polyacetylene one would also expect to see similar triplet excitations, but these have not yet been observed.

5.6 Other Conjugated Polymers

The substituents or analogues of the above-described conjugated polymers, which have a polyene backbone structure, have been synthesized and studied for their magnetic behaviour. For substituted polyacetylene, poly-(1,6-heptadiyne) has been synthesized using Ziegler–Natta catalysts and exhibits conducting behaviour on oxidation by iodine [5.275, 276]. The ESR linewidth decreases on doping from $\approx 20\,\mathrm{G}$ to $\approx 10\,\mathrm{G}$. In contrast to iodine-doped $(CH)_x$, the number of spins here increases which can be considered as being due to the formation of radical cations or polarons. A fluorinated analogue of poly-(1,6-heptadiyne) has also been synthesized and analysed for its structure using ^1H, ^{13}C and ^{19}F NMR spectroscopy [5.277]. This fluorinated hydrocarbon, in contrast to poly-(1,6-heptadiyne), does not appear to be prone to oxidation by iodine or oxygen; this may be explained by the fluorine substitutent as regards its role in stabilising the polyene against oxidation [5.277].

Polyvinyl chloride (PVC) has been degraded thermally to obtain polyacetylene-like segments [5.278]. The degraded PVC exhibits much smaller conductivity compared to Shirakawa $(CH)_x$ on iodine doping. The ESR signal with $\Delta B_{pp} \approx$ 13 G appears soon after the thermal treatment begins. The linewidth reaches a plateau at ≈ 7 G in highly degraded PVC, and the width of the conjugational defect has been deduced to be ≈ 22 G (c.f. 14 G as predicted by SSH for an ideal $(CH)_x$ chain). Polyvinyl sulphide $(CH=CH-S)_x$ (PVS) has been analysed [5.279] for the ^{13}C NMR spectrum and is consistent with the presence of alkenic carbon atoms. Polyphenylacetylene (PPA) is also an analogue of $(CH)_x$ and its ESR spectrum exhibits two lines [5.280-282]. It contains π-electron spins of the order of 2.5×10^{16} spins/gram and the spin concentration increases on doping with iodine and AsF_5, less so in PPA–AsF_5 than in PPA-iodine. The lineshapes show no Dysonian character, probably due to the smaller particle size. Most of the charge storage is thought to be due to spinless polymer cations or bipolarons. Annealing of pristine PPA above 255 K has been found to yield a much higher spin concentration, and is believed to be due to cis-trans isomerization [5.283]. A highly resolved ESR spectrum at 493 K with $g = 2.0024$, in contrast to a structureless single-line spectrum at room temperature, has been assigned to the trans-transoidal structure of PPA. With iodine doping, hyperfine coupling constants decrease, which is interpreted as being due to partial oxidation of PPA by iodine [5.283].

The substituted compounds of PPP have been more thoroughly investigated. Polyphenylene vinylene (PPV) and its analogues show increases in conductivity on doping with AsF_5 [5.284]. The symmetric ESR lineshapes of width ≈ 0.3–0.6 G indicate a defect highly delocalized along the chain direction. When the protons at the vinylic positions are deuterated, the linewidth decreases by a factor of two, suggesting again delocalization along the chain. The relatively smaller conjugation length of its analogues is reflected in the much larger ESR linewidths. The susceptibility in PPV is of the Curie type and the spin concentration on doping increases to as much as 1 spin per 20 units. The spin-lattice relaxation time decreases by a factor of 7 from $\approx 10 \mu s$ for PPV.0.0036 AsF_5 to 1.5 μs for PPV.0.54 AsF_5, which indicates increased coupling between the spin system and the dopant nuclei. A PPV film, stretched by 7 times, shows a clear anisotropy in both the g value and the ESR linewidth when the field is applied along and perpendicular to the stretch direction [5.129]. This anisotropy is quite consistent with that expected of a π-electron defect [5.129].

A fluorinated substituent of PPP (FPPP) has been synthesized [5.285] and doped with potassium to investigate its ESR behaviour. For lightly doped FPPP, a single Lorentzian line ($g = 2.0027$) with a line width of $\Delta B_{pp} \sim 1.8$ G at 6.6 K results. Spin concentration as a function of temperature, obtained from the ESR spectra, obeys the Curie law [5.285]. Heavy doping does not change the g value but the linewidth increases by about 2.5 times. As a function of temperature, the linewidth decreases by $\sim 10\%$, from 8 K to 300 K, but the g value remains the same. Spin susceptibility shows a non-Curie behaviour, similar to that plotted in Fig. 5.33 for a alkali-doped PPP, but here the spin concentration is lower by an order of magnitude compared to PPP doped heavily with K. The observations that K–FPPP with lower spin concentration is slightly more conducting than K–PPP suggests that the carriers of conductivity are bipolarons [5.285].

Another derivative of PPP which has a heteroatom (sulphur) in the polymer chain backbone and shows increased conductivity is polyparaphenylene sulphide (PPS) [5.286, 287]. Undoped PPS exhibits a weak ESR signal centred at $g \cong$ 2.004 compared to a slightly narrower signal for PPP centred at ≈ 2.003; this could be due to larger g-factor anisotropy in PPS due to spin-orbit interactions with sulphur [5.286]. Doping of PPS with SO_3, AsF_5 or SbF_5 leads to a change in the g value to ≈ 2.0074 which could be due to finite spin density on sulphur. The same is true for the oligomers of PPS which also show similar g values. The intense ESR signal observed on doping corresponds to an increase in the spin concentration by 300 and 760 in SbF_5- and SO_3-doped PPS, respectively. Whereas AsF_5-doped PPP powder gives an A/B ratio of the ESR lines of between 2 and 3, PPS film shows a ratio A/B of ≈ 1.16. This difference is intimately related to lower conductivity in doped PPS and perhaps also to smaller particle size. The static susceptibility measurements of PPS doped with $SbCl_5$ (1 mole percent) show a Curie–Weiss behaviour with a very small Curie constant [5.288]. The number of Curie spins have been found to be one tenth of the number of dopants, implying the formation of bipolarons [5.288].

NMR linewidths, second moments and spin-lattice relaxation rates have been measured in undoped PPS to study its dynamical behaviour [5.287]. The temperature dependence of the linewidth and second moments are analysed using a phenomenological expression for activated motional narrowing. Two distinct regimes are seen, one below 330 K corresponding to an activation energy of ≈ 7.5 kJ/mole and the other above 330 K corresponding to an activation energy of ≈ 14 kJ/mole. An activated process with energy ≈ 11.5 kJ/mole and which modulates the proton–proton dipole interactions accounts for the spin-lattice relaxation rates at high temperatures. T_1 at low temperatures shows only a weak temperature dependence, rather similar to that for PPP. The activated paramagnetic relaxation due to the fixed defects, corresponding to an activation energy of ≈ 1.5 kJ/mole, when combined with the process of activated proton–dipolar interactions reproduces nicely the overall behaviour of T_1^{-1}.

The azomethine $-CH=N-$ and azo $-N=N-$ linkages have been effected in PPP aromatic rings to yield poly-p-phenylene-azomethine (PPCN) and poly-p-azophenylene (PPNN), respectively [5.139, 289]. The ESR signal in PPNN is symmetrical (with $g \approx 2.0037$), unresolved and with a temperature-independent lineshape; the typical spin concentration of the spins is 1 per 100 monomers. These defects are fixed, as shown by the SSE. With progressive doping of PPNN with iodine, the conductivity increases (but remains much smaller than in $(CH)_x$), the g value increases from 2.0037 to 2.0060, the linewidth increases from 8 G to 15 G (as in iodine-doped $(CH)_x$) and the number of spins increases. This increase in the number of spins clearly demonstrates iodine addition/substitution at the N=N bonds. On the other hand, PPNN doped with AsF_5 shows no significant increase in conductivity and the ESR signal remains virtually unchanged relative to that in the virgin polymer; PPCN also behaves similarly. Iodine-doped PPCN shows instead increased conductivity along with an increase in the g factor and the ESR linewidth, but no increase in the number of spins. The last point might imply that the charge storage takes place predominantly in bipolaronic states (as in ClO_4^--doped PT [5.195]) and that, in this system, bipolarons are the intrinsic lowest energy charged configurations. It is also implied that –CH=N– bonds are more stable than –N=N– bonds.

Polyselenophene (PS) is the analogue of polythiophene and, like PT, exhibits a broad ESR line of ≈ 15 G [5.290]. On doping with ClO_4^-, the ESR line decreases to ≈ 6 G (at the highest doping levels) which is much larger than for PT, most likely due to increased spin-orbit coupling when going from S to Se. PS also gives much lower conductivities compared to doped PT. All these observations seem to indicate that in doped PS, the delocalization of the π electrons and their mobility is greatly suppressed due to shorter conjugation lengths, owing to the non-coplanarity of the heterocyclic rings effected by the Se atoms. The spin concentration, as in doped PT, increases with the dopant level. This might imply that chains with smaller conjugation lengths inhibit the conversion of induced polarons to bipolarons.

^{13}C NMR and $T_{1\rho}$ experiments have been performed in 2,5-dimethylthiophene (P3MeT) [5.92]. The measurements show a lower molecular mobility in P3MeT than in PT.

Polythienylpyrrole (PTP) may be considered to be an alternating co-polymer of pyrrole and thiophene. Four peaks observed in the ^{13}C CPMAS spectrum, assigned to the pyrrole and thiophene rings, show an interesting behaviour [5.291]. Compared with PPy and PT, the peak positions in PTP for the PPy units shift high fields whereas the peak for the B carbons of PT shifts low fields. This is explained by asserting that the pyrrole units donate electrons to PT units in PTP [5.291].

Polyazulene is an example of a "fused-ring polymer" where each azulene monomer is formed by fusing a seven-membered ring with a five-membered ring. The ESR line in polyazulene ($g = 2.0031$) has a Gaussian shape and a linewidth $\Delta B_{pp} \sim 6.5$ G [5.292]. Upon doping with ClO_4^-, ΔB_{pp} decreases to about 0.3 G. On light doping, spin concentration decreases from $\sim 1.5 \cdot 10^{20}$ spins/gm to $\sim 5 \cdot 10^{19}$ spins/gm. As doping progresses, it increases to $\sim 2.5 \cdot 10^{20}$ spins/gm before decreasing again at higher levels of doping [5.292]. These results at low-doping levels resemble those for $(CH)_x$ where the decrease in the spin concentration is due to the transformation of neutral solitons to spinless charged solitons. In the absence of a similar mechanism in polyazulene, the authors have discussed these results on the basis of an *ad-hoc* "dual band structure" model [5.292].

Highly localized π-electron defects whose concentration remains well below the ppm limit have been observed via ESR in 2,5-polythiazole [5.293]. The linewidth is $\sim 10\,G$ which is much larger than that for trans-$(CH)_x$ or PT [5.293].

A new polyconjugated system, polymetacyclophane, has recently been synthesized which, as with co-facially stacked polyphthalocyanines, has longitudinal π-electron interactions [5.294]. ^{13}C CPMAS technique has been employed to understand the structure of its monomers and the subsequent structural changes en route to polymetacyclophane during polymerization. The conductivity of this polymer when doped with H_2SO_4 vapour is, however, smaller compared to doped $(CH)_x$, PPP, PPy or PT.

Polyaniline is a potential candidate for electrodes in electrochemical batteries, and exhibits a good deal of fascinating physics. In contrast to other conjugated polymers, it undergoes insulator-metal transition through protonation, i.e. electrons on the polymer backbone remain unchanged but the number of protons increase. It is thus natural that its properties depend on the water content and the pH of the solution in which polyaniline is equilibrated. An extensive analysis based on relaxation rates has been carried out by *Nechtschein* et al. [5.295] to understand the role of water in the conduction process. *Epstein* et al. [5.296] have, on the other hand, focused on susceptibility measurements to investigate the conducting-insulating transition. Aniline oligomers, oxidized to different levels, have also been synthesized and analysed using ^{13}C and ^1H NMR spectroscopy [5.297].

5.7 Conclusions and Remarks

It should be evident from this chapter that the magnetic properties of conducting polymers, and in particular the application of magnetic resonance spectroscopy to these solids, is of fundamental importance for the understanding of their different features.

After a brief introduction to magnetic resonance spectroscopic techniques in Sect. 5.1 we have discussed in Sect. 5.2 the different possible applications of NMR to obtain information on the lattice structure and bond lengths in polymers. There the precise knowledge of the dipole–dipole interaction between nuclear spins was utilized as a structural parameter. Special two-dimensional NMR techniques have proven to be very effective for structural investigations. Even lattice dynamics can be monitored by the nuclear dipole–dipole interaction, as demonstrated in Sect. 5.2. Polyacetylene was shown to undergo no phase transition but a contraction with decreasing temperature, whereas polyparaphenylene shows a ring flipping at high temperatures which freezes in at lower temperatures.

The chemical shift tensor, especially of ^{13}C, can serve as a valuable tool for investigation of the electronic structure of polymers. It has been used in the past mainly for differentiating between different chemical species or isomers. This way *trans* and *cis* polyacetylene could be distinquished by a chemical shift difference of about 10 ppm. However, the full tensorial character of the chemical shift can be measured successfully even for polymers with advanced NMR techniques and provides subtle information on the electronic configuration of the polymer. Unfortunately, there are no simple theoretical models available for a clean theoretical calculation of the chemical shift tensor in polymers. Even highly sophisticated MO calculations cannot reproduce accurately enough the experimental values. Since it was exemplified in Sect. 5.2.3 that the chemical shift tensor does vary with chain length, a better theoretical understanding of these findings would be most welcome.

The main role in magnetic resonance spectroscopy of polymers, however, in played by electron spin. The ESR lineshape and electron spin relaxation provide direct evidence for mobile objects in polymers, such as solitons, polarons, excitons, conduction electrons and defects. The sheer existance of an ESR signal or the observation of a susceptibility often reveals important information. However, to unravel the detailed nature of the object under investigation is again a difficult task and relies very much on the correct interpretation of experimental results.

Sections 5.3–6 provided ample evidence for the fact that a rich variety of defects, excitons, solitons and polarons can give rise to similar ESR signals. It often depends therefore on one's imagination and/or further experimental support before one can draw firm conclusions. However, let us state a few established facts. Doublet ($S = 1/2$), triplet ($S = 1$) and quintet ($S = 2$) states of electronic excitations can easily be distinguished by their fine structure. The electron spin is usually coupled to the surrounding protons by hyperfine interaction. The typical hyperfine coupling constants are known, and the hyperfine structure (if observed) therefore gives some information on the delocalization of the electron. This information is more directly obtained from ENDOR investigations. If the electron spin is fixed the hyperfine structure, as observed in ENDOR and triple-ENDOR experiments, allows firm conclusions to be drawn on the "spreading" of the electron wave function. DNP experiments confirm these findings by observing the solid-state effect.

If the electron spin is mobile, a gradual averaging of the hyperfine interaction occurs, leading to a narrowing of the ESR line and the corresponding ENDOR line. The DNP reduces then to the Overhauser effect, a narrow line at the Larmor frequency of the electron spin. If the rate of motion of the electron spin is much larger than the hyperfine interaction (in $rad\,s^{-1}$), a narrow ESR line is observed with no hyperfine structure whatsoever.

Electron spin motion can be most effectively observed by spin relaxation measurements. If the correct theory is used, the correlation time for the motion can in principle be extracted. Both nuclear and electron spin relaxation should give the same answer. Because of several unknowns, and because the appropriate model is often missing, large discrepancies can arise here, as is obvious from

Sects. 5.4 and 5. A more detailed and thorough theoretical treatment of spin relaxation could possibly bridge this gap.

Finally we would like to remark that fascinating spin physics is involved in conducting polymers and further investigation along these lines will prove to be very useful for application of these solids, and will certainly be very exciting simply for scientific reasons.

Acknowledgments

Most of this work was completed while P.K. Kahol was at the University of Stuttgart on a Humboldt Fellowship, and during the visit of W.G. Clark as a Humboldt Awardee. M. Mehring would like to acknowledge financial support from the Stiftung Volkswagenwerk and in part by the Deutsche Forschungsgemeinschaft (SFB 329).

References

5.1   J.A. Pople, S.H. Walmsley: Mol. Phys. **5**, 15 (1962)
5.2   W.P. Su, J.R. Schrieffer, A.J. Heeger: Phys. Rev. Lett. **42**, 1698 (1979); Phys. Rev. B **22**, 2099 (1980)
5.3   M.J. Rice: Phys. Lett. **71A**, 152 (1979)
5.4   S.A. Brazovskii: Sov. Phys. JETP Lett. **51**, 345 (1980)
5.5   S.A. Brazovskii: Sov. Phys. JETP Lett. **78**, 677 (1980)
5.6   S.A. Brazovskii, N.N. Kirova: Sov. Phys. JETP Lett. **33**, 6 (1981)
5.7   W.P. Su, J.R. Schrieffer: Proc. Nat. Acad. of Sci. **77**, 5626 (1980)
5.8   E.J. Mele, M.J. Rice: Phys. Rev. Lett. **45**, 926 (1980)
5.9   D.K. Campbell, A.R. Bishop: Phys. Rev. B **24**, 4859 (1981)
5.10 J.L. Brédas, R.R. Chance, R. Silbey: Mol. Cryst. Liq. Cryst. **77**, 319 (1981)
5.11 Y.R. Lin-Liu, K. Maki: Phys. Rev. **22**, 5754 (1980)
5.12 S. Roth, H. Bleier: Adv. Phys. **36**, 385 (1987)
5.13 C. Kittel: *Introduction to Solid State Physics* (Wiley, New York 1966)
5.14 A. Abragam: *The Principles of Nuclear Magnetism* (Oxford Univ. Press, Oxford 1961)
5.15 C.P. Slichter: *Principles of Magnetic Resonance*, 3rd ed. Springer Ser. Solid-State Sci., Vol.1 (Springer, Berlin, Heidelberg 1989)
5.16 J. Winter: *Magnetic Resonance in Metals* (Oxford Univ. Press, Oxford 1971)
5.17 E.L. Hahn: Phys. Rev. **80**, 580 (1950)
5.18 M. Mehring: *Principles of High Resolution NMR in Solids* (Springer, Berlin, Heidelberg 1983)
5.19 U. Haeberlen: Adv. Magn. Res. Suppl. (Academic, New York 1976)
5.20 P.K. Kahol, M. Mehring: J. Phys. C **19**, 1045 (1986)
5.21 M. Mehring, J. Spengler: Phys. Rev. Lett. **53**, 2441 (1984)
5.22 M. Mehring, M. Helmle, D. Köngeter, G.G. Maresch, S. Demuth: Synth. Metals **19**, 349 (1987)
5.23 M. Mehring: In *Low Dimensional Conductors and Superconductors*, ed. by D. Jerome, L.G. Caron, NATO ASI Series B (Plenum, New York 1987) Vol. 155, p 185
5.24 (a) P.K. Kahol, M. Mehring, X. Wu: J. de Phys. **46**, 1683 (1985) (b) C. Bourbonnais: In *Low Dimensional Conductors and Superconductors*, ed. by D. Jerome, L.G. Caron, Nato ASI Series B (Plenum, New York 1987) Vol. 155, p.155
5.25 M. Nechtschein, F. Devreux, F. Genoud, M. Guglielmi, K. Holczer: Phys. Rev. B **27**, 61 (1983)

5.26   G. Soda, D. Jerome, M. Weger, J. Alizon, G. Gallice, H. Robert, J.M. Fabre, L. Giral: J. de
       Phys. **38**, 931 (1977)
5.27   A. Carrington, A.D. McLachlan: *Introduction to Magnetic Resonance* (Harper and Row, New
       York 1967)
5.28   L.R. Dalton, A.L. Kwiram: J. Chem. Phys. **57**, 1132 (1972)
5.29   H. Thomann, L.R. Dalton: In *Handbook of Conducting Polymers* ed. by T.A. Skotheim Vol.
       2 (Dekker, New York 1986) p. 1157
5.30   C.L. Young, D. Whitney, A.I. Vistnes, L.R. Dalton: Ann. Rev. Phys. Chem. **37**, 459 (1986)
5.31   H. Thomann: In *Electronic Resonance of the Solid State*, ed. by J. Weil, Canadian Chemical
       Society, Symposium Series (1986)
5.32   E.L. Hahn, S.R. Hartmann: Phys. Rev. **128**, 2042 (1962)
5.33   D.E. Kaplan, E.L. Hahn: J. Phys. Radium **19**, 821 (1958) M. Emschwiller, E.L. Hahn, D.
       Kaplan: Phys. Rev. **118**, 414 (1960)
5.34   M. Mehring, P. Höfer, A. Grupp: Phys. Rev. **A33**, 3523 (1986)
5.35   P. Höfer, A. Grupp, M. Mehring, Phys. Rev. **A33**, 3519 (1986)
5.36   A. Grupp, M. Mehring: In *Modern Pulsed and Continuous—Wave Electron Spin Resonance*,
       eds. L. Kevan, M. Bowmen, John Wiley 1990
5.37   E.R. Andrew: Prog. NMR Spectrosc. **8**, 1 (1971)
5.38   A. Pines, M.G. Gibby, J.S. Waugh: J. Am. Chem. Soc. **92**, 7222 (1970)
5.39   J.S. Waugh, L.M. Huber, U. Haeberlen: Phys. Rev. Lett. **20**, 180 (1968)
5.40   R.H. Baughman, S.L. Hsu, G.P. Pez, A.J. Signorelli: J. Chem. Phys. **68**, 5405 (1978)
5.41   R.H. Baughman, S.L. Hsu, L.R. Anderson, G.P. Pez, A.J. Signorelli: NATO Conf. Ser. **6**,
       187 (1979)
5.42   R.H. Baughman, B.E. Kohler, I.J. Levy, C. Spangler: Synth. Metals **11**, 37 (1985)
5.43   C.R. Fincher, Jr., C.E. Chen. A.J. Heeger, A.G. MacDiarmid, J.B. Hastings: Phys. Rev. Lett.
       48, 100 (1982)
5.44   K. Shimamura, F.E. Karasz, J.A. Hirsh, J.C.W. Chien: Makromol. Chem. Rapid Commun.
       2, 473 (1981)
5.45   T. Akaishi, K. Miyasaka, K. Ishikawa: Rep. Prog. Polym. Phys. Jpn. XII, 125 (1979)
5.46   C. Riekel, H.W. Hasslin, K. Menke, S. Roth: J. Chem. Phys. **77**, 4254 (1982)
5.47   G. Perego, G. Lugli, U. Pedretti, E. Cernia: J. Phys. Colloq. **44**, C3-93 (1983)
5.48   G. Lieser, G. Wegner, W. Müller, V. Enkelmann: Makromol. Chem. Rapid Commun. **1**, 621,
       627 (1980)
5.49   V. Enkelmann, M. Monkenbusch, G. Wegner: Polymer **23**, 1581 (1982)
5.50   G. Lieser, G. Wegner, R. Weizenhoffer, L. Brombacher: Polymer Preprints **25**, 221 (1984)
5.51   L. Mihaly, S. Pekker, A. Janossy: Synth. Metals **1**, 349 (1979/80)
5.52   S. Ikehata, M. Druy, T. Woerner, A.J. Heeger, A.G. MacDiarmid: Solid State Commun. **39**,
       1239 (1981)
5.53   B. Meurer, P. Spegt, G. Weill, C. Mathis, B. Francois: Solid State Commun. **44**, 201 (1982)
5.54   W.K. Rhim, A. Pines, J.S. Waugh: Phys. Rev. B **3**, 684 (1971)
5.55   M. Mehring, H. Seidel, H. Weber, G. Wegner: J. Phys. Colloq. **44**, C3-217 (1983)
5.56   H. Weber: Ph. D. Dissertation, Univ. Dortmund (1983)
5.57   H.W. Gibson, R.J. Weagley, R.A. Mosher, S. Kaplan, W.M. Prest, Jr., A.J. Epstein: Phys. Rev.
       B **31**, 2338 (1985)
5.58   M. Ziliox, B. Francois, C. Mathis, B. Meurer, G. Weill, K. Holczer: Mol. Cryst. Liq. Cryst.
       117, 483 (1985)
5.59   M. Helmle: Diploma thesis, Univ. Stuttgart (1984)
5.60   M. Stamm, J. Hocker: J. Phys. Colloq. **44**, C3-667 (1983)
5.61   H. Cailleau, A. Girad. F. Moussa, C.M.E. Zeyen: Solid State Commun. **29**, 259 (1979)
5.62   S. Schlick, B.R. McGarvey: Polymer Commun. **25**, 369 (1984)
5.63   B.J. Tabor, E.P. Magre, J. Boon: Eur. Polym. J. **7**, 1127 (1971)
5.64   C.S. Yannoni, T.C. Clarke: Phys. Rev. Letts. **51**, 1191 (1983)
5.65   M.J. Duijvestijn, A. Manenscheijn, A. Smidt, R.A. Wind: J. Magn. Res. **64**, 461 (1985)
5.66   C.S. Yannoni, T.C. Clarke: Phys. Rev. Letts. **51**, 1191 (1983)
5.67   D. Horne, R.D. Kendrick, C.S. Yannoni: J. Magn. Res. **52**, 299 (1983)
5.68   H. Kuzmany: Phys. Status Solidi B **97**, 521 (1980)
5.69   D.B. Fitchen: Mol. Cryst. Liq. Cryst. **83**, 95 (1982)
5.70   M.M. Maricq, J.S. Waugh, A.G. MacDiarmid, A.J. Heeger: J. Am. Chem. Soc. **100**, 7729 (1978)
5.71   H.A. Resing, D. Slotfeldt-Ellingsen, A.N. Garroway, T.J. Pinnavaia, K. Unger: In *Magnetic*

Resonance in Colloid and Interface Science, ed. by J.P. Fraissard, H.A. Resing (Reidel, Dordrecht 1980) p. 239

5.72 P. Bernier, F. Schue, J. Sledz, M. Rolland, L. Giral: Chemica Scripta **17**, 151 (1981)

5.73 H.W. Gibson, J.M. Pochan, S.J. Kaplan: J. Amer. Chem. Soc. **103**, 4619 (1981)

5.74 H.A. Resing, A.N. Garroway, D.C. Weber, J. Ferraris, D. Slotfeldt-Ellingsen: Pure Appl. Chem. **54**, 594 (1982)

5.75 M. Mehring, H. Weber, W. Müller, G. Wegner: Solid State Commun. **45**, 1079 (1983)

5.76 T. Terao, S. Maeda, T. Yamabe, K. Akagi, H. Shirakawa: Chem. Phys. Lett. **103**, 347 (1984)

5.77 I. Heinmaa, M. Alla, V. Vainrub, E. Lippmaa, M.L. Khidelkel. A.I. Kotov, G.I. Kozub: J. Phys. Colloq. **44**, C3-357 (1983)

5.78 T. Yamanobe, R. Chujo, I. Ando: Mol. Phys. **50**, 1231 (1983)

5.79 I. Ando, H. Saito, R. Tabeta, A. Shoji, T. Ozaki: Macromol. **17**, 457 (1984)

5.80 T. Yamanobe, I. Ando: J. Chem. Phys. **83**, 3154 (1985)

5.81 H.W. Gibson, R.J. Weagley, W.M. Prest Jr., R. Mosher, S. Kaplan: J. Phys. Colloq. **44**, C3-123 (1983)

5.82 B. Francois, M. Bernard, J.J. Andre: J. Chem. Phys. **75**, 4142 (1981)

5.83 M. Helmle, J.D. Becker, M. Mehring: In *Electronic Properties of Polymers and Related Compounds*, ed. by H. Kuzmany, M. Mehring, S. Roth, Springer Ser. Solid-State Sci. Vol. 63 (Springer, Berlin, Heidelberg 1985) p. 275

5.84 F. Genoud, M. Nechtschein, K. Holczer. F. Devreux, M. Guglielmi: Mole. Cryst. Liq. Cryst. **96**, 161 (1983)

5.85 H. Hübsch: Diploma thesis, Univ. Stuttgart (1983)

5.86 C.E. Brown, M.B. Jones, P. Kovacic: J. Poly. Sci. Poly. Lett. ed. **18**, 653 (1980)

5.87 J. Spengler: Diploma thesis, Univ. Stuttgart (1983)

5.88 F. Devreux, G. Bidan, A.A. Syed, C. Tsintavis: J. Phys. **46**, 1595 (1985)

5.89 G.B. Street, T.C. Clarke, M. Krounbi, K. Kanazawa, V. Lee, P. Pfluger, J.C. Scott, G. Weiser: Mol. Cryst. Liq. Cryst. **83**, 253 (1982)

5.90 S. Hotta, T. Hosaka, W. Shimotsuma: J. Chem. Phys. **80**, 954 (1984)

5.91 S. Hotta, T. Hosaka, M. Soga, W. Shimotsuma: Synth. Metals **10**, 95 (1984/85)

5.92 J.E. Osterholm, P. Sunila, T. Hjertberg, Synth. Metals **18**, 169 (1987)

5.93 S. Kazama, K. Arai, E. Maekawa: Synth. Metals **15**, 299 (1986)

5.94 J. Tsukamoto, S. Fukuda, K. Tanaka, T. Yamabe: Synth. Metals **17**, 673 (1987)

5.95 M. Nechtschein: J. Polym. Sci. C **1**, 1367 (1963)

5.96 T. Ito, H. Shirakawa, S. Ikeda: J. Polym. Sci. Polym. Chem. ed. **12**, 11 (1974); **13**, 1943 (1975)

5.97 I.B. Goldberg, H.R. Crowe, P.R. Newman, A.J. Heeger, A.G. MacDiarmid: J. Chem. Phys. **70**, 1132 (1979)

5.98 P. Bernier, M. Rolland, M. Galtier, A. Montaner, M. Regis, M. Candille, C. Benoit, M. Aldissi, C. Linaya, F. Schue, J. Sledz, J.M. Fabre, L. Giral: J. Phys. **40**, L-297 (1979)

5.99 B.R. Weinberger, J. Kaufer, A.J. Heeger, A. Pron, A.G. MacDiarmid: Phys. Rev. B **20**, 223 (1979)

5.100 P. Kovacic, J. Oziomek: Macromol. Synth. **2**, 23 (1966)

5.101 T. Yamamoto, A. Yamamoto: Chem. Lett. 353 (1977)

5.102 G. Froyer, F. Maurice, P. Bernier, P. McAndrew: Polymer **23**, 1103 (1982)

5.103 J.C. Scott, P. Pfluger, M.T. Krounbi, G.B. Street: Phys. Rev. B **28**, 2140 (1983)

5.104 M. Mehring, H. Seidel, W. Müller, G. Wegner: Solid State Commun. **45**, 1075 (1983)

5.105 G. Volkel, S.A. Dzuba, A. Bartl, W. Brunner, J. Frohner: Phys. Status Solidi A **85**, 257 (1984)

5.106 K. Holczer, J.P. Boucher, F. Devreux, M. Nechtschein: Phys. Rev. B **23**, 1051 (1981)

5.107 M. Nechtschein, F. Devreux, F. Genoud, M. Guglielmi, K. Holczer: Phys. Rev. B **27**, 61 (1983)

5.108 S. Kuroda, M. Tokumoto, N. Kinoshita, T. Ishiguro, H. Shirakawa: J. Phys. Colloq. **44**, C3-303 (1983)

5.109 M. Sasai, H. Fukutome: Solid State Commun. **51**, 609 (1984)

5.110 H. Thomann, L.R. Dalton, Y. Tomkiewicz, N.S. Shiren, T.C. Clarke: Mol. Cryst. Liq. Cryst. **83**, 33 (1982)

5.111 L.R. Dalton, H. Thomann, A. Morrobel-Sosa, C. Chiu, M.E. Galvin, G.E. Wnek, Y. Tomkiewicz, N.S. Shiren, B.H. Robinson, A.L. Kwiram: J. Appl. Phys. **54**, 5583 (1983)

5.112 H. Thomann, L.R. Dalton, Y. Tomkiewicz, N.S. Shiren, T.C. Clarke: Phys. Rev. Lett. **50**, 533 (1983)

5.113 J.F. Cline, H. Thomann, H. Kim, A. Morrobel-Sosa, L.R. Dalton, B.M. Hoffman: Phys. Rev. B **31**, 1605 (1985)

5.114 H. Thomann, L.R. Dalton, M. Grabowski, T.C. Clarke: Phys. Rev. B **31**, 3141 (1985)
5.115 H. Thomann, G.L. Baker: J. Am. Chem. Soc. **109**, 1569 (1987)
5.116 S. Kuroda, H. Shirakawa: Solid State Commun. **43**, 591 (1982)
5.117 S. Kuroda, H. Bando, H. Shirakawa: Solid State Commun. **52**, 893 (1984); J. Phys. Soc. Jpn. **54**, 3956 (1985)
5.118 G.L. Baker, J.B. Raynor, A. Pron: J. Chem. Phys. **80**, 5250 (1984)
5.119 G.J. Baker, J.H. Holloway, J.B Raynor, H. Selig: Synth. Metals **20**, 323 (1987)
5.120 A.J. Heeger, J.R. Schrieffer: Solid State Commun. **48**, 207 (1983)
5.121 H. Käss, P. Höfer, A. Grupp, P.K. Kahol, G. Weizenhöfer, G. Wegner, M. Mehring: Europhys. Lett. **4**, 947 (1987)
5.122 A. Grupp, P. Höfer, H. Käss, M. Mehring, R. Weizenhöfer, G. Wegner: In *Electronic Properties of Conjugated Polymers*, ed. by H. Kuzmany, M. Mehring, S. Roth, Springer Ser. Solid-State Sci. Vol. 76 (Springer, Berlin, Heidelberg 1987)
5.123 M. Mehring, A. Grupp, P. Höfer, H. Käss: Synth. Metals **28**, D 399 (1989)
5.124 S. Kuroda, H. Shirakawa: Phys. Rev. B **35**, 9380 (1987); Synth. Metals **17**, 423 (1987)
5.125 J.E. Hirsch, M.Grabowski; Phys. Rev. Lett. **52**, 1713 (1984)
5.126 K.R. Subaswamy, M. Grabowski: Phys. Rev. B **24**, 2168 (1981)
5.127 K.P. Dinse, G. Wäckerle: Private Communication
5.128 C.T. White, F.W. Kutzler, M. Cook: Phys. Rev. Lett. **56**, 252 (1986)
5.129 S. Kuroda, I. Murase, T. Ohnishi, T. Noguchi, Synth. Metals **17**, 663 (1987)
5.130 D. Davidov, F. Moraes, A.J. Heeger, F. Wudl, H. Kim, L.R. Dalton: Solid State Commun. **53**, 497 (1985)
5.131 H. Thomann, H. Kim, A. Morrobel-Sosa, C. Chiu, L.R. Dalton, B.H. Robinson: Mol. Cryst. Liq. Cryst. **117**, 455 (1985)
5.132 S. Kuroda, H. Shirakawa: Solid State Commun. **59**, 261 (1986)
5.133 M. Mehring, P. Höfer, H. Käss, A. Grupp: Europhys. Lett. **6**, 463 (1988)
5.134 K. Holczer, F. Devreux, M. Nechtschein, J.P. Travers: Solid State Commun. **39**, 881 (1981)
5.135 M. Nechtschein, F. Devreux, R.L. Greene, T.C. Clarke, G.B. Street: Phys. Rev. Letts. **44**, 356 (1980)
5.136 W.G. Clark, K. Glover, G. Mozurkewich, C.T. Murayama, J. Sanny, S. Etemad, M. Maxfield: J. Phys. Colloq. **44**, C3-239 (1983)
5.137 W.G. Clark, K. Glover, G. Mozurkewich, S. Etemad, M. Maxfield: Mol. Cryst. Liq. Cryst. **117**, 447 (1985)
5.138 F. Barbarin, G. Berthet, J.P. Blanc, C. Fabre, J.P. Germain, M. Hamdi, H. Robert: Synth. Metals **6**, 53 (1983)
5.139 F. Barbarin, G. Berthet, J.P. Blanc, M. Dugay, C. Fabre, J.P. Germain, H. Robert: J. Phys. Colloq. **44**, C3-749 (1983)
5.140 N.S. Shiren, Y. Tomkiewicz, H. Thomann, L. Dalton, T.C. Clarke: J. Phys. Colloq. **44**, C3-223 (1983)
5.141 I.J. Lowe, D. Tse: Phys. Rev. **166**, 279 (1968)
5.142 F. Masin, G. Gusman, R. Deltour: Solid State Commun. **39**, 507 (1981)
5.143 K. Kume, K. Mizuno, K. Mizoguchi, K. Nomura, J. Tanaka, M. Tanaka, H. Fujimoto: Mol. Cryst. Liq. Cryst. **83**, 49 (1982)
5.144 P.E. Sokol, J.R. Gaines, S.I. Cho, D.B. Tanner, H.W. Gibson, A.J. Epstein: Synth. Metals **10**, 43 (1984/85)
5.145 F.D. Devreux: Phys. Rev. B **25**, 6609 (1982)
5.146 J.C. Scott, T.C. Clarke: J. Phys. Colloq. **44** C3-365 (1983)
5.147 M. Ziliox, P. Spegt, C. Mathis, B. Francois, G. Weill: Solid State Commun. **51**, 393 (1984)
5.148 F. Masin, G. Gusman: Phys. Rev. B **36**, 2153 (1987)
5.149 J. Alizon, J.P. Blanc, J. Gallice, H. Robert, C. Thibaud, L. Giral, M. Aldissi, F. Schue: Solid State Commun. **39**, 169 (1981)
5.150 F. Masin, G. Gusman, R. Deltour: Solid State Commun. **40**, 415, 513 (1981)
5.151 J.B. Miller, C. Dybowski: Solid State Commun. **46**, 487 (1983)
5.152 D. Tse, S.R. Hartmann: Phys. Rev. Letts. **21**, 511 (1968)
5.153 N. Bloembergen, E.M. Purcell, R.V. Pound: Phys. Rev. **73**, 679 (1948)
5.154 K. Mizoguchi, K. Kume, H. Shirakawa: Solid State Commun. **50**, 213 (1984)
5.155 K. Mizoguchi, K. Kume, H. Shirakawa: Mol. Cryst. Liq. Cryst. **118**, 469 (1985)
5.156 K. Mizoguchi, K. Kume, H. Shirakawa: Mol. Cryst. Liq. Cryst. **117**, 459 (1985)
5.157 B.H. Robinson, J.M. Schurr, A.L. Kwiram, H. Thomann, H. Kim, A. Morrobel-Sosa, P. Bryson, L.R. Dalton: Mol. Cryst. Liq. Cryst. **117**, 421 (1985); J. Phys. Chem. **89**, 4994 (1985)

5.158 G. Volkel, W. Brunner: Phys. Status Solidi A **94**, 673 (1986)

5.159 R. Saitu, H. Kamimura: Synth. Metals **17**, 81 (1987)

5.160 K. Mizoguchi, K. Kume, S. Masubichi, H. Shirakawa: Synth. Met. **17**, 405 (1987)

5.161 J. Tang, C.P. Lin, M.K. Bowman, J.R. Norris, J. Isoya, H. Shirakawa: Phys. Rev. B **28**, 2845 (1983)

5.162 N.S. Shiren, Y. Tomkiewicz, T.G. Kazyaka, A.R. Taranko, H, Thomann, L. Dalton, T.C. Clarke: Solid State Commun. **44**, 1157 (1982)

5.163 Y. Wada, J.R. Schrieffer: Phys. Rev. B **18**, 3897 (1978)

5.164 K. Maki: J. Phys. Colloq. **44**, C3-485 (1983)

5.165 J.D. Flood, A.J. Heeger: Phys. Rev. B **28**, 2356 (1983)

5.166 R. Ball, W.P. Su, J.R. Schrieffer: J. Phys. Colloq. **44**, C3-429 (1983)

5.167 F. Moraes, Y.W. Park, A.J. Heeger: Synth. Metals **13**, 113 (1986)

5.168 L. Orenstein: In *Handbook of Conducting Polymers*, ed. by T.J. Skotheim (Dekker, New York 1986) p. 1297

5.169 C.G. Levey, D.V.Lang, E. Etemad, G.L. Baker, J. Orenstein: Synth. Metals **17**, 569 (1987)

5.170 F. Moraes, H. Schaffer, M. Kobayashi, A.J. Heeger, F. Wudl: Phys. Rev. B **30**, 2948 (1984)

5.171 Z. Vardeny, E. Ehrenfreund, O. Brafman, M. Nowak, H. Schaffer, A.J. Heeger, F. Wudl: Phys. Rev. Letts. **56**, 671 (1986)

5.172 K. Kaneto, K. Yoshino: Synth. Metals **18**, 133 (1987)

5.173 L. Robins, J. Ovenstein, R. Superfine: Phys. Rev. Lett. **56**, 1850 (1986)

5.174 M. Winter, A. Grupp, M. Mehring, H. Sixl: Chem. Phys. Lett. 133, 482 (1987)

5.175 H. Shirakawa, I. Ito and S. Ikeda: Makromol. Chem. **179**, 1565 (1978)

5.176 M. Schwoerer, U. Lauterbach, W. Müller, G. Wegner: Chem. Phys. Lett. **69**, 359 (1980)

5.177 Y. Tomkiewicz, T.D. Schultz, H.B. Brom, T.C. Clarke, G.B. Street: Phys. Rev. Lett. **43**, 1532 (1979)

5.178 A. Snow, P. Brandt, D. Weber, N.L. Yang: J. Polym. Sci. Polym. Lett. ed. **17**, 263 (1979)

5.179 B.R. Weinberger, E. Ehrenfreund, A. Pron, A.J. Heeger, A.G. MacDiarmid: J. Chem. Phys. **72**, 4749 (1980)

5.180 P.J.S. Foot, N.C. Billingham, P.D. Calvert: Synth. Metals **16**, 265 (1986)

5.181 Y. Tomkiewicz, T.D. Schultz, H.B. Brom, A.R. Taranko, T.C. Clarke, G.B. Street: Phys. Rev. B **24**, 4348 (1981)

5.182 S. Ikehata, J. Kaufer, T. Woerner, A. Pron, M.A. Druy, A. Sivak, A.J. Heeger, A.G. MacDiarmid: Phys. Rev. Lett. **45**, 1123 (1980)

5.183 J.C.W. Chien, J.M. Warakomski, F.E. Karasz, W.L. Chia, C.P. Lillya: Phys Rev. B **28**, 6937 (1983)

5.184 J.C.W. Chien, J.M. Warakomski, F.E. Karasz: J. Chem. Phys. **82**, 2118 (1985)

5.185 A.J. Heeger, A.G. MacDiarmid: Mol. Cryst. Liq. Cryst. **77**, 1 (1981)

5.186 M. Tokumoto, N. Kinoshita, S. Kuroda, T. Ishiguro, H. Shirakawa, H. Nemoto: J. Phys. Colloq. **44**, C3-299 (1983)

5.187 A.J. Epstein, R.W. Bigelow, H. Rommelmann, H.W. Gibson, R.J. Weagley, A. Feldblum, D.B. Tanner, J.P. Pouget, J.C. Pouxviel, R. Comes, P. Robin, S. Kivelson: Mol. Cryst. Liq. Cryst. **117**, 147 (1985)

5.188 T.C. Chung, F. Moraes, J.D. Flood, A.J. Heeger: Phys. Rev. B **29**, 2341 (1984)

5.189 M. Peo, H. Forster, K. Menke, J. Hocker, J.A. Gardner, S. Roth, K. Dransfeld: Solid State Commun. **38**, 467 (1981)

5.190 F. Moraes, J. Chen, T.C. Chung, A.J. Heeger: Synth. Metals **11**, 271 (1985)

5.191 K. Kume, K. Mizuno, K. Mizoguchi, K. Nomura, Y. Maniwa, J. Tanaka, M. Tanaka, A. Watanabe: Mol. Cryst. Liq. Cryst. **83**, 285 (1982)

5.192 P. Kuivalainen, H. Stubb, P. Raatikainen, C. Holmstrom: J. Phys. Colloq. **44**, C3-757 (1983)

5.193 K. Kaneto, S. Hayashi, S. Ura, K. Yoshino: J. Phys. Soc. Jpn. **54**, 1146 (1985)

5.194 S. Hayashi, K. Kaneto, K. Yoshino, R. Matshushita, T. Matsuyama: J. Phys. Soc. Jpn. **55**, 1971 (1986)

5.195 J. Chen, A.J. Heeger, F. Wudl: Solid State Commn. **58**, 251 (1986)

5.196 S. Hayashi, S. Takeda, K. Kaneto, K. Yoshino, T. Matsuyama: Synth. Metals **18**, 591 (1987)

5.197 F. Genoud, M. Guglielmi, M. Nechtschein, E. Genies, M. Salmon: Phys. Rev. Lett. **55**, 118 (1985)

5.198 F. Devereux: Synth. Metals **17**, 129 (1987); Europhys. Lett. **1**, 233 (1986)

5.199 M. Nechtschein, F. Devreux, F. Genoud, E. Vieil, J.M. Pernaut, E. Genies: Synth. Metals **15**, 59 (1986)

5.200 M. Schärli, H. Kiess, G. Harbeke, W. Berlinger, K.W. Blazey, K.A. Müller: Synth. Met. **22**, 317 (1988)
5.201 A.J. Epstein, H. Rommelmann, R. Bigelow, H.W. Gibson, D.M. Hoffman, D.B. Tanner: Phys. Rev. Lett. **50**, 1866 (1983)
5.202 K. Mizoguchi, K. Misoo, K. Kume, K. Kaneto, T. Shiraishi, K. Yoshino: Synth. Metals **18**, 195 (1987)
5.203 F. Moraes, D. Davidov, M. Kobayashi, T.C. Chung, J. Chen, A.J. Heeger, F. Wudl: Synth. Metals **10**, 169 (1985)
5.204 P. Pfluger, U.M. Gubler, G.B. Street: Solid State Commun. **49**, 911 (1984)
5.205 L.D. Kispert, J. Joseph, G.G. Miller, R.H. Baughman: Mol. Cryst. Liq. Cryst. **118**, 313 (1985)
5.206 L.D. Kispert, J. Joseph, G.G. Miller, R.H. Baughman: J. Chem. Phys. **81**, 2119 (1984)
5.207 F. Devreux, F. Genoud, M. Nechtschein, J.P. Travers, G. Bidan: J. Phys. Colloq. **44**, C3-621 (1983)
5.208 P.K. Kahol, M. Mehring: Synth. Metals **16**, 257 (1986)
5.209 D. Davidov, S. Roth, W. Neumann, H. Sixl: Z. Phys. B **51**, 145 (1983)
5.210 A. El-Khodary, P. Bernier: J. Phys. **45**, L-551 (1984)
5.211 J. Ghanbaja, C. Goulon, J.F. Marêché, D. Billaud: Solid State Commun. **64**, 69 (1987)
5.212 D. Billaud, J. Ghanbaja, J.F. Marêché, E. Mcrae, C. Goulan: Synth. Met. **24**, 23 (1988)
5.213 A. Bakhtali, J. Ghanbaja, D. Billaud, C. Goulon: Synth. Met. **24**, 47 (1988)
5.214 S. Lefrant, L.S. Lichtman, H. Temkin, D.B. Fitchen, D.C. Miller, G.E. Whitwell, J.M. Burlitch: Solid State Commun. **29**, 191 (1979)
5.215 A. Zanobi, L. D'Ilario: J. Phys. Colloq. **44**, C3-309 (1983)
5.216 Y.W. Park, A.J. Heeger, M.A. Druy, A.G. MacDiarmid: J. Chem. Phys. **73**, 946 (1980)
5.217 C. Benoit, M. Rolland, M. Aldissi, A. Rossi, M. Cadena, P. Bernier: Phys. Status Solidi A **68**, 209 (1981)
5.218 F. Rachdi, P. Bernier, E. Faulques, S. Lefrant, F. Schue: J. Phys. Colloq. **44**, C3-97 (1983)
5.219 K. Kaneto, M. Maxfield, D.P. Nairns, A.G. MacDiarmid, A.J. Heeger: J. Chem. Soc. Faraday Trans. I **78**, 3417 (1982)
5.220 F. Rachdi, P. Bernier, E. Faulques, S. Lefrant, F. Schue: J. Chem. Phys. **80**, 6285 (1984)
5.221 A. El-Khodary, P. Bernier: J. Chem. Phys. **85**, 2243 (1986)
5.222 P. Bernier, F. Rachdi, A. El-Khodary, C. Fite: Synth. Met. **24**, 31 (1988)
5.223 F.J. Dyson: Phys. Rev. **98**, 349 (1955)
5.224 G. Feher, A.F. Kip: Phys. Rev. **98**, 337 (1955)
5.225 H. Kodera: J. Phys. Soc. Jpn. **28**, 89 (1970)
5.226 F. Maurice, C. Fontaine, A. Morisson, J.Y. Goblot, G. Froyer: Mole. Cryst. Liq. Cryst. **118**, 319 (1985)
5.227 R.L. Elsenbaumer, P. Delannoy, G.G. Miller, C.E. Forbes, N.S. Murthy, H. Eckhardt, R.H. Baughman: Synth. Metals **11**, 251 (1985)
5.228 D. Billaud, I. Kulszewicz, A. Pron, P. Bernier, S. Lefrant: J. Phys. Colloq. **44**, C3-33 (1983)
5.229 A. Bartl, H.G. Doege, J. Froehner, G. Lehmann, B. Pietrass: Synth. Metals **10**, 151 (1985/85)
5.230 M. Tokumoto, H. Oyanagi, T. Ishiguro, H. Shirakawa, H. Nemoto, T. Matsushita, H. Kuroda: Mol. Cryst. Liq. Cryst. **117**, 139 (1985)
5.231 M.J. Kletter, A.G. MacDiarmid, A.J. Heeger, E. Faulques, S. Lefrant, P. Bernier, F. Barbarin, J.P. Blanc, J.P. Germain, N. Robert: Mol. Cryst. Liq. Cryst. **83**, 165 (1982)
5.232 D. Davidov, S. Roth, W. Neumann, H. Sixl: Solid State Commun. **52**, 375 (1984)
5.233 R.J. Elliott: Phys. Rev. **96**, 266 (1954)
5.234 Y. Yafet: Solid State Physics **14**, 1 (1963)
5.235 F. Rachdi, P. Bernier: Phys. Rev. B **33**, 11 (1986)
5.236 L.D. Kispert, J. Joseph, M.K. Bowman, G.H. van Brakel, J. Tang, J.R. Norris: Mol. Cryst. Liq. Cryst. **107**, 81 (1984)
5.237 L.D. Kispert, J. Joseph, J. Tang, M.K. Bowman, G.H. van Brakel, J.R. Norris: Synth. Metals **17**, 617 (1987)
5.238 J. Isoya, H. Nagasawa, H. Shirakawa, M.K. Bowman, C.P. Lin, J. Tang, J.R. Norris: Synth. Metals **17**, 215 (1987)
5.239 M. Ziliox, B. Francois, C. Mathis, B. Meurer, P. Spegt, G. Weill: J. Phys. Colloq. **44**, C3-361 (1983)
5.240 T.C. Clarke, J.C. Scott: Solid State Commun. **41**, 389 (1982)
5.241 T. Terao, S. Maeda, T. Yamabe, K. Akagi, H. Shirakawa: Solid State Commun. **49**, 829 (1984)

5.242 M. Audenaert, P. Bernier, Mol. Cryst. Liq. Cryst. **117**, 83 (1985)
5.243 L. Soderholm, C. Mathis, B. Francois, J.M. Friedt: Synth. Metals **10**, 261 (1985)
5.244 J.B. Miller, C. Dybowski: Synth. Metals **6**, 65 (1983)
5.245 S. Kazama: Synth. Metals **16**, 77 (1986)
5.246 H. Lecavelier, F. Devreux, M. Nechtschein, G. Bidan: Mol. Cryst. Liq. Cryst. **118**, 183 (1985)
5.247 G. Wegner: Z. Naturforsch. **24b**, 824 (1969)
5.248 G. Wegner: Makromol. Chem. **134**, 219 (1970); **154**, 35 (1972)
5.249 G. Wegner: Chemia **28**, 475 (1974)
5.250 G. Wegner: In *Molecular Metals*, ed. by W.E. Hatfield (Plenum, New York 1979) p. 209
5.251 a) G. Wegner: In *Chemistry and Physics of One-Dimensional Metals* ed. by H.J. Keller (Plenum, New York 1977) p. 297 b) V. Enkelmann: Acta Cryst. **B33**, 2842 (1977)
5.252 R.H. Baughmann: J. Appl. Phys. **43**, 4362 (1972)
5.253 R.H. Baughmann: J. Polym. Sci. Polym. Phys. ed. **12**, 1511 (1974)
5.254 R.H. Baughmann, K.C. Yee: J. Polym. Sci. Polym. Chem. ed. **12**, 2467 (1974)
5.255 R.H. Baughmann, R.R. Chance: In *Synthesis and Properties of Low-Dimensional Materials* ed. by J.S. Miller, A. Epstein (Academic, New York 1978) p. 705
5.256 D. Bloor: In *Developments in Crystalline Polymers*, ed. by D.C. Basset (Applied Sci, London 1982)
5.257 (a) H. Sixl: Adv. Polymer Sci. **63**, ed. by H.J. Contow (Springer, Berlin, Heidelberg 1984) (b) M. Schwoerer, H. Niederwald: Makromol. Chem. Suppl. **12**, 61 (1985)
5.258 R.J. Leyrer, G. Wegner, W. Wettling: Ber. Bunsenges. Phys. Chem. **82**, 697 (1978)
5.259 H. Bässker, R. Modong: Chem. Phys. Lett. **78**, 371 (1981)
5.260 D. Siegel, H. Sixl, V. Enkelmann, G. Weny: Chem. Phys. **72**, 201 (1982)
5.261 D. Bloor, L. Koski, G.G. Stevens, F.H. Preston, D.J. Ando: J. Mater. Sci. **10**, 1678 (1975)
5.262 (a) H. Eichele, M. Schwoerer, R. Huber, D. Bloor: Chem. Phys. **42**, 342 (1976) (b) H. Niederwald, K.H. Richter, W. Gütter, M. Schwoerer: Laser Chemistry (1983)
5.263 H. Gross, H. Sixl, C. Kröhne, V. Enkelmann: Chem. Phys. **45**, 15 (1980)
5.264 D. Bloor, D.J. Ando, F.H. Preston, G.C. Stevens: Chem. Phys. Lett. **24**, 407 (1974)
5.265 H. Eichele, M. Schwoerer, R. Huber, D. Bloor: Chem. Phys. Lett. **42**, 342 (1976)
5.266 (a) H. Eichele, M. Schwoerer, J.U. von Schütz: Chem. Phys. Lett. **56**, 208 (1978) (b) H. Niederwald, H. Eichele, M. Schwoerer: Chem. Phys. Lett, **72**, 242 (1980) (c) R. Müller-Nawrath, R. Angstl, M. Schwoerer: Chem. Phys. **108**, 121 (1986)
5.267 H. Niederwald, M. Schwoerer: Z. Naturforsch. **38a**, 749 (1983)
5.268 (a) R. Huber, M. Schwoerer: Chem. Phys. Lett. **72**, 10 (1980) (b) R. Huber, M. Schwoerer, H. Benk, H. Sixl: Chem. Phys. Lett. **78**, 416 (1981) (c) M. Schwoerer, R. Huber, W. Hartl: Chem. Phys. **55**, 97 (1981) (d) W. Hartl, M. Schwoerer: Chem. Phys. **69**, 443 (1982)
5.269 (a) W. Neumann, H. Sixl: Chem. Phys. **50**, 273 (1980); ibid. **58**, 303 (1981) (b) H. Gross, W. Neumann, H. Sixl: Chem. Phys. Lett. **92**, 584 (1983) (c) R. Huber, E. Sigmund, C. Kollmar, H. Sixl: In *Electronic Properties of Polymers and Related Compounds* ed. by H. Kuzmany, M. Mehring, S. Roth, Springer Ser. Solid-State Sci. Vol. 63 (Springer, Berlin, Heidelberg 1985) p. 249
5.270 D. Bloor, S.D.D.V. Rughooputh, D. Phillips, W. Hayes, K.S. Wong: In *Electronic Properties of Polymers and Related Compounds* ed. by H. Kuzmany, M. Mehring, S. Roth, Springer Ser. Solid-State Sci. Vol.63 (Springer, Berlin, Heidelberg 1985) p. 253
5.271 K.J. Donovan, P.D. Freeman, E.G. Wilson: In *Electronic Properties Polymers and Related Compounds* ed. by H. Kuzmany, M. Mehring, S. Roth, Springer Ser. Solid-State Sci. Vol. 63, (Springer, Berlin, Heidelberg 1985) p. 256
5.272 H. Sixl, R. Warta: In *Electronic Properties of Polymers and Related Compounds*, ed. by H. Kuzmany, M. Mehring, S. Roth, Springer Ser. Solid-State Sci., Vol. 63 (Springer, Berlin, Heidelberg 1985) p. 246; Chem. Phys. Lett. **116**, 307 (1985)
5.273 J. Orenstein, S. Etemad, G.L. Baker: J. Phys. C **17**, L 298 (1984)
5.274 Tavan, K. Schulten: J. Chem. Phys. **85**, 6602 (1986)
5.275 H.W. Gibson, A.J. Epstein, H. Rommelmann, D.B. Tanner, X.Q. Yang, J.M. Pochan: J. Phys. Colloq. **44**, C3-651 (1983)
5.276 H.W. Gibson, F.C. Bailey, A.J. Epstein, H. Rommelmann, S. Kaplan, J. Harbour, X.Q. Yang, D.B. Tanner, J.M. Pochan: J. Am. Chem. Soc. **105**, 4417 (1983)
5.277 M.M. Ahmad, W.J. Feast: Mol. Cryst. Liq. Cryst. **118**, 417 (1985)
5.278 G. Vancso, T.T. Nagy, B. Turcsanyi, T. Kelen, F. Tudos: J. Phys. Colloq. **44**, C3-701 (1983)
5.279 Y. Ikeda, M. Ozaki, T. Arakawa: Mol. Cryst. Liq. Cryst. **118**, 431 (1983)

5.280 G.M. Hobob, P. Ehrlich, R.D. Allendoerfer: Macromolecules **5**, 569 (1972)
5.281 E.T. Kang, A. Langner, P. Ehrlich: Polym. Preprints **23**, 103 (1982)
5.282 E.T. Kang, A.P. Bhatt, E. Villaroel, W.A. Anderson, P. Ehrlich: Mol. Cryst. Liq. Cryst. **83**, 1339 (1982)
5.283 M. Tabata, T. Matsura, S. Okawa, W. Yang, K. Yokota, J. Sohma: Synth. Metals **17**, 577 (1987)
5.284 J.R. Reynolds, F.E. Karasz, J.C.W. Chien, K.D. Gourley, C.P. Lillya: J. Phys. Colloq. **44**, C3-693 (1983)
5.285 L.D. Kispert, J. Joseph, R. Drobner: Synth. Metals **20**, 209 (1987)
5.286 L.D. Kispert, L.A. Files, J.E. Frommer, L.W. Shacklette, R.R. Chance: J. Chem. Phys. **78**, 4858 (1983)
5.287 S. Schlick, B.R. McGarvey: Polymer Commun. **25**, 369 (1984)
5.288 T. Sugano, M. Kinoshita: Synth. Metals **17**, 685 (1987)
5.289 F. Barbarin, J.P. Blanc, M. Dugay, C. Fabre, C. Maleysson: Synth. Metals **10**, 71 (1984/85)
5.290 K. Yoshino, Y. Kohno, T. Shiraishi, K. Kaneto, S. Inoue, K. Tsukagoshi: Synth. Metals **10**, 319 (1985)
5.291 S. Naitoh: Synth. Metals **18**, 237 (1987)
5.292 S. Hayashi, S. Nakajima, K. Kaneto, K. Yoshino: J. Phys. Soc. Jpn. **55**, 3995 (1986); Synth. Metals **17**, 667 (1987)
5.293 A. Bolognesi, M. Catellani, S. Destri, W. Porzio: Synth. Metals **18**, 129 (1987)
5.294 S. Mizogami, S. Naitoh, S. Yoshimura: Synth. Metals **18**, 479 (1987)
5.295 M. Nechtschein, C. Santier, J.P. Travers, J. Chroboczek, A. Alix, M. Ripert: Synth. Metals **18**, 311 (1987)
5.296 A.J. Epstein, J.M. Ginder, F. Zuo, R.W. Bigelow, H.-S. Woo, D.B. Tanner, A.F. Richter, W.-S. Huang, A.-G. MacDiarmid: Synth. Metals **18**, 303 (1987)
5.297 Y. Cao, S. Li, Z. Xue, D. Guo: Synth. Metals **16**, 305 (1986)

Subject Index

Springer Series in Solid-State Sciences

Editors: M. Cardona   P. Fulde   K. von Klitzing   H.-J. Queisser